SOIL, AGRICULTURE, AND ECOSYSTEM MODELING

Smart Technologies for Sustainable Solutions

SOIL, AGRICULTURE, AND ECOSYSTEM MODELING

Smart Technologies for Sustainable Solutions

Edited by

Owais Bashir, PhD

Khalid Rehman Hakeem, PhD

Surinder S. Kukal, PhD

Razeef Mohd, PhD

APPLE ACADEMIC PRESS

First edition published 2025

Apple Academic Press Inc.
1265 Goldenrod Circle, NE,
Palm Bay, FL 32905 USA

760 Laurentian Drive, Unit 19,
Burlington, ON L7N 0A4, CANADA

CRC Press
2385 NW Executive Center Drive,
Suite 320, Boca Raton FL 33431

4 Park Square, Milton Park,
Abingdon, Oxon, OX14 4RN UK

Library and Archives Canada Cataloguing in Publication

Title: Soil, agriculture, and ecosystem modeling : smart technologies for sustainable solutions / edited by Owais Bashir, PhD, Khalid Rehman Hakeem, PhD, Surinder S. Kukal, PhD, Razeef Mohd, PhD.
Names: Bashir, Owais, editor. | Hakeem, Khalid Rehman, editor. | Kukal, Surinder S., editor. | Mohd, Razeef, editor.
Description: First edition. | Includes bibliographical references and index.
Identifiers: Canadiana (print) 20240312171 | Canadiana (ebook) 2024031249X | ISBN 9781774915424 (hardcover) | ISBN 9781774915431 (softcover) | ISBN 9781032685359 (ebook)
Subjects: LCSH: Agricultural innovations. | LCSH: Crops and soils—Computer simulation. | LCSH: Artificial intelligence. | LCSH: Computer vision. | LCSH: Machine learning.
Classification: LCC S494.5.I5 S65 2024 | DDC 338.1/6—dc23

Library of Congress Cataloging-in-Publication Data

CIP data on file with US Library of Congress

ISBN: 978-1-77491-542-4 (hbk)
ISBN: 978-1-77491-543-1 (pbk)
ISBN: 978-1-03268-535-9 (ebk)

Dedication

This book is dedicated to pedologists and ecologists of the world and the Task Force on Himalayan Agriculture–National Mission for Sustaining the Himalayan Eco-System (THFA–NMSHE) of India.

About the Editors

Owais Bashir, PhD
Department of Soil Science, Sher-e-Kashmir University of Agricultural Sciences and Technology, Kashmir, India

Owais Bashir, PhD, is presently working in the Department of Soil Science at SKUAST-Kashmir, India. Dr. Bashir has been involved in the research from the last five years on soil pedology, climate change, machine learning (ML), and soil health. He is an author of more than 10 research articles (h-index 2; i-index 2; total citation 34) and two book chapters and has published one book with international publishers (Apple Academic Press). Dr. Bashir has presented and participated in numerous state, national, and international conferences, seminars, workshops, and symposiums. He has received many awards, appreciation, and recognition for his services to the science of soil analysis. He has pursued his doctorate at Sher-e-Kashmir University of Agricultural Sciences and Technology Kashmir (Division of Soil Science). He has served as an editorial board member and reviewer for reputed international journals. Currently, Dr. Bashir is experimenting with diverse capacities of soil health and climate change and its impact on plant health.

Khalid Rehman Hakeem, PhD
Professor, King Abdulaziz University, Jeddah, Saudi Arabia

Khalid Rehman Hakeem, PhD, is working as a Professor at King Abdulaziz University, Jeddah, Saudi Arabia. After completing his PhD (Botany; specialization in Plant Ecophysiology and Molecular Biology) from Jamia Hamdard, New Delhi, India in 2011, he worked as an Assistant Professor at the University of Kashmir, Srinagar for a short period. Later, he joined the University Putra Malaysia, Selangor, Malaysia, and worked there as a doctorate fellow in 2012 and fellow researcher (Associate Prof.) from 2013–2016, respectively. He joined King Abdulaziz University in August

2016 and was promoted to professorship recently. Dr. Hakeem has more than 10 years of teaching and research experience in plant eco-physiology, biotechnology, molecular biology, medicinal plant research, plant-microbe-soil interactions as well as in environmental studies. He is the recipient of several fellowships at both national and international levels, also he has served as a visiting scientist at Fatih University, Istanbul, Turkey as well as at Jinan University, Guangzhou, China. Currently, he is involved with a number of international research projects with different government organizations. So far, Dr. Hakeem has authored and edited more than 80 books with international publishers, including Springer Nature, Academic Press (Elsevier), CRC Press, etc. He has also to his credit 180 research publications in peer-reviewed international journals and 65 book chapters in edited volumes with international publishers. At present, Dr. Hakeem is serving as the editorial board member and reviewer of several high-impact international scientific journals from Elsevier, Springer Nature, Taylor, Cambridge, and Francis and John Wiley Publishers.

Surinder S. Kukal, PhD
Former Professor & Dean, Faculty of Agriculture, Punjab Agricultural University, India

Surinder S. Kukal, PhD, has worked in the field of agricultural water management at Punjab Agricultural University, India. Dr. Surinder S. Kukal, an agricultural graduate, holds a PhD in Soil Physics and Water Management. Dr. Kukal has more than 350 research publications to his credit including 150 peer-reviewed research papers in journals of international and national repute. He served as a Professor of Soil Conservation, Additional Director of Research and Dean of the Faculty of Agriculture at Punjab Agricultural University. Dr. Kukal has several awards and honors to his credit, the most prestigious being fellow, National Academy of Agricultural Sciences (NAAS); fellow, Indian Society of Soil Science; Dr. GS Khush Distinguished Professor Award; Honorary member of Watershed Management Society of Iran; 12th International Congress Commemoration Award of Indian Society of Soil Science. Dr. Kukal has been a visiting scientist at CSIRO, Australia, apart from delivering lead lectures at various platforms in countries like the USA, Australia China, Thailand, Serbia, Singapore, and Pakistan. He is an Associate Editor-in-Chief of *International Soil and Water Conservation Research.*

Razeef Mohd, PhD
Machine Learning and Artificial Intelligence Expert, Sher-e-Kashmir University of Agriculture Sciences and Technology, Kashmir, India

Razeef Mohd, PhD, is currently working as a Machine Learning and Artificial Intelligence (AI) Expert at Sher-e-Kashmir University of Agriculture Sciences and Technology, Kashmir (SKUAST-K), J&K, India. He has done MCA (Gold Medalist), MPhil, and PhD in the field of Computer Sciences from PG Department of Computer Sciences, University of Kashmir, Srinagar, J&K, India. He has more than eight years of teaching and research experience. He is an active frontline researcher in machine learning and artificial intelligence, publishing many research papers in leading peer-reviewed international journals, SCI journals, and SCOPUS journals. He has also presented papers at IEEE conferences and science congresses and guided many students in their projects and encouraged them in identifying, mentoring, nurturing innovative and entrepreneurial potential and transforming them into start-up entrepreneurs to make the innovation and venture successful. His areas of specialization are data mining, machine learning, artificial intelligence, software engineering, and cloud computing.

Contents

Contributors

Rana Afreen
Division of Soil Science, Sher-e-Kashmir University of Agricultural Sciences and Technology of Kashmir, Kashmir, India

Haseeb Ahmed
Division of Fruit Science, Sher-e-Kashmir University of Agricultural Sciences and Technology of Kashmir, Kashmir, India

Tanveer Ahmed
Division of Agronomy, Sher-e-Kashmir University of Agricultural Sciences and Technology of Kashmir, Kashmir, India

Deelak Amin
Division of Entomology, Sher-e-Kashmir University of Agricultural Sciences and Technology of Kashmir, Kashmir, India

Asmat Ara
Division of Plant Breeding and Genetics, Sher-e-Kashmir University of Agricultural Sciences and Technology of Kashmir, Kashmir, India

Sheikh Asif
Division of Entomology, Sher-e-Kashmir University of Agricultural Sciences and Technology of Kashmir, Kashmir, India

Zahoor Ahmad Baba
Division of Soil Science, Sher-e-Kashmir University of Agricultural Sciences and Technology of Kashmir, Kashmir, India

Shabir Ahmad Bangroo
Division of Soil Science, Sher-e-Kashmir University of Agricultural Sciences and Technology of Kashmir, Kashmir, India

Owais Bashir
Division of Soil Science, Sher-e-Kashmir University of Agricultural Sciences and Technology of Kashmir, Kashmir, India

Burgish Bhat
Division of Environmental Science, Sher-e-Kashmir University of Agricultural Sciences and Technology of Kashmir, Kashmir, India

M. Auyoub Bhat
Division of Soil Science, Sher-e-Kashmir University of Agricultural Sciences and Technology of Kashmir, Kashmir, India

Owais Bhat
Division of Agriculture Engineering, Sher-e-Kashmir University of Agricultural Sciences and Technology of Kashmir, Kashmir, India

Tauseef Ahmad Bhat
Division of Agronomy, Sher-e-Kashmir University of Agricultural Sciences and Technology of Kashmir, Kashmir, India

Fazil Fayaz
Division of Plant Pathology, Sher-e-Kashmir University of Agricultural Sciences and Technology of Kashmir, Kashmir, India

Nazir Ahmad Ganie
Division of Animal Genetics and Breeding, Sher-e-Kashmir University of Agricultural Sciences and Technology of Kashmir, Kashmir, India

Khalid Rehman Hakeem
Department of Biological Sciences, Faculty of Science, King Abdulaziz University, Jeddah, Saudi Arabia

Ghulam Hassan
Division of Fruit Science, Sher-e-Kashmir University of Agricultural Sciences and Technology of Kashmir, Kashmir, India

Rehana Javid
Division of Fruit Science, Sher-e-Kashmir University of Agricultural Sciences and Technology of Kashmir, Kashmir, India

Mir Waseema Jehaan
Division of Soil Science, Sher-e-Kashmir University of Agricultural Sciences and Technology of Kashmir, Kashmir, India

S. S. Kukal
Punjab Water Regulation and Development Authority, Chandigarh, Punjab, India

Abdul Rouf Malik
Division of Fruit Science, Sher-e-Kashmir University of Agricultural Sciences and Technology of Kashmir, Kashmir, India

Aamir Hassan Mir
Division of Soil Science, Sher-e-Kashmir University of Agricultural Sciences and Technology of Kashmir, Kashmir, India

Shakeel Ahmad Mir
Division of Soil Science, Sher-e-Kashmir University of Agricultural Sciences and Technology of Kashmir, Kashmir, India

Abeena Mohiuddin
Department of Computer Sciences, Mewar University Rajasthan, Chittorgarh, Rajasthan, India

Razeef Mohd.
NAHEP, Sher-e-Kashmir University of Agricultural Sciences and Technology of Kashmir, Kashmir, India

Rehana Mohiuddin
Division of Agronomy, Sher-e-Kashmir University of Agricultural Sciences and Technology of Kashmir, Kashmir, India

Nasir Bashir Naikoo
Division of Soil Science, Sher-e-Kashmir University of Agricultural Sciences and Technology of Kashmir, Kashmir, India

Zahida Rashid
Division of Agronomy, Sher-e-Kashmir University of Agricultural Sciences and Technology of Kashmir, Kashmir, India

Rehana Rasool
Division of Soil Science, Sher-e-Kashmir University of Agricultural Sciences and Technology of Kashmir, Kashmir, India

Shafeeq Ur Rehman
School of Environment and Civil Engineering, Dongguan University of Technology, Dongguan, Guangdong, China

Zohra Shabir
Division of Plant Pathology, Sher-e-Kashmir University of Agricultural Sciences and Technology of Kashmir, Kashmir, India

Divya Sharma
Division of Soil Science, Sher-e-Kashmir University of Agricultural Sciences and Technology of Jammu, Jammu, India

Abbreviations

ADV	aggregation dispersion variance
AI	artificial intelligence
ANN	artificial neural network
AOGCM	atmosphere-ocean general circulation model
APSIM	agricultural production systems simulator model
AR	augmented reality
BMP	bagging and multilayer perceptron
BPDA	benzene polycarboxylic acids
C	carbon
CAGR	compound annual growth rate
CGC	canopy growth coefficient
CNN	convolutional neural network
CO_2	carbon dioxide
CSM	crop simulation models
DEM	digital elevation model
DL	deep learning
DNN	deep neural networks
DOM	dissolving organic matter
DPM	distributed parameter models
DRLSE	distance regularization level set evolution
DSIFT	dense scale invariant features transform
DSSAT	decision support system for agro-technology transfer
DT	decision trees
ELM	extreme learning machines
ERT	extremely randomized trees
ES	ecosystem services
ESA	European Space Agency
ET	evapotranspiration
EUROSEM	European soil erosion model
FAO	Food and Agriculture Organization
FC	fully connected
FCN	fully convolutional network
FOM	fresh organic matter
GAN	generative adversarial networks
GCM	general circulation model
GCM	global climate model

GCTESOMNET	global change and terrestrial ecosystems soil organic matter network
GCV	generalized cross-validation
GDP	global gross domestic product
GHG	greenhouse gas
GIS	geographic information system
GLCM	gray level co-occurrence matrix
GLDD	grape leaf disease dataset
GMM	gaussian mixture model
HALE	high-altitude long endurance
HI	harvest index
HIS	hyper-spectral imaging
HMD	head-mounted display
HTMEs	humid tropical mountain environments
IAM	integrated assessment models
ICT	information and communication technologies
IoT	internet of things
IPM	integrated pest management
KBS	knowledge base systems
KNN	K-nearest neighbor
LAI	leaf area index
LASE	low altitude-short endurance
LF	light fraction
LIF	laser-induced fluorescence
LPMs	lumped parameter models
LR	linear regression
LSM	least-squares method
MARS	multivariate adaptive regression splines
ML	machine learning
MLP	multilayer perceptron
MLR	multiple linear regression
MMF	Morgan-Morgan and Finney
NGOs	non-governmental organizations
NM	near-mature
NMS	non-maximum suppression
NPQ	non-photochemical fluorescent quenching
NY	near-young
OM	organic matter
OR-AC-GAN	outlier removal auxiliary classifier generative adversarial nets
PCs	personal computers
POC	particle size fractionation
QNN	quantum neural network
RAINS	regional acidification information and simulation

RCP	representative concentration pathways
RMSE	root mean square error
RNNs	recurrent neural networks
ROI	region of interest
RPN	region proposal network
RUE	radiation use efficiency
SALUS	system approach to land use sustainability
SAR	synthetic aperture radar
SDGs	sustainable development goals
SLEMSA	soil loss estimation model for Southern Africa
SMC	soil moisture content
SOC	soil organic carbon
SOM	soil organic matter
SoS	system of systems
SRCNN	super-resolution convolutional neural network
SVM	support vector machine
SVR	support vector regression
SWAT	soil and water analysis tool
UAV	unmanned aerial vehicle
VIA	vulnerability, impacts, and adaptation
VR	virtual reality
VTOL	vertical take-off and landing

About the Book

This book explores computer vision, a non-contact as well as non-invasive method that entails the creation of conceptual and algorithmic techniques for automatic visual comprehension and appreciation, which has huge applications throughout ecosystem services (ES) and agricultural production. In addition to this, it discusses how the use of machine learning (ML) strategies to computer vision algorithms is helping improve productivity in agriculture by fostering the development of more accurate systems.

The combination of computer vision and machine learning (ML) aids in the diagnosis of plant diseases, the monitoring of agricultural conditions, and the assessment of environmental risks, all of which are important in preventing the loss of yield and quality as well as sustenance of the human population. Significant scientific and technological advancements have been made in the areas of defect assessment, quality grading, soil characterization disease recognition, pests, and insect identification, fruit, and vegetable type identification, and evaluation of a wide variety of agricultural plants, crops, leaves, and fruits.

This book examines intelligent robots that have been constructed with a touch of ML that may assist farmers in doing a variety of jobs. Plant, leaf, and fruit disease detection, crop health monitoring, uses of robots in agriculture, precision farming, evaluation of product quality and flaws, and recognition of pests, insects, fruits, and vegetables are some of the important subjects that are discussed in this book.

Preface

Soil, Agriculture, and Ecosystem Modeling is a book about the scientific knowledge of artificial intelligence (AI) and its use for the advancement of agriculture and creation of an improved ecosystem. AI has grown rapidly as a multidisciplinary discipline in the recent years. This inspired us to write a book on AI as a formal, persuasive theory and a boisterous experimental wing. In this book, we have tried to strike a balance between experiment and theory and demonstrate how to connect the two closely. Together with its agricultural applications, we advance AI science.

There is nothing more practical than a sound theory. The saying "Everything should be made as simple as possible, but not simpler" summarizes our philosophy. The need of the hour is to develop basic models for our ecosystems and agriculture in science. The underlying structures and building blocks should be straightforward, notwithstanding the complexity of the resulting systems.

The integration of machine learning methodologies within computer vision algorithms has been instrumental in transforming agricultural methods, leading to a substantial increase in productivity. This fusion of machine learning with computer vision techniques plays a pivotal role in the identification of plant diseases, monitoring agricultural conditions, and assessing environmental risks. These advancements are critical in preventing losses in yield and quality, thereby playing a vital role in sustaining the global population.

Significant progress has been achieved today in the scientific and technological spheres encompassing defect assessment, quality grading, soil characterization, disease detection, pest and insect identification, as well as the categorization of diverse fruits and vegetables. Moreover, this technology facilitates a comprehensive evaluation of an array of agricultural plants, crops, leaves, and fruits. Therefore, a course in agriculture, soil, and ecosystem modeling should be used for all studies focusing on agriculture systems, pedological study, environmental management, environmental sciences, and ecology.

Building models for soil, agriculture, and ecosystems involves a practical computer lab component and an introduction to the fundamental theory. The creation of models is crucial for managing and comprehending ecological processes. A vital job is figuring out the intricate connections between ecological patterns and processes.

This book provides various examples of agriculture and ecological modeling problems or assignments, and the answer is then provided in detail to ease the exercises. Each chapter ends with various tasks, but the greatest ones require the students to create their own models based on issues they have posed and are particularly interested in solving.

This book covers the fundamental concepts of soil agriculture and ecosystem modeling, "how to go modeling," and an overview of the numerous model types used for ecological modeling. The creation and use of the many model types in agriculture modeling are covered in depth in the book including: environmental models, pathological models, agronomic models, and structurally dynamic models.

Thorough and instructive examples are also provided. The examples were chosen to give the readers enough knowledge to make it easier for them to create similar models independently.

—Editors

CHAPTER 1

State-of-the-Art Modeling in Soil and Ecosystem Services Under the Current Scenario

OWAIS BASHIR,[1] DIVYA SHARMA,[2] SHABIR AHMAD BANGROO,[1] NASIR BASHIR NAIKOO,[1] NAZIR AHMAD GANIE,[3] S. S. KUKAL,[4] ZAHOOR AHMAD BABA,[1] ABDUL ROUF MALIK,[5] REHANA RASOOL,[1] SHAFEEQ UR REHMAN,[6] and KHALID REHMAN HAKEEM[7]

[1]*Division of Soil Science, Sher-e-Kashmir University of Agricultural Sciences and Technology of Kashmir, Kashmir, India*

[2]*Division of Soil Science, Sher-e-Kashmir University of Agricultural Sciences and Technology of Jammu, Jammu, India*

[3]*Division of Animal Genetics and Breeding, Sher-e-Kashmir University of Agricultural Sciences and Technology of Kashmir, Kashmir, India*

[4]*Punjab Water Regulation and Development Authority, Chandigarh, Punjab, India*

[5]*Division of Fruit Science, Sher-e-Kashmir University of Agricultural Sciences and Technology of Kashmir, Kashmir, India*

[6]*School of Environment and Civil Engineering, Dongguan University of Technology, Dongguan, Guangdong, China*

[7]*Department of Biological Sciences, Faculty of Science, King Abdulaziz University, Jeddah, Saudi Arabia*

Soil, Agriculture, and Ecosystem Modelling: Smart Technologies for Sustainable Solutions.
Owais Bashir, Khalid Rehman Hakeem, Surinder S. Kukal, and Razeef Mohd (Eds.)

ABSTRACT

The ecosystem services (ES) and processes are facilitated by ecosystems and biodiversity supporting human existence. These ES have emerged as useful tools for connecting ecosystem function to the well-being of humans. ES is not only an environmental issue but also an economic and social one, impacting the economy and individual health. However, because ES is holistic and interact, anthropogenic pressures frequently affect multiple services. Scientific research has focused more on modeling the demand for ES, which is an effective tool for conservation, land use, and management. However, different people have different ideas about ES demand, which impacts how and where ES demand is modeled and mapped. The ecosystem service modeling includes various processes, including SOM, climate change, nutrients, land use, and land cover modeling. Simulation models play an important role in SOM research because they provide a mathematical framework for integrating, examining, and testing the awareness of SOM dynamics. Various models exist for predicting changes in SOC over time and space, which is useful for land use planning and soil management. Models that describe and assign changes in land use and land cover are crucial components of integrated land change research. We have examined existing ecosystem service mapping technologies and previous studies to assess the current state of the art. We provide our early findings and describe the techniques utilized to model various soil and ecosystem models for the major ES. We end by discussing the potential applications of our methodology and the inherent uncertainty in any attempt to map such ES.

1.1 INTRODUCTION

The term "ecosystem services (ES)" popped up in the 1970s and gained steam over the next few decades. On the other hand, the notion that ES benefits well-being is much older. Plato, in 400 BCE, documented the link between deforestation and water supply. The ES includes processes and instances facilitated by ecosystem and biodiversity to improve human existence (Gomiero, 2016; Mazzoni et al., 2017; Padilla, 2021). Processes contributing to the production of consumable goods (e.g., food and timber), processes that control and sustain life (i.e., crop pollination, storm surge protection, and carbon sequestration), and conditions that improve life's recreational, esthetic, and spiritual values, as well as undiscovered medicinal benefits from plants, are all examples of ES (Butler et al., 2019; Levine & Chan, 2013; Maes et al., 2011; Mulder et al., 2015). The ecosystem performs a variety of functions. Provisioning services provide people with food, materials, and energy (Heidenstrøm & Hebrok, 2022; Watz et al., 2022). Regulating services are ecosystems that regulate other environmental media or processes (Naselli-Flores & Padisák, 2022; Urlich et al., 2022; Yin et al., 2022). Services that meet people's spiritual or cultural needs are cultural (Milner et al., 2020; Riechers et al., 2019). Supporting services are activities and functions in the ecosystem that aid the other three types of services (Das et al., 2022; Jones et al., 2022). All these demand modeling for ES ensure demand does not exceed supply. Thus, modeling may communicate and assist decision-making by showing and quantifying the geographical mismatch between ecosystem service supply and demand. Multiple ES supply has recently been modeled at global and provisional levels (Zhao et al., 2022). Recently published frameworks integrated ecological and economic value aspects of ES to compute monetary values (Bostian & Lundgren, 2022; Grizzetti et al., 2016; Kapsalis et al., 2019). These tools show how to model and map ES at various sizes. As a result, both what is modeled and how it is modeled are unknown. As part of a larger sustainability strategy, it is necessary to understand better where and what services are supplied by a particular piece of land, landscape, area, state, continent, and worldwide (Canedoli et al., 2022; Marando et al., 2022; Pettorelli et al., 2018; Tulbure et al., 2022; Xu et al., 2022). The ecosystem service evaluations should be based on an extensible, conceptual framework. For practical purposes, it urges increased modeling of:

- Service flow;
- Ecosystem services, including cultural and regulatory functions, to better understand trade-offs;
- Biodiversity and ultimate benefit.

There is an obvious need to create a design and set of standards for modeling ecosystem service stocks, fluxes, and supply and demand. We examine existing ecosystem service mapping technologies and previous studies to assess the current state of the art.

The ES concerning the soil includes modeling land use and land cover, soil structural complexities in foretelling the soil's buffering or rather filtering capacity; modeling of soil habitat and biodiversity in connection with soil function models; simulating soil development processes over prolonged timeframes; modeling soil nutrients, upscale activities of the rhizosphere, and soil–root interactions; integrate new insights on structural dynamics of the ES (Barker & Fisher, 2019; Höök et al., 2020; Mason-Jones et al., 2022; Morris et al., 2022; Schwartz, 2020; Sokol et al., 2022; Yang et al., 2022c). Soil provides ES such as biomass production for food, fiber, energy, habitat, and physical support (Löbmann et al., 2022; Petsch et al., 2022). It is possible to investigate the chronological scales of climate control, i.e., short-term storage of soil water impacts the climatic patterns (e.g., rainfall, droughts, and heat waves), whereas soil carbon sequestration in the long-term impacts GHG emissions (Wehrli et al., 2019). The influence of soil on plant communities that affect climate, land use patterns, surface albedo, and nutrient dynamics are examples of mechanistic feedback linked to soil regulatory function (Berdugo et al., 2022; Heijmans et al., 2022). In the segment on soil formation, the interaction of soil and climate was highlighted and quantified in several soil formation models. According to new research, new research, droughts, floods, and heat waves are all caused by soil processes (water content). The role of the soil either as a carbon source or sink and other greenhouse gases (GHGs) over decadal to millennial time scales is critical (Nottingham et al., 2022). Estimated soil carbon reserves vary in quantity and stability depending on the function and characteristics of soil (with climate and land use). Because of the fast melting of permafrost soils in temperate zones, enormous land areas may quickly become substantial GHG producers. Weathering rocks and soil formation sustain flora, accumulate carbon, and thus regulate global temperature during geologic periods. Many human activities, such as tillage and land clearing (grasslands and forests), have increased CO_2 emissions, with much of the emissions mediated by soil microbes (Hartmann et al., 2013; Navarro-Pedreño et al., 2021).

Furthermore, increased fertilizer use has led to significant atmospheric nitrous oxide emissions, limiting nutrient use efficiency and contributing directly to global warming (Pires et al., 2015; Tian et al., 2020). Scientists must study the interconnected soil system, integrating upcoming climatic patterns with physical, chemical, and biological processes, and include information on state-of-the-art soil processes into existing and functioning models of the terrestrial biosphere so that models can be used to assess the GHG emissions affected by management, and land-use practices (Brüggemann et al., 2011; Dang & Lovell, 2016; Lappalainen et al., 2022; Li et al., 2022; Sajjad et al., 2022; Wen et al., 2022). However, current models lack adequate parameterization and an understanding of the underlying mechanisms. Advanced soil modeling platforms employ existing information and feedback for improved performance and forecast of integrated soil systems. Our study examines the techniques used to model and quantify ES. We provide our early findings and describe the methods utilized to model various soil and ecosystem models for the major ES (Table 1.1). We end by discussing the potential applications of our methodology and the inherent uncertainty in any attempt to map such ES.

TABLE 1.1 Various Models for Soil and Ecosystem Services

Tool and the Author	Description
The soil plant atmosphere system model (SPASMO) (Hardie et al., 2022)	• It uses a process-based method to describe soil dynamics, crop production, and farming techniques. • SPASMO allows for the assessment of soil service provision by investigating the impact of management practices on growth of plant, soil processes, and its properties. • The model's core consists of supporting and degradation processes.
The modeling soil ecosystem services (MOSES) (Rawat et al., 2022)	• A soil model based on processes providing information well about soil profile. • MOSES focuses on functions related to soil ecosystem services, such as biomass productivity and carbon sequestration. • The results are used to judge how effectively soil ecosystem services are being delivered.
The nature value explorer (Castellar et al., 2021)	• A web-based application was used to calculate both cultural and regulating ecosystem services. • Ecosystem services can be estimated using this tool, which can be used to determine how land use change affects both the regulating and cultural functions provided by ecosystems.
Toolkit for ecosystem service at site-based assessment (TESSA) (Peh et al., 2014)	• Measurement and monitoring of ecosystem services at the site level with low-cost tools. • With the use of this tool, users may choose which ES should be evaluated, what information is required to quantify them, what sources or methodologies to utilize in various situations, and how to present the findings.
Artificial intelligence for ecosystem services (ARIES) (Hamann et al., 2021)	• Web application and ES modeling methodology for the evaluation of environmental effects and the illumination of their values to assist environmental decision-making. • Using model integration technologies makes it possible to assemble customized models from an expanding model base, which is beneficial to the methodology.
The integrated valuation of ecosystem services and tradeoffs (InVEST) (Yu et al., 2022)	• Multiple software models for mapping and valuing natural resources. • In these models, the ecosystem's structure and function are used to explain how the flows and values of ecosystem services are affected by the ecosystem's production functions. Models for terrestrial, freshwater, and marine ecosystems are currently available in the toolbox.
The multiscale integrated models of ecosystem services (MIMES) (Oliveira et al., 2022)	• Land use change and its long-term effects on ecosystem services can be studied at the global, regional, and local levels using a set of models. • Formalizing material exchanges between natural, built, and social capitals can be done using MIMES. This information is then used to create a system model that predicts the ES and human use dynamics in various scenarios.

1.2 MODELING LAND USE AND LAND COVER CHANGES

It is estimated that over a 1,000 LULC models are reported in the literature in domains such as landscape ecology and geographic information science (Singleton et al., 2022; Zaman-Ul-Haq et al., 2022). Disparities in academic viewpoints, methodological methods, data availability, and modeling aims hamper efforts to classify models. Several evaluations of LULC models have been published recently, each with a unique approach and resulting typologies (Anderle et al., 2022; Jatayu et al., 2022; Jozdani et al., 2022; Souza et al., 2022). Models of land use change predict potential changes in land use according to historical patterns and causes (Leta et al., 2021). Early land use change modeling aimed to anticipate tropical deforestation and its numerous effects. An interdisciplinary approach to change, including social, ecological, and information sciences, is critical to understand the ecological and human elements of land use and land cover change (Chen et al., 2022c; Magness et al., 2022). Models of land use change that describe and assign changes in land use and land cover are a key component of integrated land change research (Borrelli et al., 2017; Li et al., 2017; Ren et al., 2022; Wang & Wang, 2022; Zope et al., 2016). Because there is a dynamic linking of land use and land cover, such models are among the most potent tools for combining human and biophysical subsystems. Land transformation models are categorized by application or size, utilizing an array of data sources (Pandey et al., 2021). Models for Land use change have traditionally utilized survey and census data but also consist of spatial data on land management actions and socio-economic variables. Such geographical data is frequently utilized in geographic information systems (GIS) (software systems that store, retrieve, modify, analyze, and display georeferenced data). Data is being collected throughout time to help determine land change trajectories. They usually provide explanations of historical and current land usage and predictions of future land use.

Non-iterative equations are used in the relatively simple land change models to meet a particular solution which describes represented system as either static or in equilibrium (Tassi et al., 2021; Zhao et al., 2021). System models typically characterize relevant details, information, or energy stocks and fluxes as configurations of differential calculus connected by data and functions (Kuznik et al., 2022; Meunier et al., 2022; Temesgen et al., 2015). Time moves in small increments when differential equations are numerically solved, enabling dynamic modeling of variable feedback. Statistical models may undervalue the importance of human and institutional decisions if not theoretically grounded (Smith et al., 2022a). Expert models communicate qualitative information quantitatively, typically to predict where certain land will emerge (Gunawardana et al., 2022; Mallampalli et al., 2016). Expert and rule-based knowledge systems use logical rules and data to enable models to deal with new situations. Using an independent ecological model, we can evaluate the effects of potential land changes. Probabilities for these branching rules are not calculated statistically but rather by interviewing many land managers and synthesizing their responses. Artificial neural networks (ANNs) that are imparted to correlate performance with various parameters, such as population density or distance to water bodies, are currently the most promising models.

An increasing number of land use and land cover change models utilize cellular modeling techniques operating on a grid of congruent cells. Cells in the two-dimensional grid can only have one of a few states, each representing a different land use. Future states are determined by transition rules depending on the condition of the nearby neighborhood. In Markov modeling, the cells' states in the lattice are based on initial cell states in a probabilistic but simple way (Ghalehteimouri et al., 2022). Cellular models can simulate a land change in human-environment systems (Rienow et al., 2022). Agent-based models are a newer land change modeling approach. They are groups of adaptive autonomous entities (such as farmers, road builders, or village elders) that use information from the surroundings to guide their actions, such as perception, planning, and learning. Agent-based models are typically used in conjunction with other land change models, and their strength lies in their ability to represent a wide range of agents, including agent interaction and communication (Stieler et al., 2022; Yang et al., 2022a). However, agent-based models may be challenging to apply since they frequently provide findings that are not generalizable. Many of these models are still under-evaluated; thus, more effort has to be made to create standard modeling platforms and validate agent-based models. Likewise, these are frequently employed at smaller spatial scales and must be enlarged for complete ecological evaluation. Convergent and time-dependent connections are challenging to describe using simultaneous equations as they require familiar parameters throughout time scales and equilibrium assumptions, while system models may effectively describe dynamic systems (Bettencourt, 2020; Milano et al., 2018; Williams & Johnson, 2013). However, system models have limitations due to the difficulty of expressing parameter complexity in linked equations, and unless the system is represented at a small size or in-depth, equations may restrict system models to statistically idealized flows and stocks (Burchard, 2002). Because important system dynamics are often glossed over or assumed away, the modeler usually makes assumptions about the aggregate consequences of behavior. Despite this, assumptions like observation independence must be carefully checked, particularly in light of spatial or temporal autocorrelation and data aggregation, which combine data from various sources or scales. Researchers are now focusing on spatial data statistical issues, reducing parameter estimation bias and inefficiencies, and utilizing spatial and temporal autocorrelation to guide model development (Krisztin et al., 2022; Song et al., 2014; Staab et al., 2022). Expert models can convert qualitative expert domain knowledge into quantitative forms. Although these models work because of their logical underpinnings, it may be difficult to incorporate all elements of the issue domain, leading to discrepancies in model findings.

Land use and land cover change may be predicted using evolutionary models, and their main purpose is to find patterns and connections in high-dimensional, noisy, and stochastic settings. Statistical techniques, for example, make it easier to identify causation and correlation structures than evolutionary methods, which may quickly build complex computer systems that achieve excellent outcomes without explaining how or why. In summary, land use change modeling ranges from simple equation-based models to complex and computationally expensive models. These techniques are increasingly being integrated and hybridized to address multidisciplinary problems, spatiotemporal scale concerns, and land change complexity. GEOMOD2 uses statistical modeling, systems thinking, and expert decision-making to forecast regional land change. CLUE family of models simulates land change and related processes at regional sizes. Finally, another example is integrated assessment models (IAM), which connect continental and global terrestrial environments.

1.3 MODELING CHANGES IN SOIL ORGANIC MATTER (SOM)

Organic matter (OM) in the soil is a valuable natural resource essential for soil and ecosystem functions on various scales, from soil fertility and water-holding capacity to biogeochemical cycles globally (Osborne et al., 2022; Sokol et al., 2022). Furthermore, it is a highly complex substance that is highly vulnerable to direct and indirect human impacts. Simulation models are useful in soil organic matter (SOM) research as they provide a mathematical framework for integrating, examining, and testing the understanding of SOM dynamics (Bashir et al., 2016, 2021; Gerber, 2016; Liang et al., 2017; Mohi-Ud-Din et al., 2017; Rahman et al., 2022). SOM simulation models are progressively being used in "applied" settings to assess the impacts of humans on ecosystem functions as well as for the management of SOM for GHG mitigation, enhanced soil health, and long-term resource use (Hessel et al., 2022; Libutti et al., 2021). In this regard, it's critical to keep a strong link between scientific advances in SOM modeling approaches and applications of the SOM model. The level of model hypotheses and usage in SOM dynamics must be carefully considered, as different scales impose different limitations. Developing models at one scale and applying them at another can produce erroneous results, which is a major consideration in policy and land management decision-making, in which data supporting model evaluations may be unavailable at the scale of a decision being examined (Höchtl et al., 2016; Renschler & Harbor, 2002). The 'toolbox' of mathematical approaches in SOM modeling has recently been seen to expand slightly at any scale. Instead, advances are largely the result of a new fundamental understanding of SOM dynamics, though they can also be attributed to increased data availability or advancements in computational systems (Mouloodi et al., 2022). Many SOM models have multiple pools for mass loss and use first-order decay kinetics. The organic C equilibrium in the soil is determined by the balance of C inputs from plant roots and C outputs from microbial decomposition.

For land use and soil management planning, a number of models exists for predicting changes in SOC over time and space. Simplistic models consider plant productivity, climate, and soil texture for decomposition turnover and stabilization (Manning et al., 2015). Roth C, DAISY, and CENTURY models are such examples. There are no models for inorganic C in soil that can be used to guide policy and decision-making. CASPER (Carbonate Accumulation in Soils through the Prediction of Elemental Release) is a different model from SOC in that it stimulates the one-way assimilation of inorganic C irreversibly from CO_2 when calcium-bearing precursors are present. As model pools do not always depict quantifiable fractions in the field, it isn't easy to initialize them with measured data. Measurements or subjective-based initialization can result in state variable values that are significantly different from those obtained by a model running with any good consolidation of data and input variables for estimating current trends (Demetriou, 2017; Poore et al., 2016). The consequences can be simulations having incredibly high carbon loss or gain (even chaotic trajectory can be seen in some models). Many models avoid such behavior by simulating previous land use and management using survey data and expert knowledge. To achieve more reasonable iterations of state variables, historical data are essential for measuring recent trends and relatively brief differences in soil carbon, especially in applications where the assumption of steady-state equilibrium is violated. Table 1.2 illustrates the model-specific effects on the decomposition process.

1.4 MODELING IMPACTS ON COASTAL ECOSYSTEMS

Climate change is believed to have a significant influence on ecosystems and their components. Coastal ecosystem predictions in the short-term, such as forecasting hazardous algal bloom landfall, are comparable to weather and hurricane forecasts, which can impact human health (Ross et al., 2020). Contrarily, macroeconomic forecasting is comparable to predicting large-scale, long-term ecological changes using expert judgment and analysis and numerical simulations and predictions (Hammad & Falchetta, 2022; Sitar-Taut & Mican, 2022). Forecasting broad-based, long-term impacts are critical because persistent, subtle influences can cause several of the most drastic and long-lasting influences on ecosystems (Smith & Mair, 2022). One aspect of coastal forecasting seems to have a long history due to the widespread use of modeling in assessing fisheries stock. Climate change, social change, and coastal governance systems should predict long-term shifts in the ecosystem's functions. The decision maker's values and the data from the model's physical, chemical, biological, spatial, and temporal models will impact the model's outcomes. The main direct drivers of coastal ecosystems are habitat modification, introduction of non-native species, hydrologic, and hydrodynamic disruption, resource exploitation, eutrophication, global climate change, shoreline erosion, pathogens, and toxins (Elko et al., 2022; Gómez et al., 2022; Lepage et al., 2022). A meta-data gateway and an analytical system will be included in the new coastal forecasting model's meta-data gateway. It is an algorithm that compiles information from several sources into a format that the analysis technique can use. Archives, satellites, and on-site measurement devices can contribute to the data. It is necessary to analyze the physical, chemical, and biological data to determine the root causes using an array of interactive deterministic multi-modal (Abreu et al., 2022; Brunton et al., 2016).

Interactive or linked multi-modals comprise a collection of stand-alone models that interact with each other to forecast the probability of a result (e.g., a harmful bloom of algae). The ultimate result is typically a GIS or mapping module visual presentation. Empirical models based on local experience and

TABLE 1.2 Impact of Environmental Factors on Modeled Decomposition

Environmental Factors	Modeled Decomposition					
	Yasso	**ROMUL**	**SOILN**	**Roth C**	**Forest-DNDC**	**CENTURY**
Nitrogen	–	The variability of N is determined in conjunction with the dynamics of SOM (soil C).	C and N models which are coupled dynamically.	–	Explicit N model, with a DNDC-inspired algorithm.	C and N models which are coupled dynamically.
Temperature and moisture	Empirical regression models developed on basis of litter bag experiments are used to attune D-rates.	Empirical regression models based on laboratory data.	Empirical functions and simulated daily soil temperature and moisture content is used to derive D-rates.	Rate modifier multiplying the D-rates of active compartments is affected by temperature and soil moisture deficit.	Potential D, rates of nitrification or denitrification are modified using a combined moisture and temperature factor.	The effects of soil moisture and temperature on D were modeled using the DEFAC relationship (multiplier with combined temperature and moisture effects).
Nutrients (excl. N)	–	–	–	–	–	P, S optional sub-models.
Texture	–	In SCLISS, it has an effect on the stable humus mineralization rate and soil moisture.	Affects decomposition rate, N leaching, denitrification.	Affects the stabilization of SOM and the retention of soil water.	Aeration, water movement, soil moisture, ammonium absorption, decomposition, and other factors are all affected.	Affects moisture availability and decomposition rates.

theory-based mechanistic models describe events in strictly physical terms. Common but restricted usage empirical models are not explored further. Instead, we need models that can accept human decision-making value judgments. The current state of the art does not give coastal managers enough time to act to avoid possible negative long-term consequences. That is not the purpose of this evaluation. It seeks to provide state of the art regarding construction scenarios influencing coastal managers' choices for long-term planning.

1.5 MODELING CLIMATE CHANGE (GLOBAL CLIMATE MODELS: GCMS)

General circulation models (GCMs) are the main source of regional information on possible future climates. Due to advancements in the resolution of models, simulation of important processes for regional change, and the broadening set of available simulations, a better view of the vital factors of regional climate change is evolving (Chen et al., 2020; Gao & O'Neill, 2020). The development of probabilistic information at regional scales from GCM simulations has progressed, but these methods are still in the exploratory stage. Less progress is there for applying this to downscaled regional data. To better understand and predict the climate's behavior on various time scales, climatic models are a useful tool (Rolnick et al., 2022). There is an investigation into whether the reported climate changes result from natural variability, human activity, or both. Using their findings and projections, local, regional, and national decision-makers can make more informed choices concerning water resource management, agriculture, transportation, and urban planning (Bouska et al., 2022; Dalu & Wasserman, 2021; Feiner et al., 2022). Most experts consider future climates to be linked to changing global precipitation. Though such changes have an impact at all spatial scales, variations in daily precipitation characteristics (including extreme events) are crucial for impact assessment at local and regional scales.

The climate model treats the following major components of the climate system:

1. **Atmospheric:** Clouds and aerosols, or rather the global transfer of heat and water, are simulated by this component.
2. **Land Surface Component:** Surface features, such as vegetation, soil moisture, snow cover, rivers, and carbon sequestration, are simulated using this model.

3. **Ocean Component:** It predicts the movement and mixing of current and its biogeochemistry, as the climate system's primary heat and carbon reservoir is the ocean. The sea ice component controls the absorption of solar radiation and heat and water exchanges between the air and the sea.

The climate models divide the world into a 3D grid of cells corresponding to specific geographic locations and elevations. For climatic factors like temperature, precipitation, etc., equations are calculated on the global grid by every component, including the atmosphere, land surface, sea ice and ocean (Sein et al., 2015). The various parts of the model swap heat, water, and momentum fluxes in addition to the computation of how they have been gradually changing over time. Their interaction with each another is a coupled system. In today's climate models, two types of processes are used: simulated and parameterized (Bakhshi et al., 2022; Chakraborty et al., 2022; Liang et al., 2022; Pant et al., 2022). Simulated processes are based on well-established scientific principles (i.e., conservation of energy, mass, and momentum) and are larger than grid-scale. A simulated process representing tropical cyclones and storm occurrence is one such example. Parameterized processes use simpler processes to correspond to complex processes smaller than grid-scale and thus can't be represented physically (Xia et al., 2022). The representation of the composition of clouds and aerosols is an example of a parameterized process.

GCMs are the most effective tool for estimating precipitation for climate change scenarios, but they do not remediate small spatial scales. Numerous organizations utilize regional climatic models to simulate the climate in almost all areas worldwide (Hamed et al., 2022; Yuan et al., 2022). Regional climate models can be used in conjunction with global climate models (GCMs), which provide greater clarity, more extensive modeling runs, and the development of local climate change models (Golub et al., 2022; Rolnick et al., 2022). Increased resolution allows a more accurate portrayal of landscape elements like hills, estuaries, lakes, and other surface characteristics. GCMs attempt to shed light on the ebb and flow of the climate system over timescales ranging from months to centuries (He, 2022). In addition, their outputs have contributed to various evaluations of the effects of large-scale climatic variation, changes in natural resources, human health, infrastructure, and commerce (Awan, 2022; St-Laurent et al., 2022). It was decided to develop RCMs to downscale climate fields derived from GCMs with coarse resolution so that data at finer, sub-GCMs grid scales could be collected and applied to VIA (vulnerability, impacts, and adaptation) evaluations more effectively. Regional modeling, also known as dynamical downscaling, accomplishes this by utilizing physically based models with a limited spatial scope. It is a process that incorporates local physical factors like topography to simulate high spatial resolution climatic information.

Due to advancements in modeling and consideration of physical processes of the climate system, we now have highly reliable regional climate change projections for various regions of the globe. While AOGCMs (atmosphere-ocean general circulation models) continue to be the foundation for projections, downscaling techniques provide further valuable detail (Robison & Scanlon, 2018). AAOGCMs cannot provide information at finer scales than their computational grid, and the processes at unresolved scales are crucial. Higher resolution in dynamical models or empirical statistical downscaling can be used to provide information at finer scales. The development of downscaling methodologies is still a priority. Only recently have downscaled climate change projections designed for specific needs become available. Downscaling methods have been designed to generate regional or local scale information on the climate from GCMs in overcoming the resolution disparity among GCMs and impact assessment models (Fiedler et al., 2021; Hernanz et al., 2022; Jaiswal et al., 2022). Variations in large-scale climate variables are transferred to regional or local scale meteorological parameters (predictands) using dynamical or statistical downscaling techniques. Dynamical downscaling is the dynamic extrapolation of the impact on regional or local scales of interest by large-scale climate processes using regional simulations of high-resolution (Martinez-García et al., 2021; Yu & Liu, 2021). Such relationships are then applied to GCM results for the transformation of climate model outputs to products that are statistically refined and are frequently considered to suit better for usage as a study of the impact of input on the regional or local climate.

Regional climate modeling initiatives have served as a bridge between hydrology, glaciers, energy, agriculture, forestry, spatial planning, and air quality research in several individual studies and within coordinated programs (Orr et al., 2022; Teo et al., 2019). Stakeholder engagement may help make RCM findings more relevant in impact research and adaptation initiatives. But the constraints of RCM in terms of sampling (number of runs), systemic biases, and so forth must be disclosed and acknowledged. Additionally, the distinction between modeled (gridded) and point-wise data of observed stations is problematic since various consumers of climate data are acquainted with the latter (McNally et al., 2022). The granularity of RCMs enables the use of more realistic climatic data to constrain models of regional or local impact. As previously stated, any systemic biases must still be considered. As a result, differences deduced from RCM between a modeled future period and a present-day period are now used in more notable transient runs, not only in regard to time-slice runs (Mottram et al., 2019; Paeth et al., 2011; Zampieri et al., 2012). The advantages are self-evident, but so is the risk of being trapped in a previous trajectory of fluctuation and extremes. This may be refined further if the RCM bias structure is established. Rescaling or otherwise minimizing systematic biases and directly using time series from RCMs is an emerging option for delta-changing. This solves the problem of implementing simulated variability and extremes when regional climate and impact models are interfacing. RCMs provide information to findings of global modeling or other global data and are essential in situations where fine-scale processes must be accounted for in impact models.

1.5.1 BIAS CORRECTION PERFORMANCE INDICATORS FOR APPROPRIATE CLIMATE MODEL SELECTION

Several performance indicators, like the coefficient of determination (R2) and root mean square error (RMSE) between historical GCMs data and observed ground data of respective GCMs (after bias correction), are used to evaluate the effectiveness of the bias correction method. R2 ranges between 0 and 1, where values near 1 indicate better agreement between the two sets of data. R2 greater than 0.5 is usually considered to be a fair approximation. The RMSE is the most frequently used test of error statistics, where a value near zero signifies that the model is performing better. The best models are thought to have as much variance as the observation, the minimum RMSE, and the highest correlation. GCM with a significantly greater agreement is selected based on the degree of agreement on performance indicators.

Representative concentration pathways (RCP) scenarios are the climate projection scenarios used in the recent IPCC 5th Assessment Report (AR5). RCP2.6, RCP4.5, RCP6.0, and RCP 8.5 are four distinct climate change projection scenarios that consider the impact of variability in GHG emissions, atmospheric concentrations of aerosols, land-use change, population changes, GDP changes, and overall technological advancement. The RCP' primary goal is to provide atmospheric GHG concentration projections in a time-dependent manner. Climate variables from the scenarios RCP2.6, RCP4.5, and RCP8.5 were obtained for use in climate projections for all

GCMs to encompass the low, medium, and high ends of future climate projections (Ashrafzadeh et al., 2022; Jia et al., 2022; Li et al., 2018; Lu et al., 2017; Mottram et al., 2019; Pan et al., 2020; Sok et al., 2022; Tenfie et al., 2022). The scenarios of RCP depict how the planet is likely to advance in terms of GHG emissions in the future. RCPs are the fundamental concept, and they are measured in watts per square meter of radioactive forcing W/m^2. RCP 2.6 is a low or peak-and-decline scenario wherein radiative forcing peaks in the mid-21st century and declines gradually to a sensible range of 2.6 W/m^2. RCP4.5 is a stabilization scenario leading to moderate GHG emissions levels (Hope, 2020; Rao, 2019). RCP8.5 is a high scenario that describes radiative forcing all through the 21st century before actually peaking at around 8.5 W/m^2 by the end of the century.

1.6 MODELING BIODIVERSITY

For many, the ethical imperative is to protect the world's biological diversity (genes, species, and ecosystems). 'ES, or the processes and operations which preserve and improve human well-being, are also supported by biodiversity. ES like pollination, biological control of pests in agroecosystems, and food and feed can be provided by biodiversity (Maslo et al., 2022; Steward et al., 2014; Thiesen et al., 2022). Cropping systems can remain productive while reducing their reliance on external inputs that contribute to harmful environmental impacts. Producers' farms, moderately natural habitats, human transportation systems (e.g., roads), and occasional wildlife ecosystems provide a broad spectrum of services, such as food and fiber, carbon capture, and recycling of nutrients (Cavender-Bares et al., 2018; Feeley et al., 2017; Frouz & Frouzová, 2022; Scialabba, 2022; Shugart et al., 2022). This means a landscape that is used for agriculture can be as little as a few acres or as vast as a whole continent. Ecological, economic, and social benefits may be gained by preserving biodiversity in agricultural landscapes (Macfadyen et al., 2012; Montagnini et al., 2022). However, the landscape's advantages are dwindling due to the widespread reduction in species and persons of huge sizes. One of the main causes of the decline in landscape biodiversity in agriculture. Agricultural land expansion, loss of landscape heterogeneity, elevated pesticide and fertilizer use, and the shift to low-diversity crop systems have all had significant impacts on global biodiversity in the last few decades (Isbell et al., 2017; Jafari et al., 2022; Nelson et al., 2022; Van Der Werf & Bianchi, 2022). As a result of these modifications, fewer services are being offered to the public. Due to agricultural intensification, farmland birds are undergoing a massive drop in all parts of the globe. Because of the drastic decline in farmland bird populations, such as the skylark, the ES they provide, such as reducing the number of weed seeds, have suffered (Priya et al., 2022). Pesticides have reduced the diversity of aquatic invertebrates, like stream invertebrates, by as much as 42% in Europe and Australia (Liess & Ohe, 2005). Associated with global drivers, like market requirements for more cost-effective commodities and the latest policy to produce bioenergy on agricultural land, the processes deleterious to species diversity are the consequence of augmentation of production (Bowles et al., 2018; Gabrielle et al., 2014). Farm landscapes can host a wide range of wild biodiversity while not negatively impacting agricultural production or livelihoods. Landscape-level techniques to stabilize manufacturing and preservation have been fiercely disputed, with the most predominant discussion concentrating on land sharing versus preservation. Biodiversity safeguard regions such as preservation agreements or indigenous reservoirs can be established globally or at field scales ranging from local to global. Studies of biodiversity at a larger scale are necessary because agriculture has ecological impacts on biodiversity that are more than just at the field scale (Batáry et al., 2017; Pe'er et al., 2022). An important consideration in landscape design is the distribution of agricultural and natural features relative to one another and their spatial configuration. Planners need to consider the socioeconomic aspects of agricultural landscapes when examining the future of agricultural landscapes for biodiversity considerations. Farm landscapes are multi-stakeholder adaptive socio-ecological systems, with farmers, local and regional government agencies, and non-governmental organizations (NGOs) all having a role to play (NGOs) (Amin & Shammin, 2022; Dumont et al., 2019; Sorge et al., 2022). As a result, both the difficulty of dealing with biodiversity and the number of factors influencing landscape characteristics are exacerbated. Socioeconomic, political, technological, natural, and cultural factors can all play a role (Agnoletti, 2014; Blanco et al., 2022). These factors are constantly changing and are neither fixed in space nor time. The state of biodiversity in landscapes is altered due to shifts in vehicle drivers. European Common Policy's agri-environmental schemes for biodiversity protection are one example (CAP). Scientists are working on ways to simultaneously monitor changes in biodiversity and ES in agricultural landscapes to keep track of changes in biodiversity (e.g., production of food, esthetic value) (Chen et al., 2022b; Gordon et al., 2010; Reboud et al., 2022). Scenarios can be an effective tool for predicting how biodiversity will respond to various future paths of human development and policy choices in the future. Different modeling approaches for agricultural landscape design and biodiversity assessment simplify the landscape and process that influence biodiversity in models (Lin & Fuller, 2013; Perović et al., 2022; Santos et al., 2016). The key characteristics of approaches, such as their spatial characteristics and landscape depiction, the kind of biodiversity investigated, scenario approaches, and the kind of modeling techniques used to design and evaluate landscapes, can provide more insight into the nature and sophistication of approaches.

1.7 MODELING IMPACTS ON HUMAN HEALTH

It's difficult to predict the effects of future ecosystem change on human health globally and for the next century. This job is so ambiguous that it could be considered impossible or, at the very least, unscientific. On the other hand, uncertainty is not infinite, and many boundary conditions can be discovered. This would include not only the continuity of anatomy, physiology, and pathophysiology but also the fact that human health will continue to be affected and altered by social and technological influences as it had been for millennia (Bordoni & Zanier, 2014; Bragazzi et al., 2019; Damari & Ehsani Chimeh, 2017). While there is much ambiguity regarding the features of even current global health, a great deal is known. The traditional epidemiological and bioclimatic malarial and even animal illnesses exemplify one dominating method of simulating human health and disease (Li et al., 2021). These methods usually begin with a thorough examination of a single illness and one host, progressing to forecasts of epidemic critical thresholds and their consequences for public policy (Gomes et al., 2022; Leach et al., 2022; Marion et al., 2022). Because temperature variations affect the survival and behavior of vectors, their habitats may be established or destroyed by land use changes, so land use changes or climatic changes are incredibly significant when vectors carry illnesses such as malaria (transmitted by mosquitoes). Even from the perspective of classical epidemiology, there is currently a trend toward simulating multiple species simultaneously, e.g., there is growing recognition that climate and habitat disruption may play a role in the emergence of many novel human illnesses, even though no models of this process exist at this time (Robbins, 2019). Due to the limits of single-species illness models, a more phenomenological or aggregate approach has been emphasized for global evaluations, emphasizing the links between population demography, social behavior, and poverty. Because three critical "black box" factors crucial to this job are uncertain (Matthiopoulos et al., 2020), these boxes stand for inputs, outputs, and moderating factors, wherein input stands for ecosystems, output for health, and moderators for cultural, technical, and political variables that can either reduce or increase the impact of ecosystem degradation on health (Bangroo et al., 2021; Bashir et al., 2021; Bou Zeineddine & Pratto, 2017; Fiksel, 2015; Rather et al., 2019; Sedlacko et al., 2014; Wang et al., 2021; Wurth et al., 2022). In reality, these two groups are not entirely distinct. These three groups are constantly interacting, but the causal connections are frequently subtle, underappreciated, and in some cases, insignificant when they are not near thresholds. There are infinite connections and cascade interactions between ecological services, human health, and the cultures and institutions that modify and impact both health and the ecosystem (Table 1.3).

One approach has been suggested for categorizing possible adverse reactions into four main categories: "direct," "mediated," "modulated," and "systems failure" impacts. Direct health impacts are defined in this category as those caused by the failure of a beneficial ecosystem function, like the supply of adequate food, clean water, fertile soil, or the prevention of floods and erosion (Crawford et al., 2022; Hoekstra et al., 2018). Such impacts are caused by physical factors and do not contain infections in and of themselves. They are perhaps the easiest of the four impacts to comprehend. Compared to direct impacts, mediated effects have a greater degree of causal intricacy and in certain instances, including pathogens. Certain types may be associated with a significant risk of morbidity and death. Additionally, there is a wider gap between ecological change and resulting health than there is for direct impacts. This area includes a variety of infectious and some chronic illnesses. In many of these instances, illness developed as a consequence of ecosystems' enhanced ability to produce food (a provisioning ecosystem function), as a result of animal domestication, irrigation, dams, and other intensive agricultural techniques (Giagnocavo et al., 2022; Millhauser & Earle, 2022; Rohr et al., 2019). The unanticipated rise in the frequency and prevalence of many infectious illnesses has been a trade-off. As large-scale, more delayed, and plausibly complicated negative repercussions of ecosystem change "modulated," and "systems failure" effects have been discovered. Periods of state failure or eventual or actual large-scale social and economic collapse are associated with modulated impacts. The term "systems failure" refers to the social and economic breakdown occurring at the supranational level due to converging, modulated impacts. This

TABLE 1.3 Ecosystem Services and Their Biophysical Structures

Ecosystem Service	Ecosystem Function	Biophysical Structure/ Process of Greenspace	Example Health Outcome	References
Climate regulation	Solar radiation reflection, evaporation, and transpiration, and carbon sequestration.	Shrub, tree, and plantation areas.	Reduced rates of mortality in cities.	Anderson & Gough (2022); Bashir et al. (2021); Kumar et al. (2022a)
Waste services	Hair or foliage wax traps particulates on the leaf surface. CO_2 uptake during photosynthesis.	Shrub, tree, and plantation areas.	Decrease of respiratory and cardiovascular diseases.	Ram et al. (2015); Sharma & Saxena (2022)
Prevention of floods	Water retention by the canopy, rainfall retention by the soil, reduced runoff, and discharge.	Multi-level grassland, soil, and vegetation.	Improved water quality; decreased mental health issues and financial costs related to exposure to flooding.	Newman et al. (2022); Richards et al. (2022)
Recreation	Biodiversity and primary productivity.	Biodiverse; processes; accommodation areas.	Possibilities for leisure and exercise; reducing stress; lowering of blood pressure, obesity, and diabetes.	Smith et al. (2022b); Von Hertzen et al. (2015)

categorization system may be imaginatively applied to various ecological scenarios. For instance, if an area suffers a significant loss of provisioning services, immediate detrimental consequences are expected, such as flood interruption or increased hunger rates. Reduced biodiversity and increased interaction between people and non-domesticated species may increase mediated effects, such as new infectious illnesses (Albery et al., 2022). Enhanced intensification of animal husbandry could help spread recombinant versions of some familiar diseases like influenza. More accurate and meaningful forecasts of the impact of ecosystem change on health will necessitate a better understanding of, or at the very least an attempt to anticipate, the social, institutional, and technical variables influencing health — the 3rd of the black boxes mentioned previously (Abelson, 1986; He & Harris, 2020; Hoque & Sorwar, 2017). In some cases, like when there is a regulation of health-related impacts, health can adversely affect the quality of these human services. Modulated effects are less likely to occur in civilizations with strong institutions, rational government, and advanced technology, to be sure. However, adverse health consequences of ecosystem change, and decreased ES may overwhelm already fragile civilizations in certain instances, pushing them beyond a "tipping point" beyond which collapse is very probable (Haldon et al., 2018; Marques, 2020; Obbard, 2010). Although much less probable, it is conceivable to envision paths and causal webs that put civilizations' social, economic, and health fabric that seems virtually impervious to adverse environmental change under strain. Traditionally, scenario theorists amplify and extend current patterns into the future, envisioning and simulating various outcomes. With the scenario method, future assumptions are usually "bundled," as in "TechnoGarden," where global problems are anticipated to be addressed by worldwide technological innovation. But a future characterized by fast technological advancement may harm human health if other kinds of advancement do not precede it. Vaccines, surgery, and medicines may be unable to recompense for bad health entirely if advanced technology also dehumanizes society (Batelaan, 2022). Second, technology, environment, and social systems influence human health. Predicting how they'll interact without theory and data is difficult. The notion that climate change would alter income distribution and destabilize civilizations is not incorporated into these health projections (Zhu & Zhang, 2022). Reductionist science seeks to simplify and understand reality by focusing on individual components. This method has been extremely effective in numerous areas and has greatly aided our time's scientific, technical, and material development (Mobus, 2022). While financial development does not guarantee human health or happiness, it contributes to both. But reductionist approaches have limitations. Their main flaw is that they obscure the importance of threshold effects and more importantly, the notion of thresholds. A small change is insignificant from a linear, reductionist perspective. If an experimenter determines that a change causes an undesirable reaction, linear thinking suggests merely removing the change (Craney & Surles, 2002; Grummon et al., 2022). Reversal is difficult or excessively costly beyond many thresholds in ecosystems, society, and human health. While this expensive or impossible reversibility is increasingly acknowledged in ecology, little is known about how it affects health and society (Hook et al., 2014). Critical study topics include connecting the various strands that influence human health. We still lack fundamental knowledge on how the environment affects illness epidemiology. Human systems react to health risks, but global health forecasts do not predict potential adaptive responses. Social system breakdown may negatively impact human health (Figure 1.1). Our capacity to foresee system breakdowns, which impose pressure on healthcare systems, is limited. Many possibilities exist to translate particular epidemiological model achievements into genuine global evaluations. The rise of ecosystem research and biodiversity inspired new models with important ecological foundations. Attempts to use epidemiological models to scale up and aggregate across illnesses might improve global health assessments.

1.8 INTEGRATED ASSESSMENT MODELS (IAMS)

IAMs is a term used to describe a type of scientific modeling that attempts to connect the key features of economy and society with the biodiversity and atmosphere into a single modeling framework (Krumm et al., 2022; Pyka et al., 2022). When making informed decisions about the future of human and social development, integrated assessment modeling is a useful tool (Purvis et al., 2022; Sacchi et al., 2022). Many environmental issues include environmental, social, and economic aspects. Because disciplinary methods cannot adequately explain these webs, interest in integrated assessment models (IAMs) has grown. They define IAMs as modeling frameworks for organizing and structuring scientific information to evaluate cause-effect connections. The analyzes should cover various issues and incorporate interactions with other issues. These models describe the socioeconomic system (economic activity, human behavior, population dynamics, resource usage) and also how it interacts with the environmental system (regional air

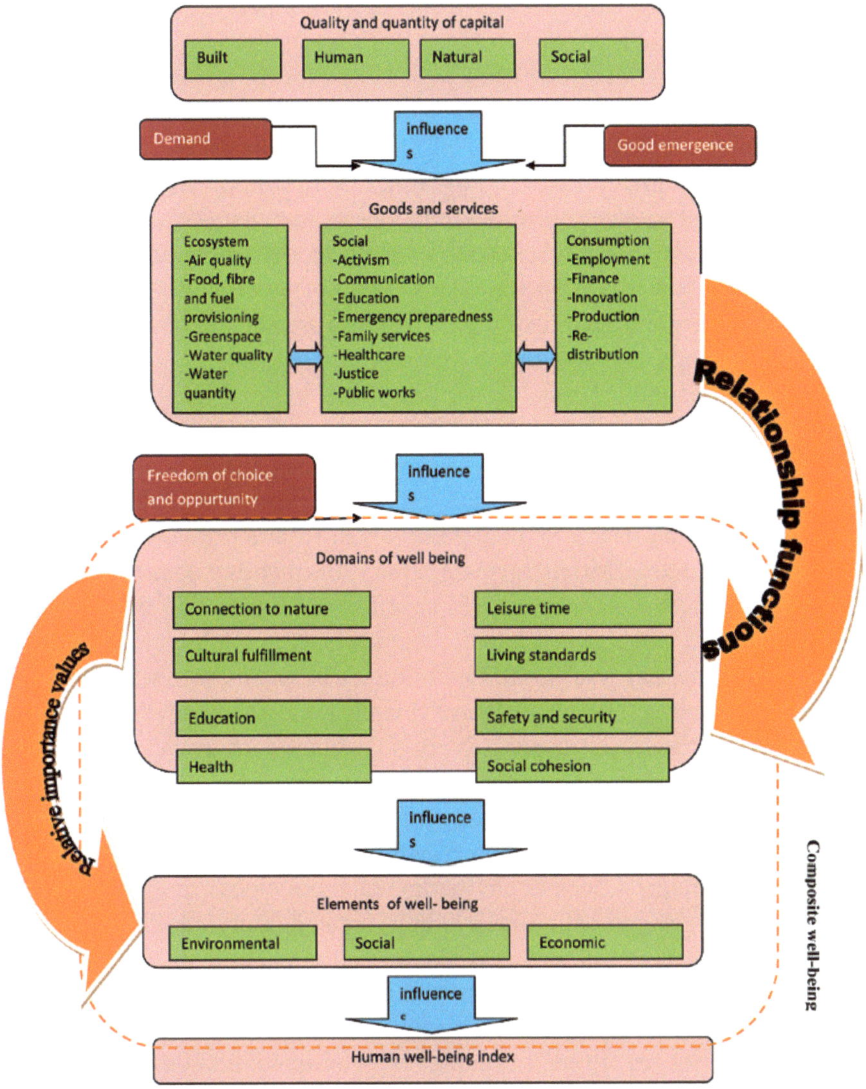

FIGURE 1.1 Ecosystem functioning and its various governing factors.

pollution, climatic system, land cover/use, etc.) (Hritonenko & Yatsenko, 2022; Karutz et al., 2022; Kumar et al., 2022b; Liu et al., 2021, 2022; Muriithi, 2016). IAMs are used to organize and synthesize scientific knowledge from many fields. IAMs are distinguished by their emphasis on policy support and evaluation rather than pure scientific research (Tavoni & Valente, 2022).

They are used for policy assessment, early warning systems, and direct assistance for public decision-making and negotiation.

IAMs are classified in several ways. The prevailing modeling paradigm is first class. IAMs may be simulated or optimized. Most simulation IAMs use differential equations (Brinkerink et al., 2022; Yang et al., 2022b). Partial optimization is also utilized. They utilize optimization methods to assess the lowest cost or other objective functions of various paths. IAMs like this are dynamic cost-benefit assessments. Both modeling frameworks employ deterministic and stochastic methods, though the former dominates. The degree to which external restrictions or goals are satisfied may be differentiated inside optimization IAMs (Stern & Stiglitz, 2022; Żebrowski et al., 2022). The second category is horizontal and vertical integration. Vertical integration is the model's ability to encompass the whole chain of cause-effect of a problem, from driving factors to pressures, state changes, effects, and potential solutions (Chakori et al., 2022; Kayikci et al., 2022; Kosamu et al., 2022). Environmental policy has evolved from a pressure-state-impact-response paradigm. Horizontal integration is the merging of several elements of a study object. This may be very specific, like integrating water and land cover/land use, or very broad, like integrating demographic/health problems with environmental issues (Adhikari et al., 2022). Another way to classify them is by subject. This typically reflects the policy's intended applicability. Many IAMs have been used to study acidifying contaminants and GHGs. Cost modules and allocation algorithms may make IAMs policy-relevant in addressing climate change (Marangoni et al., 2017). Significant uncertainties are population and economic growth, socio-political decisions affecting human development (such as environmental regulations), and discount rates. Uncertainty management tools have been created. Storyline-based scenarios are a useful tool for creating a set of consistent assumptions. Others include conventional uncertainty analysis and model result qualifying. The area of IA modeling is, by definition, vast and ambiguous. Because it is fresh, universal norms and principles have not yet been solidified. This section's summary is thus eclectic and focused on particular instances. We do not aim to cover all possible models. Comprehensive overviews exist for numerous applications, such as integrated climate change assessment (Kasperson et al., 2022; Rolnick et al., 2022). Regarding air pollution and climate change, IAMs were the first to catch on. Many models developed in this area in the late 1980s and 1990s can be used to assess global environmental change or sustainable development because of their long-time horizons and global perspectives. Global Environment Outlook (United Nations Environment

Program with a Foreword by Kofi Annan, 2002) and IPCC Special Report on Emissions Scenarios have used these models. In response to the concerns about global trends, the first recognized IAM was founded in the early 1970s, expressed by the Club of Rome, a group of businessmen and government officials (Schmelzer, 2012, 2017). On a global basis, the computer simulation model World defined links between the world system's main demographic, resource, and economic components (Montanari et al., 2013; Wei et al., 2015). It made use of the Massachusetts Institute of Technology's technique of system dynamics (developed from the science of electrical engineering). Its main goal was to raise understanding of the nature of exponential growth in a constrained environment, as well as the systemic nature of observed and projected trends, due to the multitude of connections between what were once thought to be separate processes (Arancibia, 2013; Capellán-Pérez et al., 2020; Sterman, 2018). It emphasized the dangers of continuing on business-as-usual growth paths to the point where only the doomsday message was heard. This model and subsequent attempts at more disaggregated models aided in developing a comprehensive set of derivative models. It also ignited a heated debate about the nature of market mechanisms, something many economists believe would solve most problems even before the emergence of a disaster. The second generation of IAMs focused more narrowly on a specific environmental issue and policy options for addressing it (Babbitt et al., 2022; Fremstad & Paul, 2022; Stern, 2022). This was partly in response to the first-generation model's aggregate structure, which made meaningful validation and policy support implausible. In this regard, the regional acidification information and simulation (RAINS) acidification model developed in Europe in the 1980s—is an outstanding example. It continued to play a significant role in European international air pollution negotiations. In the late 1970s, IAMs took their first steps in addressing the causes and effects of climate change (Zenios, 2022). Since then, over 50 IAMs related to climate change have been formed. To search for cost-effective mitigation strategies, economists have combined the neoclassical economic growth model and a simple climate system model (Campiglio and van der Ploeg, 2022). Among the new features are a more complex energy system, a profound treatment of technological dynamics, and its integration with impact modules. Climate change or environmental models gradually shift to global change or sustainable development models in these third-generation IAMs (Capellán-Pérez et al., 2020). For discussing IAMs in MA scenarios, three evaluative criteria were established: Ecosystems interact with other elements of the system (water, land, and atmosphere), as well as people and the economy.

We grouped three IAM models into three categories. One category includes models developed specifically to provide comprehensive cognizance of various environmental, social and economic elements of sustainability. The other category includes models focusing on economic development, energy, and climate interplay. Models in the third category are a subset of those in the first. These models started as energy-environment models, but they've since evolved into a third category: global change models. The IAM models on regional air pollution management were excluded due to their limited emphasis. It's worth noting that the RAINS modeling team is currently working on expanding their framework to include not only acidification but also eutrophication and ground-level ozone.

1.8.1 MODELS OF SUSTAINABLE DEVELOPMENT

This IAM category has the most integrated social, economic, and environmental problems. They utilize a lot of aggregation to prevent too much complexity. A low degree of geographical or regional disaggregation generates expert model-derived meta-level descriptions of underlying processes. World3, TARGETS, Threshold 21, Polestar system and GUMBO, are also part of this group. To avoid these trade-offs, most of these models highlight potential trade-offs between economic growth and ecological functioning but do not provide specific solutions or policy recommendations (Kirchner et al., 2015). In the MA setting, these models benefit from high integration and much feedback. However, they frequently describe complex (ecological) processes in simple terms. The TARGETS model created three coherent views on sustainable development (D'Amore et al., 2022). One of its five sub-models mimics important biogeochemical cycles. The dynamics of food, water, energy supply and demand are aggregated. The model's high aggregate level allows for obvious links to ES. Unlike most previous global models, TARGETS incorporated a complete connection from environmental change to demographic (including health) and economic changes (Awan, 2022; Chen et al., 2022a; Pradhan et al., 2017). This model lacks geographical disaggregation and like other IAMs in this category, is not linked to a particular decision-making process. The Stockholm Environment Institute's Boston Centre created Polestar, whose most famous use was with the Global Scenarios Group. The model is built on a large data collection of social, economic, and environmental factors. As it has few connections between its variables, it's more of an accounting framework for testing different assumptions.

Polestar has assisted user-driven scenario creation processes, including scenarios that explain ecological processes (Wissert, 2022). It is more suited to investigating potential futures and the effect of specific policy actions than providing comprehensive and balanced solutions to political decision-making questions (Gordon et al., 2022; Hunt et al., 2012). The GUMBO (Global Unified Met model of the Biosphere) met model was created due to the characterization of ecological services and their future potential progress under various assumptions (Jackson et al., 2013). It synthesizes and simplifies both current dynamic global models in scientific and social sciences. It considers versatile interactions between human technology, economic output, and ecologic products and services. Carbon, water, and nutrient flow in environmental and biological systems are all simulated. GUMBO links 11 biomes (coastal ocean, Open Ocean, grasslands, wetlands, woods, lakes/rivers, tundra, deserts, ice/rock, and farmland) geologically to one another. The World Bank's approach to sustainable development and measures of sustainable social welfare include the varied types of capital (human, built, social, and natural). IAMs had great success fighting climate change. There are several IAMs, and their findings are routinely presented and debated with decision-makers. Their impact on policy has been apparent. Over 50 IAMs now exist that focus on the intersection of economic growth, energy consumption, and climate change (Stern, 2022). Few of them mention associated problems like air pollution and resource depletion. These models are often called 3E models (Economy, Energy, and Environment) (de Carvalho et al., 2015; Laimon et al., 2022; Oliveira et al., 2016). There is a significant distinction between basic macroeconomic models and models containing comprehensive energy sector descriptions in this category. The DICE and FUND models are examples of aggregated meta-models attempting to link the economic drivers of climate change and their effects on conducting a cost-benefit analysis (Callicott et al., 2006). However, these model models' descriptions of ecological roles are vague and abstract. DICE presents climate change as a global restricted set of temperature equations with fewer damaging functions. The FUND model has evolved to include additional aspects of climate change effects, such as disease transmission. The latest IPCC Special Report on Emission Scenarios included most of these models (ASF, MESSAGEMACRO, MARIA, MiniCAM, AIM, and IMAGE). Other notable IAM models are ICAM and MIT's Integrated Assessment Model. The ICAM model stresses uncertainty. The MIT model is presumably the most advanced in the field and is a collection of state-of-the-art disciplinary models.

1.9 KEY GAPS IN OUR MODELING ABILITIES

Many models flaws stem from data or model constraints. For example, we often fail to integrate spatially explicit data. In many instances, there are no crucial feedback. Almost every area of forecasting has examples. Food supply affects land use patterns, affecting ES, climate change, and future food supplies. Changes in land usage affect the climate, which modifies the kinds of plants that may be grown on any given piece of land. This kind of input is missing. This implies predictions are better made sooner rather than later before feedback can reverberate. The MA predictions are for 50 years, which may be too short to allow for important feedback. If this feedback are significant, our models may be far off. It also lacks models and theories that predict thresholds and the ones that cause major changes in the system or even its collapse when crossed. It is well known that a species can be hunted or fished to extinction. A short prediction time frame does not protect us against thresholds since we may be near some thresholds now. As a result, improving models should focus on anticipating thresholds. Finally, model openness and decision-makers usage of models are issued. Although modelers will often get the technical aspects of the modeling process right, they overlook the fact that evaluation tools must be integrated into existing social and political processes (Fletcher et al., 2022; Fodstad et al., 2022; Kyker-Snowman et al., 2022; McKenna et al., 2022; Süsser et al., 2022). Because most global ecosystems have so many nonlinearities, time lags, and feedback, more research is needed on how to ask modeling questions relevant to policy and other processes. Table 1.4 enlists the pros and cons of modeling.

Several IAMs have predicted environmental services. Ecological services, such as food, water, and biogeochemical cycles, are regulated and delivered this way. On the other hand, only a few models have been developed to provide information about "sustainable development." Existing models only cover a few services. Using the criteria above, we may draw various conclusions. While most models have strong integration for the specific (environmental) issues they were designed for, their integration for the MA's goal is still poor. There is ecological feedback on socioeconomic forces. For global evaluations, better-describing connections and feedback from a sustainable development viewpoint (rather than a single problem focus) would be beneficial. A worldwide effort to clarify these links and prioritize them within a "sustainability science" framework may be useful. Simultaneously extensive and detailed models are unrealistic. Thus, layered methods may significantly enhance current work. More comprehensive yet aggregated models drive dedicated (and therefore more detailed) models. Aggregation

TABLE 1.4 Pros and Cons of Ecosystem Modeling

Environmental Goals	Pros	Cons
An agricultural landscape with a wide range of crops.	Models of crop growth models are available for scenario modeling along with multiple crop types.	These simulations are often complex, requiring a lot of parameterizations, and the results must be interpreted carefully.
Thriving wetlands	There have been established soil models designed specifically for simulating peat soils.	The status of peat land is hard to ascertain for these models' baseline data, that has a significant impact on the results' accuracy.
Good-quality groundwater	Various processes related to erosion and chemical transport are incorporated into catchment-scale soil models, which are flexible and comparatively accurate while operating at low spatial and temporal resolution.	Models pertinent to this goal do not consider all probable concerning chemicals (though a wide range is covered), and soil hydrology models do not typically include snow and ice implementations.
Decreased impact of climate	There is a great variety of soil carbon and nitrogen models, as well as active frameworks for integrating soil models into global circulation models.	Climate models are already at the simple end of the spectrum, and they don't do a good job of incorporating some critical processes (such as soil hydrology).

Level Ecosystems and ES are best understood as nested dynamical processes happening at different scales. Models must span several spatial and temporal dimensions to understand their reaction to direct and indirect disturbances induced by humans (Tong et al., 2022). Given the MA's aim, models must account for regional/local differences. With the introduction of new GIS tools and an abundance of field and satellite data, a strong trend toward investing in models of better spatial resolution arises (Radočaj et al., 2022; Zahawi et al., 2015). It comes at the expense, as less global or regional coverage is usually the result. Moreover, as globalization continues, many economic activities may have ever-wider implications for ecological disruption. Understanding regional disparities and interregional connections, particularly commerce, will be a key study topic for integrated modeling in the future years.

Ecosystem models are widely used in scenario development and analysis. Models should be used solely to create and arrange quantitative predictions. As shown in the IPCC's SRES scenario study, descriptive narratives may effectively communicate a scenario's wider importance. Finally, given their complexity and decision-making emphasis, ecosystem models are uncertain. Uncertainties include parameter variability, model specification inaccuracies, and unknown model limits. Although uncertainty analysis is incorporated in many IAMs, it is often either partly or not. Again, a layered approach to IAM may be beneficial, with global models guiding regional (ecological) models. If IAMs are to be utilized for policy-relevant evaluation, they should include as many elements of sustainable development as possible. Inevitably, certain aspects of ecological change go unmentioned. Most IAMs lack or weaken links between environmental change and human growth in broad economic and socio-political terms. Institutionally, they are famously weak. This reflects the increasing interconnectedness of economic and social systems. No consensus currently exists on how to link the different levels of a vertically integrated paradigm. In particular, sophisticated simulation models of socio-ecological systems may point the way ahead. Another approach is to incorporate qualitative tales and back them with quantitative models. For example, marine habitats and coastal zones have a dearth of IAMs. This is true for many ecological services, including regulatory and cultural functions.

1.10 CONCLUSION

As part of a larger sustainability strategy, it becomes necessary to understand better where and what services are supplied by a particular piece of land, landscape, area, state, continent, and worldwide. Modeling communicates and assists decision-making by showing and quantifying the geographical mismatch between ecosystem service supply and demand. We have assessed the current state of the art by examining existing ecosystem service mapping technologies and previous studies. We reviewed modeling to assess land use and land cover changes, changes in SOM, impact on coastal ecosystems, climate change, biodiversity, human health, etc. Researchers should delve into the interconnected soil ecosystem, integrating emerging climate patterns with physical, chemical, and biological processes. It's essential to incorporate cutting-edge information on soil processes into current operational models of the terrestrial biosphere. This integration is crucial for using these models to evaluate the greenhouse gas emissions impacted by management and land use practices. Advancements in understanding soil dynamics and their interaction with climate change are fundamental in accurately assessing and mitigating the environmental impacts of human activities on the planet. Land use and land cover change models are among the most potent tools for combining human and biophysical subsystems where non-iterative equations are used in the relatively simple land change models to meet a particular solution. These may be predicted using evolutionary models, and their main purpose is to find patterns and connections in high-dimensional, noisy, and stochastic settings. An increasing number of land use and land cover change models utilize cellular modeling techniques operating on a grid of congruent cells. SOM simulation models have been progressively used in "applied" settings to assess humans' impacts on ecosystem functions and to manage SOM for GHG mitigation, enhanced soil health, and long-term resource use. GCMs are the main source of regional information on possible future climates. The major components of climate models include atmosphere, land surface, ocean, and sea ice. Climate models employ a three-dimensional grid system that divides the world into cells representing precise geographic locations and elevations. These models calculate equations for various climatic factors such as temperature, precipitation, and more across the global grid. Each component, encompassing the atmosphere, land surface, sea ice, and oceans, contributes to the computation of these factors. This grid-based approach allows for a comprehensive analysis of the interactions. Regional climate models, when used in conjunction with GCMs, provide greater clarity, more extensive modeling runs, and the development of local climate change models. RCP scenarios are the climate projection scenarios used in the recent IPCC 5th Assessment Report, where the primary goal is to provide projections of atmospheric.

GHG concentration in a time-dependent manner. The scenarios of RCP depict how the planet is likely to advance in terms of GHG emissions in the future. Forecasting the impact of forthcoming alterations in ecosystems on global human health for the next century is a formidable task. The influence stems from the intricate relationship between temperature fluctuations, vector survival, and behavior. Changes in land use or climate can potentially establish or disrupt habitats crucial for vectors, like mosquitoes, responsible for transmitting diseases such as malaria. Hence, the significance of these alterations, whether in land use or climate, cannot be overstated when considering their role in disease transmission by vectors. There are infinite connections and cascade interactions between ecological services, human health, and the cultures and institutions that modify and impact health and the ecosystem, which modeling approaches can examine. However, the best models are those that have as much variance as the observation, the minimum RMSE, and the highest correlation.

KEYWORDS

- **bias correction performance indicators**
- **biodiversity**
- **global climate models**
- **human health**
- **integrated assessment models**
- **soil organic matter**
- **sustainable development**

REFERENCES

Abelson, M. A., (1986). Strategic management of turnover: A model for the health service administrator. *Health Care Management Review*, 61–71.

Abreu, C. M., Costa-Silva, B., Reis, R. L., Kundu, S. C., & Caballero, D., (2022). Microfluidic platforms for extracellular vesicle isolation, analysis, and therapy in cancer. *Lab on a Chip, 22*, 1093–1125.

Adhikari, B., Mishra, S. R., & Schwarz, R., (2022). Transforming Nepal's primary health care delivery system in global health era: Addressing historical and current implementation challenges. *Globalization and Health, 18*, 1–12.

Agnoletti, M., (2014). Rural landscape, nature conservation and culture: Some notes on research trends and management approaches from a (Southern) European perspective. *Landscape and Urban Planning, 126*, 66–73.

Albery, G. F., Carlson, C. J., Cohen, L. E., Eskew, E. A., Gibb, R., Ryan, S. J., Sweeny, A. R., & Becker, D. J., (2022). Urban-adapted mammal species have more known pathogens. *Nature Ecology & Evolution*, 1–8.

Amin, R., & Shammin, M. R., (2022). A resilience framework for climate adaptation: The Shyamnagar experience. *Climate Change and Community Resilience* (pp. 69–84). Springer, Singapore.

Anderle, M., Paniccia, C., Brambilla, M., Hilpold, A., Volani, S., Tasser, E., Seeber, J., & Tappeiner, U., (2022). The contribution of landscape features, climate, and topography in shaping taxonomical and functional diversity of avian communities in a heterogeneous Alpine region. *Oecologia*, 1–14.

Anderson, V., & Gough, W. A., (2022). Nature-based cooling potential: A multi-type green infrastructure evaluation in Toronto, Ontario, Canada. *International Journal of Biometeorology, 66*, 397–410.

Arancibia, F., (2013). Challenging the bioeconomy: The dynamics of collective action in Argentina. *Technology in Society, 35,* 79–92.

Ashrafzadeh, M. R., Khosravi, R., Mohammadi, A., Naghipour, A. A., Khoshnamvand, H., Haidarian, M., & Penteriani, V., (2022). Modeling climate change impacts on the distribution of an endangered brown bear population in its critical habitat in Iran. *Science of The Total Environment, 837*, 155753.

Awan, U., (2022). Industrial ecology in support of sustainable development goals. *Responsible Consumption and Production* (pp. 370–380). Springer.

Babbitt, C. W., Neff, R. A., Roe, B. E., Siddiqui, S., Chavis, C., & Trabold, T. A., (2022). Transforming wasted food will require systemic and sustainable infrastructure innovations. *Current Opinion in Environmental Sustainability, 54*, 101151.

Bakhshi, S., Chenaghlou, M. R., Rahimian, F. P., Edwards, D. J., & Dawood, N., (2022). Integrated BIM and DfMA parametric and algorithmic design-based collaboration for supporting client engagement within offsite construction. *Automation in Construction, 133*, 104015.

Bangroo, S. A., Sofi, J. A., Bhat, M. I., Mir, S. A., Mubarak, T., & Bashir, O., (2021). Quantifying spatial variability of soil properties in apple orchards of Kashmir, India, using geospatial techniques. *Arabian Journal of Geosciences, 14*, 1–10.

Barker, T., & Fisher, J., (2019). Ecosystem health as the basis for human health. In: Selendy, J. M. H., Farmer, P., & Fawzi, W., (eds), *Water and Sanitation Related Diseases and the Changing Environment: Challenges, Interventions, and Preventive Measures* (pp. 245–270).

Bashir, O., Ali, T., Baba, Z. A., Rather, G., Bangroo, S., Mukhtar, S. D., Naik, N., et al., (2021). Soil organic matter and its impact on soil properties and nutrient status. *Microbiota and Biofertilizers* (Vol 2, pp. 129–159). Springer.

Bashir, O., Khan, K., Hakeem, K. R., Mir, N. A., Rather, G. H., & Mohiuddin, R., (2016). Soil microbe diversity and root exudates as important aspects of rhizosphere ecosystem. *Plant, Soil and Microbes* (pp. 337–357). Springer.

Batáry, P., Gallé, R., Riesch, F., Fischer, C., Dormann, C. F., Mußhoff, O., Császár, P., et al., (2017). The former iron curtain still drives biodiversity–profit trade-offs in German agriculture. *Nature Ecology & Evolution, 1*, 1279–1284.

Batelaan, K., (2022). 'It's not the science we distrust; it's the scientists': Reframing the anti-vaccination movement within black communities. *Global Public Health, 17*, 1099–1112.

Berdugo, M., Vidiella, B., Solé, R. V., & Maestre, F. T., (2022). Ecological mechanisms underlying aridity thresholds in global drylands. *Functional Ecology, 36*, 4–23.

Bettencourt, L. M., (2020). Urban growth and the emergent statistics of cities. *Science Advances, 6*, eaat8812.

Blanco, J., Bellón, B., Barthelemy, L., Camus, B., Jaffre, L., Masson, A. S., Masure, A., et al., (2022). A novel ecosystem (dis) service cascade model to navigate sustainability problems and its application in a changing agricultural landscape in Brazil. *Sustainability Science, 17*, 105–119.

Bordoni, B., & Zanier, E., (2014). Skin, fascias, and scars: Symptoms and systemic connections. *Journal of Multidisciplinary Healthcare, 7*, 11.

Borrelli, P., Robinson, D. A., Fleischer, L. R., Lugato, E., Ballabio, C., Alewell, C., Meusburger, K., et al., (2017). An assessment of the global impact of 21st century land use change on soil erosion. *Nature Communications, 8*, 1–13.

Bostian, M., & Lundgren, T., (2022). Valuing ecosystem services for agricultural TFP: A review of best practices, challenges, and recommendations. *Sustainability, 14*, 3035.

Bou, Z. F., & Pratto, F., (2017). The need for power and the power of need: An ecological approach for political psychology. *Political Psychology, 38*, 3–35.

Bouska, K. L., De Jager, N. R., & Houser, J. N., (2022). Resisting-accepting-directing: Ecosystem management guided by an ecological resilience assessment. *Environmental Management*, 1–20.

Bowles, T. M., Atallah, S. S., Campbell, E. E., Gaudin, A., Wieder, W. R., & Grandy, A. S., (2018). Addressing agricultural nitrogen losses in a changing climate. *Nature Sustainability, 1*, 399–408.

Bragazzi, N. L., Sellami, M., Salem, I., Conic, R., Kimak, M., Pigatto, P. D. M., & Damiani, G., (2019). Fasting and its impact on skin anatomy, physiology, and physiopathology: A comprehensive review of the literature. *Nutrients, 11*, 249.

Brinkerink, M., Zakeri, B., Huppmann, D., Glynn, J., Gallachóir, B. Ó., & Deane, P., (2022). Assessing global climate change mitigation scenarios from a power system perspective using a novel multi-model framework. *Environmental Modeling & Software, 150*, 105336.

Brüggemann, N., Gessler, A., Kayler, Z., Keel, S., Badeck, F., Barthel, M., Boeckx, P., et al., (2011). Carbon allocation and carbon isotope fluxes in the plant-soil-atmosphere continuum: A review. *Biogeosciences, 8*, 3457–3489.

Brunton, S. L., Proctor, J. L., & Kutz, J. N., (2016). Discovering governing equations from data by sparse identification of nonlinear dynamical systems. *Proceedings of the National Academy of Sciences, 113*, 3932–3937.

Burchard, H., (2002). *Applied Turbulence Modeling in Marine Waters*. Springer Science & Business Media.

Butler, C. D., Higgs, K., & McFarlane, R. A., (2019). Environmental health, planetary boundaries, and limits to growth. *Encyclopedia of Environmental Health*, 533.

Callicott, J. B., Acevedo, M., Gunter, P., Harcombe, P., Lindquist, C., & Monticino, M., (2006). Biocomplexity in the big thicket. *Ethics, Place and Environment, 9*, 21–45.

Campiglio, E., & Van, D. P. F., (2022). Macrofinancial risks of the transition to a low-carbon economy. *Review of Environmental Economics and Policy, 16*, 000–000.

Canedoli, C., Ficetola, G. F., Corengia, D., Tognini, P., Ferrario, A., & Padoa-Schioppa, E., (2022). Integrating landscape ecology and the assessment of ecosystem services in the study of karst areas. *Landscape Ecology, 37*, 347–365.

Capellán-Pérez, I., De Blas, I., Nieto, J., De Castro, C., Miguel, L. J., Carpintero, O., Mediavilla, M., et al., (2020). MEDEAS: A new modeling framework integrating global biophysical and socioeconomic constraints. *Energy & Environmental Science, 13*, 986–1017.

Castellar, J. A., Popartan, L. A., Pueyo-Ros, J., Atanasova, N., Langergraber, G., Säumel, I., Corominas, L., et al., (2021). Nature-based solutions in the urban context: Terminology, classification and scoring for urban challenges and ecosystem services. *Science of The Total Environment, 779*, 146237.

Cavender-Bares, J., Arroyo, M., Abell, R., Ackerly, D., Ackerman, D., Arim, M., Belnap, J., et al., (2018). *Status and Trends of Biodiversity and Ecosystem Functions Underpinning Nature's Benefit to People* (pp. 207–363). Secretariat of the intergovernmental science-policy platform on biodiversity and ecosystem services. https://www.ipbes.net/system/tdf/ipbes-6-inf-4-rev.1.pdf?file=1&type=node&id=16517 (accessed on 30 November 2023).

Chakori, S., Richards, R., Smith, C., Hudson, N. J., & Aziz, A. A., (2022). Taking a whole-of-system approach to food packaging reduction. *Journal of Cleaner Production, 338*, 130632.

Chakraborty, T., Pattnaik, S., Baisya, H., & Vishwakarma, V., (2022). Investigation of ocean sub-surface processes in tropical cyclone Phailin using a coupled modeling framework: Sensitivity to ocean conditions. *Oceans* (pp. 364–388). Multidisciplinary Digital Publishing Institute.

Chen, D., Zhao, Q., Jiang, P., & Li, M., (2022a). Incorporating ecosystem services to assess progress towards sustainable development goals: A case study of the Yangtze River economic belt, China. *Science of the Total Environment, 806*, 151277.

Chen, G., Li, X., Liu, X., Chen, Y., Liang, X., Leng, J., Xu, X., et al., (2020). Global projections of future urban land expansion under shared socioeconomic pathways. *Nature Communications, 11*, 1–12.

Chen, H., Fleskens, L., Schild, J., Moolenaar, S., Wang, F., & Ritsema, C., (2022b). Impacts of large-scale landscape restoration on spatio-temporal dynamics of ecosystem services in the Chinese loess plateau. *Landscape Ecology, 37*, 329–346.

Chen, J., John, R., Yuan, J., Mack, E. A., Groisman, P., Allington, G., Wu, J., et al., (2022c). Sustainability challenges for the social-environmental systems across the Asian drylands belt. *Environmental Research Letters, 17*, 023001.

Craney, T. A., & Surles, J. G., (2002). Model-dependent variance inflation factor cutoff values. *Quality Engineering, 14*, 391–403.

Crawford, S. E., Brinkmann, M., Ouellet, J. D., Lehmkuhl, F., Reicherter, K., Schwarzbauer, J., Bellanova, P., et al., (2022). Remobilization of pollutants during extreme flood events poses severe risks to human and environmental health. *Journal of Hazardous Materials, 421*, 126691.

D'Amore, G., Di Vaio, A., Balsalobre-Lorente, D., & Boccia, F., (2022). Artificial intelligence in the water–energy–food model: A holistic approach towards sustainable development goals. *Sustainability, 14*, 867.

Dalu, T., & Wasserman, R. J., (2021). *Fundamentals of Tropical Freshwater Wetlands: From Ecology to Conservation Management*. Elsevier.

Damari, B., & Ehsani, C. E., (2017). Public health activist skills pyramid: A model for implementing health in all policies. *Social Work in Public Health, 32*, 407–420.

Dang, H., & Lovell, C. R., (2016). Microbial surface colonization and biofilm development in marine environments. *Microbiology and Molecular Biology Reviews, 80*, 91–138.
Das, S., Kumar, S. P., Bera, B., & Adhikary, P. P., (2022). Effect of urbanization on the dynamics of ecosystem services: An analysis for decision making in Kolkata urban agglomeration. *Urban Ecosystems*, 1–19.
De Carvalho, A. L., Antunes, C. H., Freire, F., & Henriques, C. O., (2015). A hybrid input–output multi-objective model to assess economic–energy–environment trade-offs in Brazil. *Energy, 82*, 769–785.
Demetriou, D., (2017). A spatially based artificial neural network mass valuation model for land consolidation. *Environment and Planning B: Urban Analytics and City Science, 44*, 864–883.
Dumont, E. S., Bonhomme, S., Pagella, T. F., & Sinclair, F. L., (2019). Structured stakeholder engagement leads to development of more diverse and inclusive agroforestry options. *Experimental Agriculture, 55*, 252–274.
Elko, N., Foster, D., Kleinheinz, G., Raubenheimer, B., Brander, S., Kinzelman, J., Kritzer, J. P., et al., (2022). Human and ecosystem health in coastal systems. *Shore & Beach, 90*, 64.
Feeley, H. B., Bruen, M., Bullock, C., Christie, M., Kelly, F., & Kelly-Quinn, M., (2017). *ESManage Program: Irish Freshwater Resources and Assessment of Ecosystem Services Provision.* Environmental Protection Agency.
Feiner, Z. S., Shultz, A. D., Sass, G. G., Trudeau, A., Mitro, M. G., Dassow, C. J., Latzka, A. W., et al., (2022). Resist-accept-direct (RAD) considerations for climate change adaptation in fisheries: The Wisconsin experience. *Fisheries Management and Ecology.*
Fiedler, T., Pitman, A. J., Mackenzie, K., Wood, N., Jakob, C., & Perkins-Kirkpatrick, S. E., (2021). Business risk and the emergence of climate analytics. *Nature Climate Change, 11*, 87–94.
Fiksel, J., (2015). *Resilient by Design: Creating Businesses that Adapt and Flourish in a Changing World.* Island Press.
Fletcher, S., Hadjimichael, A., Quinn, J., Osman, K., Giuliani, M., Gold, D., Figueroa, A. J., & Gordon, B., (2022). Equity in water resources planning: A path forward for decision support modelers. *Journal of Water Resources Planning and Management, 148*, 02522005.
Fodstad, M., Del Granado, P. C., Hellemo, L., Knudsen, B. R., Pisciella, P., Silvast, A., Bordin, C., et al., (2022). Next frontiers in energy system modeling: A review on challenges and the state of the art. *Renewable and Sustainable Energy Reviews, 160*, 112246.
Fremstad, A., & Paul, M., (2022). Neoliberalism and climate change: How the free-market myth has prevented climate action. *Ecological Economics, 197*, 107353.
Frouz, J., & Frouzová, J., (2022). *Forestry and Hunting, Applied Ecology* (pp. 221–314). Springer.
Gabrielle, B., Bamière, L., Caldes, N., De Cara, S., Decocq, G., Ferchaud, F., Loyce, C., et al., (2014). Paving the way for sustainable bioenergy in Europe: Technological options and research avenues for large-scale biomass feedstock supply. *Renewable and Sustainable Energy Reviews, 33,* 11–25.
Gao, J., & O'Neill, B. C., (2020). Mapping global urban land for the 21st century with data-driven simulations and shared socioeconomic pathways. *Nature Communications, 11*, 1–12.
Gerber, A., (2016). Short-term success versus long-term failure: A simulation-based approach for understanding the potential of Zambia's fertilizer subsidy program in enhancing maize availability. *Sustainability, 8*, 1036.

Ghalehteimouri, K. J., Shamsoddini, A., Mousavi, M. N., Ros, F. B. C., & Khedmatzadeh, A., (2022). Predicting spatial and decadal of land use and land cover change using integrated cellular automata Markov chain model-based scenarios (2019–2049) Zarriné-Rūd River Basin in Iran. *Environmental Challenges, 6*, 100399.

Giagnocavo, C., De Cara-García, M., González, M., Juan, M., Marín-Guirao, J. I., Mehrabi, S., Rodríguez, E., et al., (2022). Reconnecting farmers with nature through agroecological transitions: Interacting niches and experimentation and the role of agricultural knowledge and innovation systems. *Agriculture, 12*, 137.

Golub, M., Thiery, W., Marcé, R., Pierson, D., Vanderkelen, I., Mercado-Bettin, D., Woolway, R. I., et al., (2022). A framework for ensemble modeling of climate change impacts on lakes worldwide: The ISIMIP Lake sector. *Geoscientific Model Development, 15*, 4597–4623.

Gomes, M. G. M., Ferreira, M. U., Corder, R. M., King, J. G., Souto-Maior, C., Penha-Gonçalves, C., Gonçalves, G., et al., (2022). Individual variation in susceptibility or exposure to SARS-CoV-2 lowers the herd immunity threshold. *Journal of Theoretical Biology, 540*, 111063.

Gómez, I., Silva, R., Lithgow, D., Rodríguez, J., Banaszak, A. T., & Van, T. B., (2022). A review of disturbances to the ecosystems of the Mexican Caribbean, their causes, and consequences. *Journal of Marine Science and Engineering, 10*, 644.

Gomiero, T., (2016). Soil degradation, land scarcity and food security: Reviewing a complex challenge. *Sustainability, 8*, 281.

Gordon, G., Rieder, B., & Sileno, G., (2022). On mapping values in AI governance. *Computer Law & Security Review, 46*, 105712.

Gordon, L. J., Finlayson, C. M., & Falkenmark, M., (2010). Managing water in agriculture for food production and other ecosystem services. *Agricultural Water Management, 97*, 512–519.

Grizzetti, B., Lanzanova, D., Liquete, C., Reynaud, A., & Cardoso, A., (2016). Assessing water ecosystem services for water resource management. *Environmental Science & Policy, 61*, 194–203.

Grummon, A. H., Hall, M. G., Mitchell, C. G., Pulido, M., Sheldon, J. M., Noar, S. M., Ribisl, K. M., & Brewer, N. T., (2022). Reactions to messages about smoking, vaping, and COVID-19: Two national experiments. *Tobacco Control, 31*, 402–410.

Gunawardana, A., Shani, G., & Yogev, S., (2022). Evaluating recommender systems. *Recommender Systems Handbook* (pp. 547–601). Springer.

Haldon, J., Mordechai, L., Newfield, T. P., Chase, A. F., Izdebski, A., Guzowski, P., Labuhn, I., & Roberts, N., (2018). History meets palaeoscience: Consilience and collaboration in studying past societal responses to environmental change. *Proceedings of the National Academy of Sciences, 115*, 3210–3218.

Hamann, M., Johnson, J. A., Chaigneau, T., Chaplin-Kramer, R., Mandle, L., & Rieb, J. T., (2021). Ecosystem service modeling. *The Routledge Handbook of Research Methods for Social-Ecological Systems* (pp. 426–439). Routledge.

Hamed, M. M., Nashwan, M. S., Shahid, S., Bin, I. T., Wang, X. J., Dewan, A., & Asaduzzaman, M., (2022). Inconsistency in historical simulations and future projections of temperature and rainfall: A comparison of CMIP5 and CMIP6 models over Southeast Asia. *Atmospheric Research, 265*, 105927.

Hammad, A. T., & Falchetta, G., (2022). Probabilistic forecasting of remotely sensed cropland vegetation health and its relevance for food security. *Science of the Total Environment*, 156157.

Hardie, M., Green, S., Oliver, G., Swarts, N., Clothier, B., Gentile, R., & Close, D., (2022). Measuring and modeling nitrate fluxes in a mature commercial apple orchard. *Agricultural Water Management, 263*, 107410.

Hartmann, J., West, A. J., Renforth, P., Köhler, P., De La Rocha, C. L., Wolf-Gladrow, D. A., Dürr, H. H., & Scheffran, J., (2013). Enhanced chemical weathering as a geoengineering strategy to reduce atmospheric carbon dioxide, supply nutrients, and mitigate ocean acidification. *Reviews of Geophysics, 51*, 113–149.

He, H., & Harris, L., (2020). The impact of COVID-19 pandemic on corporate social responsibility and marketing philosophy. *Journal of Business Research, 116*, 176–182.

He, M., (2022). Assessing changes in 21st century mean and extreme climate of the Sacramento–San Joaquin Delta in California. *Climate, 10*, 16.

Heidenstrøm, N., & Hebrok, M., (2022). Towards realizing the sustainability potential within digital food provisioning platforms: The case of meal box schemes and online grocery shopping in Norway. *Sustainable Production and Consumption, 29*, 831–850.

Heijmans, M. M., Magnússon, R. Í., Lara, M. J., Frost, G. V., Myers-Smith, I. H., Van, H. J., Jorgenson, M. T., et al., (2022). Tundra vegetation change and impacts on permafrost. *Nature Reviews Earth & Environment, 3*, 68–84.

Hernanz, A., García-Valero, J. A., Domínguez, M., Ramos-Calzado, P., Pastor-Saavedra, M. A., & Rodríguez-Camino, E., (2022). Evaluation of statistical downscaling methods for climate change projections over Spain: Present conditions with perfect predictors. *International Journal of Climatology, 42*, 762–776.

Hessel, R., Wyseure, G., Panagea, I. S., Alaoui, A., Reed, M. S., Van, D. H., Muro, M., et al., (2022). Soil-improving cropping systems for sustainable and profitable farming in Europe. *Land, 11*, 780.

Höchtl, J., Parycek, P., & Schöllhammer, R., (2016). Big data in the policy cycle: Policy decision making in the digital era. *Journal of Organizational Computing and Electronic Commerce, 26*, 147–169.

Hoekstra, A. Y., Buurman, J., & Van, G. K. C., (2018). Urban water security: A review. *Environmental Research Letters, 13*, 053002.

Hook, S. E., Gallagher, E. P., & Batley, G. E., (2014). The role of biomarkers in the assessment of aquatic ecosystem health. *Integrated Environmental Assessment and Management, 10*, 327–341.

Höök, T. O., Foley, C. J., Collingsworth, P., Dorworth, L., Fisher, B., Hoverman, J. T., LaRue, E., et al., (2020). An assessment of the potential impacts of climate change on freshwater habitats and biota of Indiana, USA. *Climatic Change, 163*, 1897–1916.

Hope, A. P., (2020). *Quantification of the Past and Future Anthropogenic Effect on Climate Change Using the Empirical Model of Global Climate, an Energy Balance Multiple Linear Regression Model*. University of Maryland, College Park.

Hoque, R., & Sorwar, G., (2017). Understanding factors influencing the adoption of mHealth by the elderly: An extension of the UTAUT model. *International Journal of Medical Informatics, 101*, 75–84.

Hritonenko, V., & Yatsenko, Y., (2022). Sustainable adaptation and mitigation in regions and cities: Review of decision-support methods. *Resources, Conservation & Recycling Advances*, 200066.

Hunt, D. V., Lombardi, D. R., Atkinson, S., Barber, A. R., Barnes, M., Boyko, C. T., Brown, J., et al., (2012). Scenario archetypes: Converging rather than diverging themes. *Sustainability, 4*, 740–772.

Isbell, F., Adler, P. R., Eisenhauer, N., Fornara, D., Kimmel, K., Kremen, C., Letourneau, D. K., et al., (2017). Benefits of increasing plant diversity in sustainable agroecosystems. *Journal of Ecology, 105*, 871–879.

Jackson, B., Pagella, T., Sinclair, F., Orellana, B., Henshaw, A., Reynolds, B., Mcintyre, N., et al., (2013). Polyscape: A GIS mapping framework providing efficient and spatially explicit landscape-scale valuation of multiple ecosystem services. *Landscape and Urban Planning, 112*, 74–88.

Jafari, L., Asadi, S., & Asgari, A., (2022). Temporal and regional shifts of crop species diversity in rainfed and irrigated cropland in Iran. *Plos One, 17*, e0264702.

Jaiswal, R., Mall, R., Singh, N., Kumar, T. L., & Niyogi, D., (2022). Evaluation of bias correction methods for regional climate models: Downscaled rainfall analysis over diverse agroclimatic zones of India. *Earth and Space Science, 9*.

Jatayu, A., Saizen, I., Rustiadi, E., Pribadi, D. O., & Juanda, B., (2022). Urban form dynamics and modeling towards sustainable hinterland development in North Cianjur, Jakarta–Bandung mega-urban region. *Sustainability, 14*, 907.

Jia, D., Wang, C., Han, Y., Huang, H., & Xiao, H., (2022). Impact of climate change on the yield and water footprint of winter wheat in the Haihe River basin, China. *Atmosphere, 13*, 630.

Jones, L., Boeri, M., Christie, M., Durance, I., Evans, K. L., Fletcher, D., Harrison, L., et al., (2022). Can we model cultural ecosystem services, and are we measuring the right things? *People and Nature, 4*, 166–179.

Jozdani, S., Chen, D., Pouliot, D., & Johnson, B. A., (2022). A review and meta-analysis of generative adversarial networks and their applications in remote sensing. *International Journal of Applied Earth Observation and Geoinformation, 108*, 102734.

Kapsalis, V. C., Kyriakopoulos, G. L., & Aravossis, K. G., (2019). Investigation of ecosystem services and circular economy interactions under an inter-organizational framework. *Energies, 12*, 1734.

Karutz, R., Omann, I., Gorelick, S. M., Klassert, C. J., Zozmann, H., Zhu, Y., Kabisch, S., et al., (2022). Capturing stakeholders' challenges of the food–water–energy nexus—A participatory approach for Pune and the Bhima basin, India. *Sustainability, 14*, 5323.

Kasperson, J. X., Kasperson, R. E., Turner, B., Hsieh, W., & Schiller, A., (2022). Vulnerability to global environmental change. *The Social Contours of Risk* (pp. 245–285). Routledge.

Kayikci, Y., Kazancoglu, Y., Gozacan-Chase, N., & Lafci, C., (2022). Analyzing the drivers of smart sustainable circular supply chain for sustainable development goals through stakeholder theory. *Business Strategy and the Environment.*

Kirchner, M., Schmidt, J., Kindermann, G., Kulmer, V., Mitter, H., Prettenthaler, F., Rüdisser, J., et al., (2015). Ecosystem services and economic development in Austrian agricultural landscapes—The impact of policy and climate change scenarios on trade-offs and synergies. *Ecological Economics, 109*, 161–174.

Kofi, A. U. S. G., (2002). Perspectivas del Medio Ambiente Mundial. United Nations Environment Program. Global environment outlook (GEO)-3: Past, present, and future perspectives. UNEP. Fecha de consulta. *Environmental Management and Health, 13*, 560, 561. http://www.unep.org/geo/geo3/spanish/402.htm (accessed on 30 November 2023).

Kosamu, I. B. M., Makwinja, R., Kaonga, C. C., Mengistou, S., Kaunda, E., Alamirew, T., & Njaya, F., (2022). Application of DPSIR and tobit models in assessing freshwater ecosystems: The case of Lake Malombe, Malawi. *Water, 14*, 619.

Krisztin, T., Piribauer, P., & Wögerer, M., (2022). A spatial multinomial logit model for analyzing urban expansion. *Spatial Economic Analysis, 17*, 223–244.

Krumm, A., Süsser, D., & Blechinger, P., (2022). Modeling social aspects of the energy transition: What is the current representation of social factors in energy models? *Energy, 239*, 121706.

Kumar, A., Kumari, R., & Saikia, P., (2022a). Avenue plantation as a viable carbon confiscation tool: A blueprint for eco-smart cities. *Climate Change Alleviation for Sustainable Progression* (pp. 333–346). CRC Press.

Kumar, S., Meena, R. S., Sheoran, S., Jangir, C. K., Jhariya, M. K., Banerjee, A., & Raj, A., (2022b). Remote sensing for agriculture and resource management. *Natural Resources Conservation and Advances for Sustainability* (pp. 91–135). Elsevier.

Kuznik, F., Frayssinet, L., Roux, J. J., & Merlier, L., (2022). Calculation of heating and cooling energy loads at the district scale: Development of MoDEM, a modular and technologically explicit platform. *Sustainable Cities and Society*, 103901.

Kyker-Snowman, E., Lombardozzi, D. L., Bonan, G. B., Cheng, S. J., Dukes, J. S., Frey, S. D., Jacobs, E. M., et al., (2022). *Increasing the Spatial and Temporal Impact of Ecological Research: A Roadmap for Integrating a Novel Terrestrial Process into an Earth System Model*. Wiley Online Library.

Laimon, M., Mai, T., Goh, S., & Yusaf, T., (2022). System dynamics modeling to assess the impact of renewable energy systems and energy efficiency on the performance of the energy sector. *Renewable Energy*.

Lappalainen, H. K., Petäjä, T., Vihma, T., Räisänen, J., Baklanov, A., Chalov, S., Esau, I., et al., (2022). Overview: Recent advances in the understanding of the northern Eurasian environments and of the urban air quality in China–a Pan-Eurasian experiment (PEEX) program perspective. *Atmospheric Chemistry and Physics, 22*, 4413–4469.

Leach, M., MacGregor, H., Ripoll, S., Scoones, I., & Wilkinson, A., (2022). Rethinking disease preparedness: Incertitude and the politics of knowledge. *Critical Public Health, 32*, 82–96.

Lepage, M., Capderrey, C., Elliott, M., & Meire, P., (2022). Estuarine degradation and rehabilitation. *Fish and Fisheries in Estuaries: A Global Perspective, 1*, 458–552.

Leta, M. K., Demissie, T. A., & Tränckner, J., (2021). Modeling and prediction of land use land cover change dynamics based on land change modeler (LCM) in Nashe watershed, upper Blue Nile basin, Ethiopia. *Sustainability, 13*, 3740.

Levine, J., & Chan, K. M., (2013). Global human dependence on ecosystem services. *Ecosystem Services and Global Trade of Natural Resources* (pp. 28–45). Routledge.

Li, L., Stewart, B., Zhi, W., Sadayappan, K., Ramesh, S., Kerins, D., Sterle, G., et al., (2022). Climate controls on river chemistry. *Earth's Future, 10*, e2021EF002603.

Li, X., Chen, G., Liu, X., Liang, X., Wang, S., Chen, Y., Pei, F., & Xu, X., (2017). A new global land-use and land-cover change product at a 1-km resolution for 2010 to 2100 based on human–environment interactions. *Annals of the American Association of Geographers, 107*, 1040–1059.

Li, X., Stringer, L. C., & Dallimer, M., (2021). The spatial and temporal characteristics of urban heat island intensity: Implications for East Africa's urban development. *Climate, 9*, 51.

Li, Y., Ren, T., Kinney, P. L., Joyner, A., & Zhang, W., (2018). Projecting future climate change impacts on heat-related mortality in large urban areas in China. *Environmental Research, 163*, 171–185.

Liang, C., Schimel, J. P., & Jastrow, J. D., (2017). The importance of anabolism in microbial control over soil carbon storage. *Nature Microbiology, 2*, 1–6.

Liang, J. H., Yuan, J., Wan, X., Liu, J., Liu, B., Jang, H., & Tyagi, M., (2022). Exploring the use of machine learning to parameterize vertical mixing in the ocean surface boundary layer. *Ocean Modeling, 176*, 102059.

Libutti, A., Cammerino, A. R. B., & Monteleone, M., (2021). Management of residues from fruit tree pruning: A trade-off between soil quality and energy use. *Agronomy, 11*, 236.

Liess, M., & Ohe, P. C. V. D., (2005). Analyzing effects of pesticides on invertebrate communities in streams. *Environmental Toxicology and Chemistry: An International Journal, 24*, 954–965.

Lin, B. B., & Fuller, R. A., (2013). Sharing or sparing? How should we grow the world's cities? *Journal of Applied Ecology, 50*, 1161–1168.

Liu, F., Hou, H., & Murayama, Y., (2021). Spatial interconnections of land surface temperatures with land cover/use: A case study of Tokyo. *Remote Sensing, 13*, 610.

Liu, S., Shen, J., Liu, G., Wu, Y., & Shi, K., (2022). Exploring the effect of urban spatial development pattern on carbon dioxide emissions in China: A socioeconomic density distribution approach based on remotely sensed nighttime light data. *Computers, Environment and Urban Systems, 96*, 101847.

Löbmann, M. T., Maring, L., Prokop, G., Brils, J., Bender, J., Bispo, A., & Helming, K., (2022). Systems knowledge for sustainable soil and land management. *Science of the Total Environment, 822*, 153389.

Lu, Q., Zhao, D., & Wu, S., (2017). Simulated responses of permafrost distribution to climate change on the Qinghai–Tibet Plateau. *Scientific Reports, 7*, 1–13.

Macfadyen, S., Cunningham, S. A., Costamagna, A. C., & Schellhorn, N. A., (2012). Managing ecosystem services and biodiversity conservation in agricultural landscapes: Are the solutions the same? *Journal of Applied Ecology, 49*, 690–694.

Maes, J., Paracchini, M. L., & Zulian, G., (2011). *A European Assessment of the Provision of Ecosystem Services. Towards and Atlas of Ecosystem Services*. ISPRA: Joint Research Centre, IES.

Magness, D. R., Hoang, L., Belote, R. T., Brennan, J., Carr, W., Stuart, C. III. F., Clifford, K., et al., (2022). Management foundations for navigating ecological transformation by resisting, accepting, or directing social–ecological change. *BioScience, 72*, 30–44.

Mallampalli, V. R., Mavrommati, G., Thompson, J., Duveneck, M., Meyer, S., Ligmann-Zielinska, A., Druschke, C. G., et al., (2016). Methods for translating narrative scenarios into quantitative assessments of land use change. *Environmental Modeling & Software, 82*, 7–20.

Manning, P., De Vries, F. T., Tallowin, J. R., Smith, R., Mortimer, S. R., Pilgrim, E. S., Harrison, K. A., et al., (2015). Simple measures of climate, soil properties and plant traits predict national-scale grassland soil carbon stocks. *Journal of Applied Ecology, 52*, 1188–1196.

Marando, F., Heris, M. P., Zulian, G., Udías, A., Mentaschi, L., Chrysoulakis, N., Parastatidis, D., & Maes, J., (2022). Urban heat island mitigation by green infrastructure in European functional Urban Areas. *Sustainable Cities and Society, 77*, 103564.

Marangoni, G., Tavoni, M., Bosetti, V., Borgonovo, E., Capros, P., Fricko, O., Gernaat, D. E., et al., (2017). Sensitivity of projected long-term CO_2 emissions across the shared socioeconomic PATHWAYS. *Nature Climate Change, 7*, 113–117.

Marion, G., Hadley, L., Isham, V., Mollison, D., Panovska-Griffiths, J., Pellis, L., Tomba, G. S., et al., (2022). Modeling: Understanding pandemics and how to control them. *Epidemics*, 100588.

Marques, L., (2020). Decrease and degradation of forests. *Capitalism and Environmental Collapse* (pp. 41–64). Springer.

Martinez-García, F. P., Contreras-de-Villar, A., & Muñoz-Perez, J. J., (2021). Review of wind models at a local scale: Advantages and disadvantages. *Journal of Marine Science and Engineering, 9*, 318.

Maslo, B., Mau, R. L., Kerwin, K., McDonough, R., McHale, E., & Foster, J. T., (2022). Bats provide a critical ecosystem service by consuming a large diversity of agricultural pest insects. *Agriculture, Ecosystems & Environment, 324*, 107722.

Mason-Jones, K., Robinson, S. L., Veen, G., Manzoni, S., & Van, D. P. W. H., (2022). Microbial storage and its implications for soil ecology. *The ISME Journal, 16*, 617–629.

Matthiopoulos, J., Fieberg, J. R., & Aarts, G., (2020). *Species-Habitat Associations: Spatial Data, Predictive Models, and Ecological Insights*. University of Minnesota Libraries Publishing.

Mazzoni, S., Mariano, P. M., & Lippi, D., (2017). Unexplored treasures: The university of Florence libraries on display. *Unexplored Treasures*, 1–198.

McKenna, R., Pfenninger, S., Heinrichs, H., Schmidt, J., Staffell, I., Bauer, C., Gruber, K., et al., (2022). High-resolution large-scale onshore wind energy assessments: A review of potential definitions, methodologies, and future research needs. *Renewable Energy, 182*, 659–684.

McNally, A., Jacob, J., Arsenault, K., Slinski, K., Sarmiento, D. P., Hoell, A., Pervez, S., et al., (2022). A central Asia hydrologic monitoring dataset for food and water security applications in Afghanistan. *Earth System Science Data, 14*, 3115–3135.

Meunier, F., Krishna, M. S. M., Peaucelle, M., Calders, K., Terryn, L., Verbruggen, W., Liu, C., et al., (2022). Using terrestrial laser scanning to constrain forest ecosystem structure and functions in the ecosystem demography model (ED2. 2). *Geoscientific Model Development, 15*, 4783–4803.

Milano, F., Dörfler, F., Hug, G., Hill, D. J., & Verbič, G., (2018). Foundations and challenges of low-inertia systems. In: *2018 Power Systems Computation Conference (PSCC)* (pp. 1–25). IEEE.

Millhauser, J. K., & Earle, T. K., (2022). Biodiversity and the human past: Lessons for conservation biology. *Biological Conservation, 272*, 109599.

Milner, K., Crawford, P., Edgley, A., Hare-Duke, L., & Slade, M., (2020). The experiences of spirituality among adults with mental health difficulties: A qualitative systematic review. *Epidemiology and Psychiatric Sciences*, 29.

Mobus, G. E., (2022). Principles of systems science. *Systems Science: Theory, Analysis, Modeling, and Design* (pp. 41–87). Springer.

Mohi-ud-Din, R., Dar, N., Lal, S., Rukhsana, J., Khan, K., Owais, B., & Asmat, A., (2017). Agronomic practices influencing yield attributes of sweet corn (*Zea mays saccharata* L.). *Vegetos, 30*, 459–462.

Montagnini, F., Levin, B., & Berg, K. E., (2022). Introduction. Biodiversity Islands: Strategies for conservation in human-dominated environments. *Biodiversity Islands: Strategies for Conservation in Human-Dominated Environments* (pp. 3–37). Springer.

Montanari, A., Young, G., Savenije, H., Hughes, D., Wagener, T., Ren, L., Koutsoyiannis, D., et al., (2013). "Panta Rhei—Everything flows": Change in hydrology and society—The IAHS scientific decade 2013–2022. *Hydrological Sciences Journal, 58*, 1256–1275.
Morris, R., Davis, S., Grelet, G. A., & Gregorini, P., (2022). Multiscapes and urbanization: The case for spatial agroecology. *Sustainability, 14*, 1352.
Mottram, R., Simonsen, S. B., Høyer, S. S., Barletta, V. R., Sandberg, S. L., Nagler, T., Wuite, J., et al., (2019). An integrated view of Greenland ice sheet mass changes based on models and satellite observations. *Remote Sensing, 11*, 1407.
Mouloodi, S., Rahmanpanah, H., Martin, C., Gohari, S., & Davies, H., (2022). How artificial intelligence and machine learning is assisting us to extract meaning from data on bone mechanics? *Biomedical Visualization*, 195–221.
Mulder, C., Bennett, E. M., Bohan, D. A., Bonkowski, M., Carpenter, S. R., Chalmers, R., Cramer, W., et al., (2015). 10 years later: Revisiting priorities for science and society a decade after the millennium ecosystem assessment. *Advances in Ecological Research* (pp. 1–53). Elsevier.
Muriithi, F. K., (2016). Land use and land cover (LULC) changes in semi-arid sub-watersheds of Laikipia and Athi River basins, Kenya, as influenced by expanding intensive commercial horticulture. *Remote Sensing Applications: Society and Environment, 3*, 73–88.
Naselli-Flores, L., & Padisák, J., (2022). Ecosystem services provided by marine and freshwater phytoplankton. *Hydrobiologia*, 1–16.
Navarro-Pedreño, J., Almendro-Candel, M. B., & Zorpas, A. A., (2021). The increase of soil organic matter reduces global warming, myth or reality? *Sci., 3*, 18.
Nelson, K. S., Patalee, B., & Yao, B., (2022). Higher landscape diversity associated with improved crop production resilience in Kansas-USA. *Environmental Research Letters, 17*, 084011.
Newman, G., Sansom, G. T., Yu, S., Kirsch, K. R., Li, D., Kim, Y., Horney, J. A., et al., (2022). A framework for evaluating the effects of green infrastructure in mitigating pollutant transferal and flood events in Sunnyside, Houston, TX. *Sustainability, 14*, 4247.
Nottingham, A. T., Gloor, E., Bååth, E., & Meir, P., (2022). Soil carbon and microbes in the warming tropics. *Functional Ecology*.
Obbard, J., (2010). Climate change—Living in the Anthropocene. *Crucial Issues in Climate Change and the Kyoto Protocol: Asia and the World* (pp. 63–96). World Scientific.
Oliveira, B. M., Boumans, R., Fath, B. D., Othoniel, B., Liu, W., & Harari, J., (2022). Prototype of social-ecological system's resilience analysis using a dynamic index. *Ecological Indicators, 141*, 109113.
Oliveira, C., Coelho, D., & Antunes, C. H., (2016). Coupling input–output analysis with multiobjective linear programming models for the study of economy–energy–environment–social (E3S) trade-offs: A review. *Annals of Operations Research, 247*, 471–502.
Orr, A., Ahmad, B., Alam, U., Appadurai, A., Bharucha, Z. P., Biemans, H., Bolch, T., et al., (2022). Knowledge priorities on climate change and water in the upper Indus basin: A horizon scanning exercise to identify the top 100 research questions in social and natural sciences. *Earth's Future, 10*, e2021EF002619.
Osborne, B. B., Roybal, C. M., Reibold, R., Collier, C. D., Geiger, E., Phillips, M. L., Weintraub, M. N., & Reed, S. C., (2022). Biogeochemical and ecosystem properties in three adjacent semi-arid grasslands are resistant to nitrogen deposition but sensitive to edaphic variability. *Journal of Ecology, 110*, 1615–1631.

Padilla, L. A., (2021). The Anthropocene: Are we in the midst of the sixth mass extinction? *Sustainable Development in the Anthropocene* (pp. 93–167). Springer.

Paeth, H., Hall, N. M., Gaertner, M. A., Alonso, M. D., Moumouni, S., Polcher, J., Ruti, P. M., et al., (2011). Progress in regional downscaling of West African precipitation. *Atmospheric Science Letters, 12*, 75–82.

Pan, X., Zhang, L., & Huang, C., (2020). Future climate projection in Northwest China with RegCM4. 6. *Earth and Space Science, 7*, e2019EA000819.

Pandey, P. C., Koutsias, N., Petropoulos, G. P., Srivastava, P. K., & Ben, D. E., (2021). Land use/land cover in view of earth observation: Data sources, input dimensions, and classifiers—A review of the state of the art. *Geocarto International, 36*, 957–988.

Pant, M., Ghosh, S., Verma, S., Sinha, P., Mall, R., & Bhatla, R., (2022). Simulation of an extreme rainfall event over Mumbai using a regional climate model: A case study. *Meteorology and Atmospheric Physics, 134*, 1–17.

Pe'er, G., Finn, J. A., Díaz, M., Birkenstock, M., Lakner, S., Röder, N., Kazakova, Y., et al., (2022). How can the European common agricultural policy help halt biodiversity loss? Recommendations by over 300 experts. *Conservation Letters*, e12901.

Peh, K. S.-H., Balmford, A. P., Bradbury, R. B., Brown, C., Butchart, S. H. M., Hughes, F. M. R., MacDonald, M. A., Stattersfield, A. J., Thomas, D. H. L., Trevelyan, R. J., Walpole, M., & Merriman, J. C., (2017). *Toolkit for Ecosystem Service Site-based Assessment (TESSA)*. Version 2.0 Cambridge, UK. https://www.niney.org/showcase/rain/downloads/TESSAToolkit-V1_1-20130927.pdf (accessed on 30 November 2023).

Perović, D. J., Gámez-Virués, S., Landis, D. A., Tscharntke, T., Zalucki, M. P., Saura, S., Furlong, M. J., et al., (2022). Broadening the scope of empirical studies to answer persistent questions in landscape-moderated effects on biodiversity and ecosystem functioning. *Advances in Ecological Research* (pp. 109–131). Elsevier.

Petsch, D. K., Cionek, V. D. M., Thomaz, S. M., & Dos, S. N. C. L., (2022). Ecosystem services provided by river-floodplain ecosystems. *Hydrobiologia*, 1–22.

Pettorelli, N., Barlow, J., Stephens, P. A., Durant, S. M., Connor, B., Schulte, T. B. H., Sandom, C. J., Wentworth, J., & Du Toit, J. T., (2018). Making rewilding fit for policy. *Journal of Applied Ecology*, *55*(3), 1114–1125. https://besjournals.onlinelibrary.wiley.com/doi/abs/10.1111/1365-2664.13082 (accessed on 30 November 2023).

Pires, M. V., Da Cunha, D. A., De Matos, C. S., & Costa, M. H., (2015). Nitrogen-use efficiency, nitrous oxide emissions, and cereal production in Brazil: Current trends and forecasts. *PloS One, 10*, e0135234.

Poore, A. B., Aristoff, J. M., Horwood, J. T., Armellin, R., Cerven, W. T., Cheng, Y., Cox, C. M., et al., (2016). *Covariance and Uncertainty Realism in Space Surveillance and Tracking*. Numerica Corporation Fort Collins United States.

Pradhan, P., Costa, L., Rybski, D., Lucht, W., & Kropp, J. P., (2017). A systematic study of sustainable development goal (SDG) interactions. *Earth's Future, 5*, 1169–1179.

Priya, D. S., Delu, V., Yodha, K., Dahiya, T., Kour, A., & Punia, N., (2022). Role of birds in agroecosystem: A review on agricultural and economic ornithology. *The Pharma Innovation Journal, 11*(7), 2300–2314. www.thepharmajournal.com (accessed on 30 November 2023).

Purvis, B., Mao, Y., & Robinson, D., (2022). A multi-scale integrated assessment model to support urban sustainability. *Sustainability Science, 17*, 151–169.

Pyka, A., Cardellini, G., Van, M. H., & Verkerk, P., (2022). Modeling the bioeconomy: Emerging approaches to address policy needs. *Journal of Cleaner Production, 330*, 129801.

Radočaj, D., Jurišić, M., & Gašparović, M., (2022). The role of remote sensing data and methods in a modern approach to fertilization in precision agriculture. *Remote Sensing, 14*, 778.
Rahman, S. U., Wang, X., Shahzad, M., Bashir, O., Li, Y., & Cheng, H., (2022). A review of the influence of nanoparticles on the physiological and biochemical attributes of plants with a focus on the absorption and translocation of toxic trace elements. *Environmental Pollution*, 119916.
Ram, S., Majumder, S., Chaudhuri, P., Chanda, S., Santra, S., Chakraborty, A., & Sudarshan, M., (2015). A review on air pollution monitoring and management using plants with special reference to foliar dust adsorption and physiological stress responses. *Critical Reviews in Environmental Science and Technology, 45*, 2489–2522.
Rao, K., (2019). *Environmental Impacts in Relation to Wind Energy, Wind Energy for Power Generation* (pp. 1073–1273). Springer.
Rather, G., Bansal, S. K., Bashir, O., & Waida, U., (2019). Impact of potassium nutrition on fruit yield and physicochemical characteristics of apple cultivar red delicious. *Indian Journal of Fertilizers, 15*, 790–797.
Rawat, M., Singh, R., Sharma, J., Saklani, H., Chand, T., Bhatt, I. D., & Pandey, R., (2022). An overview of the functioning of temperate forest ecosystems with particular reference Tt Himalayan temperate forest. *Trees, Forests and People*, 100230.
Reboud, X., Poggi, S., & Bohan, D. A., (2022). Effective biodiversity monitoring could be facilitated by networks of simple sensors and a shift to incentivizing results. *Advances in Ecological Research* (pp. 339–365). Elsevier.
Ren, Y., Li, Z., Li, J., Ding, Y., & Miao, X., (2022). Analysis of land use/cover change and driving forces in the Selenga River Basin. *Sensors, 22*, 1041.
Renschler, C. S., & Harbor, J., (2002). Soil erosion assessment tools from point to regional scales—The role of geomorphologists in land management research and implementation. *Geomorphology, 47*, 189–209.
Richards, D. R., Belcher, R. N., Carrasco, L. R., Edwards, P. J., Fatichi, S., Hamel, P., Masoudi, M., et al., (2022). Global variation in contributions to human well-being from urban vegetation ecosystem services. *One Earth, 5*, 522–533.
Riechers, M., Strack, M., Barkmann, J., & Tscharntke, T., (2019). Cultural ecosystem services provided by urban green change along an urban-periurban gradient. *Sustainability, 11*, 645.
Rienow, A., Kantakumar, L. N., Ghazaryan, G., Dröge-Rothaar, A., Sticksel, S., Trampnau, B., & Thonfeld, F., (2022). Modeling the spatial impact of regional planning and climate change prevention strategies on land consumption in the Rhine-Ruhr metropolitan area 2017–2030. *Landscape and Urban Planning, 217*, 104284.
Robbins, P., (2019). *Political Ecology: A Critical Introduction*. John Wiley & Sons.
Robison, A. L., & Scanlon, T. M., (2018). Climate change to offset improvements in watershed acid-base status provided by clean air act and amendments: A model application in Shenandoah national park, Virginia. *Journal of Geophysical Research: Biogeosciences, 123*, 2863–2877.
Rohr, J. R., Barrett, C. B., Civitello, D. J., Craft, M. E., Delius, B., DeLeo, G. A., Hudson, P. J., et al., (2019). Emerging human infectious diseases and the links to global food production. *Nature Sustainability, 2*, 445–456.
Rolnick, D., Donti, P. L., Kaack, L. H., Kochanski, K., Lacoste, A., Sankaran, K., Ross, A. S., et al., (2022). Tackling climate change with machine learning. *ACM Computing Surveys (CSUR), 55*, 1–96.

Ross, A. C., Stock, C. A., Dixon, K. W., Friedrichs, M. A., Hood, R. R., Li, M., Pegion, K., et al., (2020). Estuarine forecasts at daily weather to subseasonal time scales. *Earth and Space Science, 7*, e2020EA001179.

Sacchi, R., Terlouw, T., Siala, K., Dirnaichner, A., Bauer, C., Cox, B., Mutel, C., et al., (2022). PRospective EnvironMental Impact asSEment (premise): A streamlined approach to producing databases for prospective life cycle assessment using integrated assessment models. *Renewable and Sustainable Energy Reviews, 160*, 112311.

Sajjad, W., Ilahi, N., Kang, S., Bahadur, A., Zada, S., & Iqbal, A., (2022). Endolithic microbes of rocks, their community, function, and survival strategies. *International Biodeterioration & Biodegradation, 169*, 105387.

Santos, M., Ferreira, D., Bastos, R., Vicente, J., Honrado, J., Kueffer, C., Kull, C. A., et al., (2016). Linking landscape futures with biodiversity conservation strategies in northwest Iberia—A simulation study combining surrogates with a spatio-temporal modeling approach. *Ecological Informatics, 33*, 85–100.

Schmelzer, M., (2012). The crisis before the crisis: The 'problems of modern society' and the OECD, 1968–74. *European Review of History: Revue Europeenne D'histoire, 19*, 999–1020.

Schmelzer, M., (2017). 'Born in the corridors of the OECD': The forgotten origins of the Club of Rome, transnational networks, and the 1970s in global history. *Journal of Global History, 12*, 26–48.

Schwartz, J. D., (2020). *The Reindeer Chronicles: And Other Inspiring Stories of Working with Nature to Heal the Earth.* Chelsea Green Publishing.

Scialabba, N. E. H., (2022). Experiences of low-external-input livestock systems. *Managing Health Livestock Production and Consumption* (pp. 165–219). Elsevier.

Sedlacko, M., Martinuzzi, A., Røpke, I., Videira, N., & Antunes, P., (2014). Participatory systems mapping for sustainable consumption: Discussion of a method promoting systemic insights. *Ecological Economics, 106*, 33–43.

Sein, D. V., Mikolajewicz, U., Gröger, M., Fast, I., Cabos, W., Pinto, J. G., Hagemann, S., et al., (2015). Regionally coupled atmosphere-ocean-sea ice-marine biogeochemistry model ROM: 1. Description and validation. *Journal of Advances in Modeling Earth Systems, 7*, 268–304.

Sharma, P., & Saxena, P., (2022). Impacts and responses of particulate matter pollution on vegetation. *Airborne Particulate Matter*, 229–264.

Shugart, H., White, P., Saatchi, S., & Chave, J., (2022). *The World Atlas of Trees and Forests: Exploring Earth's Forest Ecosystems*. Princeton University Press.

Singleton, A., Arribas-Bel, D., Murray, J., & Fleischmann, M., (2022). Estimating generalized measures of local neighborhood context from multispectral satellite images using a convolutional neural network. *Computers, Environment and Urban Systems, 95*, 101802.

Sitar-Taut, D. A., & Mican, D., (2022). *Social Media Exposure Assessment: Influence on Attitudes Toward Generic Vaccination During the COVID-19 Pandemic*. Online Information Review.

Smith, A., & Mair, J., (2022). *Events and Sustainability: Can Events Make Places More Inclusive, Resilient and Sustainable?* Taylor & Francis.

Smith, C. E., Matthews, R. A., Mills, M. J., Hong, Y. H., & Sim, S., (2022a). Organizational benefits of onboarding contingent workers: An anchoring model approach. *Journal of Business and Psychology, 37*, 525–541.

Smith, N., Georgiou, M., King, A. C., Tieges, Z., & Chastin, S., (2022b). Factors influencing usage of urban blue spaces: A systems-based approach to identify leverage points. *Health & Place, 73*, 102735.

Sok, T., Ich, I., Tes, D., Chan, R., Try, S., Song, L., Ket, P., Khem, S., & Oeurng, C., (2022). Change in hydrological regimes and extremes from the impact of climate change in the largest tributary of the Tonle Sap Lake Basin. *Water, 14*, 1426.

Sokol, N. W., Slessarev, E., Marschmann, G. L., Nicolas, A., Blazewicz, S. J., Brodie, E. L., Firestone, M. K., et al., (2022). Life and death in the soil microbiome: How ecological processes influence biogeochemistry. *Nature Reviews Microbiology*, 1–16.

Song, J., Du, S., Feng, X., & Guo, L., (2014). The relationships between landscape compositions and land surface temperature: Quantifying their resolution sensitivity with spatial regression models. *Landscape and Urban Planning, 123*, 145–157.

Sorge, S., Mann, C., Schleyer, C., Loft, L., Spacek, M., Hernández-Morcillo, M., & Kluvankova, T., (2022). Understanding dynamics of forest ecosystem services governance: A socio-ecological-technical-analytical framework. *Ecosystem Services, 55*, 101427.

Souza, J. M. D., Morgado, P., Costa, E. M. D., & Vianna, L. F. D. N., (2022). Modeling of land use and land cover (LULC) change based on artificial neural networks for the Chapecó River Ecological Corridor, Santa Catarina/Brazil. *Sustainability, 14*, 4038.

Staab, J., Schady, A., Weigand, M., Lakes, T., & Taubenböck, H., (2022). Predicting traffic noise using land-use regression—A scalable approach. *Journal of Exposure Science & Environmental Epidemiology, 32*, 232–243.

Sterman, J., (2018). *System Dynamics at Sixty: The Path Forward* (pp. 5–47). Wiley Online Library.

Stern, N., & Stiglitz, J., (2022). The economics of immense risk, urgent action, and radical change: Towards new approaches to the economics of climate change. *Journal of Economic Methodology,* 1–36.

Stern, N., (2022). A time for action on climate change and a time for change in economics. *The Economic Journal, 132*, 1259–1289.

Steward, P. R., Shackelford, G., Carvalheiro, L. G., Benton, T. G., Garibaldi, L. A., & Sait, S. M., (2014). Pollination and biological control research: Are we neglecting two billion smallholders. *Agriculture & Food Security, 3*, 1–13.

Stieler, D., Schwinn, T., Leder, S., Maierhofer, M., Kannenberg, F., & Menges, A., (2022). Agent-based modeling and simulation in architecture. *Automation in Construction, 141*, 104426.

St-Laurent, G. P., Oakes, L. E., Cross, M., & Hagerman, S., (2022). Flexible and comprehensive criteria for evaluating climate change adaptation success for biodiversity and natural resource conservation. *Environmental Science & Policy, 127*, 87–97.

Süsser, D., Gaschnig, H., Ceglarz, A., Stavrakas, V., Flamos, A., & Lilliestam, J., (2022). Better suited or just more complex? On the fit between user needs and modeler-driven improvements of energy system models. *Energy, 239*, 121909.

Tassi, A., Gigante, D., Modica, G., Di Martino, L., & Vizzari, M., (2021). Pixel-vs. Object-based Landsat 8 data classification in google earth engine using random forest: The case study of Maiella national park. *Remote Sensing, 13*, 2299.

Tavoni, M., & Valente, G., (2022). Uncertainty in integrated assessment modeling of climate change. *Perspectives on Science, 30*, 321–351.

Temesgen, H., Affleck, D., Poudel, K., Gray, A., & Sessions, J., (2015). A review of the challenges and opportunities in estimating above ground forest biomass using tree-level models. *Scandinavian Journal of Forest Research, 30*, 326–335.

Tenfie, H. W., Saathoff, F., Hailu, D., & Gebissa, A., (2022). Selection of representative general circulation models for climate change study using advanced envelope-based and past performance approach on transboundary river basin, a case of upper Blue Nile Basin, Ethiopia. *Sustainability, 14*, 2140.

Teo, H. C., Lechner, A. M., Walton, G. W., Chan, F. K. S., Cheshmehzangi, A., Tan-Mullins, M., Chan, H. K., et al., (2019). Environmental impacts of infrastructure development under the belt and road initiative. *Environments, 6*, 72.

Thiesen, T., Bhat, M. G., Liu, H., & Rovira, R., (2022). An ecosystem service approach to assessing agro-ecosystems in urban landscapes. *Land, 11*, 469.

Tian, H., Xu, R., Canadell, J. G., Thompson, R. L., Winiwarter, W., Suntharalingam, P., Davidson, E. A., et al., (2020). A comprehensive quantification of global nitrous oxide sources and sinks. *Nature, 586*, 248–256.

Tong, X., Mohapatra, S., Zhang, J., Tran, N. H., You, L., He, Y., & Gin, K. Y. H., (2022). Source, fate, transport, and modeling of selected emerging contaminants in the aquatic environment: Current status and future perspectives. *Water Research*, 118418.

Tulbure, M. G., Hostert, P., Kuemmerle, T., & Broich, M., (2022). Regional matters: On the usefulness of regional land-cover datasets in times of global change. *Remote Sensing in Ecology and Conservation, 8*, 272–283.

Urlich, S. C., White, F. R., & Rennie, H. G., (2022). Characterizing the regulatory seascape in Aotearoa New Zealand: Bridging local, regional, and national scales for marine ecosystem-based management. *Ocean & Coastal Management, 224*, 106193.

Van, D. W. W., & Bianchi, F., (2022). Options for diversifying agricultural systems to reduce pesticide use: Can we learn from nature? *Outlook on Agriculture, 51*, 105–113.

Von, H. L., Beutler, B., Bienenstock, J., Blaser, M., Cani, P. D., Eriksson, J., Färkkilä, M., Haahtela, T., Hanski, I., Jenmalm, M. C., et al., (2015). Helsinki alert of biodiversity and health. *Annals of Medicine, 47*, 218–225.

Wang, H., Yang, G., Ouyang, X., Tand, Z., Long, X., & Yue, Z., (2021). Horizontal ecological compensation mechanism and technological progress: Theory and empirical study of Xin'an river ecological compensation gambling agreement. *Journal of Environmental Planning and Management*, 1–23.

Wang, Q., & Wang, H., (2022). Spatiotemporal dynamics and evolution relationships between land-use/land cover change and landscape pattern in response to rapid urban sprawl process: A case study in Wuhan, China. *Ecological Engineering, 182*, 106716.

Watz, J., Aldvén, D., Andreasson, P., Aziz, K., Blixt, M., Calles, O., Lund, B. K., et al., (2022). Atlantic salmon in regulated rivers: Understanding River management through the ecosystem services lens. *Fish and Fisheries, 23*, 478–491.

Wehrli, K., Guillod, B. P., Hauser, M., Leclair, M., & Seneviratne, S. I., (2019). Identifying key driving processes of major recent heat waves. *Journal of Geophysical Research: Atmospheres, 124*, 11746–11765.

Wei, Y., Huang, C., Lam, P. T., & Yuan, Z., (2015). Sustainable urban development: A review on urban carrying capacity assessment. *Habitat International, 46*, 64–71.

Wen, H., Sullivan, P. L., Billings, S. A., Ajami, H., Cueva, A., Flores, A., Hirmas, D. R., et al., (2022). From soils to streams: Connecting terrestrial carbon transformation, chemical weathering, and solute export across hydrological regimes. *Water Resources Research, 58*, e2022WR032314.

Williams, B. K., & Johnson, F. A., (2013). Confronting dynamics and uncertainty in optimal decision making for conservation. *Environmental Research Letters, 8*, 025004.

Wissert, L. P., (2022). *Servicing or Buying New? Estimating the Environmentally Optimal Time for Car Replacement in Sweden–an LCA Approach.* https://www.diva-portal.org/smash/get/diva2:1667648/FULLTEXT01.pdf (accessed on 30 November 2023).

Wurth, B., Stam, E., & Spigel, B., (2022). Toward an entrepreneurial ecosystem research program. *Entrepreneurship Theory and Practice, 46*, 729–778.

Xia, M., Wang, Z., Lu, M., & Pan, L., (2022). MFAGCN: A new framework for identifying power grid branch parameters. *Electric Power Systems Research, 207*, 107855.

Xu, J., Renaud, F. G., & Barrett, B., (2022). Modeling land system evolution and dynamics of terrestrial carbon stocks in the Luanhe River Basin, China: A scenario analysis of trade-offs and synergies between sustainable development goals. *Sustainability Science, 17*, 1323–1345.

Yang, P., Cai, X., Hu, X., Zhao, Q., Lee, Y., Khanna, M., Cortés-Peña, Y. R., et al., (2022a). An agent-based modeling tool supporting bioenergy and bio-product community communication regarding cellulosic bioeconomy development. *Renewable and Sustainable Energy Reviews, 167*, 112745.

Yang, Q., Liu, G., Casazza, M., Dumontet, S., & Yang, Z., (2022b). Ecosystem restoration programs challenges under climate and land use change. *Science of The Total Environment, 807*, 150527.

Yang, Y., Knops, J. M. H., & Brassil, C. E., (2022c). Pocket gopher disturbance slows soil carbon accumulation in abandoned agricultural lands. *Ecology, 103*, e3627.

Yin, C., Zhao, W., & Pereira, P., (2022). Soil conservation service underpins sustainable development goals. *Global Ecology and Conservation, 33*, e01974.

Yu, H., Yang, J., Sun, D., Li, T., & Liu, Y., (2022). Spatial responses of ecosystem service value during the development of urban agglomerations. *Land, 11*, 165.

Yu, M., & Liu, Q., (2021). Deep learning-based downscaling of tropospheric nitrogen dioxide using ground-level and satellite observations. *Science of The Total Environment, 773*, 145145.

Yuan, J., Lin, Q., Chen, S., Zhao, H., Xie, X., Cai, Z., Zhang, J., et al., (2022). Influence of global warming and urbanization on regional climate of Megacity: A case study of Chengdu, China. *Urban Climate, 44*, 101227.

Zahawi, R. A., Dandois, J. P., Holl, K. D., Nadwodny, D., Reid, J. L., & Ellis, E. C., (2015). Using lightweight unmanned aerial vehicles to monitor tropical forest recovery. *Biological Conservation, 186*, 287–295.

Zaman-ul-Haq, M., Saqib, Z., Kanwal, A., Naseer, S., Shafiq, M., Akhtar, N., Bokhari, S. A., et al., (2022). The trajectories, trends, and opportunities for assessing urban ecosystem services: A systematic review of geospatial methods. *Sustainability, 14*, 1471.

Zampieri, M., Giorgi, F., Lionello, P., & Nikulin, G., (2012). Regional climate change in the Northern Adriatic. *Physics and Chemistry of the Earth, Parts A/B/C 40*, 32–46.

Żebrowski, P., Dieckmann, U., Brännström, Å., Franklin, O., & Rovenskaya, E., (2022). Sharing the burdens of climate mitigation and adaptation: Incorporating fairness perspectives into policy optimization models. *Sustainability, 14*, 3737.

Zenios, S. A., (2022). The risks from climate change to sovereign debt. *Climatic Change, 172*, 1–19.

Zhao, C., Xiao, P., Qian, P., Xu, J., Yang, L., & Wu, Y., (2022). Spatiotemporal differentiation and balance pattern of ecosystem service supply and demand in the Yangtze River Economic Belt. International *Journal of Environmental Research and Public Health, 19*, 7223.

Zhao, X., Heng, Z., Jiang, X., Yang, Q., Li, Y., Yang, Y., & Gao, Z., (2021). Evaluation of the effect of stability schemes on the simulation of land surface processes at a Western Tibetan Site. *Land, 10*, 253.

Zhu, M., & Zhang, Y., (2022). Medical and public health instructors' perceptions of online teaching: A qualitative study using the technology acceptance model 2. *Education and Information Technologies, 27,* 2385–2405.

Zope, P., Eldho, T., & Jothiprakash, V., (2016). Impacts of land use–land cover change and urbanization on flooding: A case study of Oshiwara River Basin in Mumbai, India. *Catena, 145,* 142–154.

CHAPTER 2

Predicting and Modeling of Soil Organic Carbon in Various Land Uses

NASIR BASHIR NAIKOO,[1] OWAIS BASHIR,[1] SHABIR AHMAD BANGROO,[1] AAMIR HASSAN MIR,[1] M. AUYOUB BHAT,[1] SHAKEEL AHMAD MIR,[1] TAUSEEF AHMAD BHAT,[2] SHAFEEQ UR REHMAN,[3] and KHALID REHMAN HAKEEM[4]

[1]*Division of Soil Science, Sher-e-Kashmir University of Agricultural Sciences and Technology of Kashmir, Kashmir, India*

[2]*Division of Agronomy, Sher-e-Kashmir University of Agricultural Sciences and Technology of Kashmir, Kashmir, India*

[3]*School of Environment and Civil Engineering, Dongguan University of Technology, Dongguan, Guangdong, China*

[4]*Department of Biological Sciences, Faculty of Science, King Abdulaziz University, Jeddah, Saudi Arabia*

ABSTRACT

Soil organic carbon (SOC) stocks are the main ambiguity in global carbon cycle projections as they consist of more carbon than crops or even the atmosphere. Degradation and stabilization processes modulate SOC cycling; various experimental studies are being used to evaluate process-based microbiota models. It will take long-term soil C input measurements to rectify simulated and identified trends. Consolidating model-experiment divergence will allow more accurate predictions of SOC responses to climate change. The inability to measure SOC stocks directly can be due to various factors, including difficulty accessing representative sampling points, a lack

Soil, Agriculture, and Ecosystem Modelling: Smart Technologies for Sustainable Solutions.
Owais Bashir, Khalid Rehman Hakeem, Surinder S. Kukal, and Razeef Mohd (Eds.)

of equipment, or the number of samples needed to represent a certain area of interest exceeding those financially feasible.

For this reason, it is not always possible to answer all pertinent questions about SOC stock and dynamics using only the data obtained through direct measurements. In the global carbon cycle, this redistribution has a significant impact. SOC profile is depicted by the type of soil. From millimeter to km, spatial/spectral scale mathematical models have been developed in the last eight decades to characterize soil biochemical processes. In this chapter, we provide an overview of soil carbon models, which are frequently applied to simulate soil carbon stocks.

2.1 INTRODUCTION

Soil organic carbon (SOC) is a complex and diverse combination of elements ranging from quickly decomposing plant parts to microbes that is a necessary component of soil (Khatoon et al., 2017; Lützow et al., 2006; Ramesh et al., 2019). The study of SOC dynamics is becoming more relevant since it has a significant effect on soil productivity and sustainability. The quantity of SOC is a balance between the accumulation of new plants and animal materials and the continuous losses caused by the decomposition of carbon (C) and the separation of components into mineral nutrients and gases or by washing or leaching away (Assmann et al., 2014; Mukherjee et al., 2018; Rees et al., 2005). Positive SOC accumulation is feasible when there is a lot of water, nutrients, and sunshine, while prolonged fallow favors SOC loss. Five key pools govern the carbon cycle on our planet viz., oceanic (38,000 Pg), geologic (5,000 Pg), soil (2,500 Pg), atmospheric (760 Pg), and biotic (560 Pg) (Kumar et al., 2017; Senapati et al., 2014a). About 1,500 Pg of organic carbon are found in soils worldwide, which is more than three times the amount of carbon found in plants and nearly two times the concentration of CO_2 in the atmosphere. While the SOC pool is tiny in comparison to the seas, it has the potential to be considerably more labile in the near term. Around 1% change in the amount of carbon stored in soils corresponds to an approximately eight ppm change in the atmospheric CO_2 concentration, assuming that all other carbon cycle parameters remain constant. Thus, even a modest alteration in the flow of carbon into or out of the soil carbon pool may have a significant effect on a global scale (Abrams et al., 2016; Shevliakova et al., 2009). Land use change, climate, soil, and management practices are the primary determinants of the size and mobility of the SOC pool. Tillage, disturbance, aeration, and mineralization contribute to agricultural soils' carbon depletion. However, there is significant potential to adsorb and store more carbon in such soils by changing soil management techniques (Bojko & Kabala, 2017; Wiesmeier et al., 2019). Accurate measurement of C inputs and outputs from the soil is critical for determining the potential for C sequestration and variations in SOC over time. These data will aid in deciphering the processes governing C storage and developing realistic models of soil C dynamics and turnover (Smith et al., 2020; Xu et al., 2021). SOC turnover models can simulate SOC dynamics across various land uses, soil management practices, and technological advancements (varietal improvement, improved machinery, agronomical practices, and climatic conditions). As a result, they can help us better understand how SOC turnover processes change under various conditions and also help us study changes in SOC dynamics (Mandal et al., 2022; Paustian et al., 2019). The terrestrial biosphere is a significant source of carbon, with the majority of it stored in the soil. Carbon dioxide (CO_2) is exchanged between the atmosphere and soil, resulting in CO_2 sequestration in soil. Additionally, SOC is associated with soil fertility, plant development, soil and water conservation, aggregate stability, and soil nutrient cycling capacity. Soil type, climate, and terrain, as well as land use and farming methods, all affect the composition of SOC. However, soil carbon storage is limited, and as SOC storage increases, the soil carbon concentration will eventually approach saturation. As a result, the content of SOC must be monitored. However, the conventional technique of sampling sample sites is inefficient and expensive, making it unsuitable for large-scale monitoring. To address these issues, researchers started to investigate the use of robust and cost-effective methods for SOC content prediction (Hasan et al., 2020; Ramesh et al., 2019; Wang et al., 2021a). Recent research has mostly relied on laboratory-based soil reflectance spectroscopy to estimate SOC levels. For prediction, they used various spectral bands (visible, mid-infrared, near-infrared, and short-wave infrared spectroscopy). Although these techniques have high predictive accuracy, their coefficient of determination is very low (R2) of about 0.83; the laboratory spectrum has significant limitations; specifically, it can only give spectral data on the point scale and not at the landscape size. Remote sensing data are readily available, cover a large area, and may be utilized for year-round ground monitoring. As a result, it is essential to investigate how distant sensing data contribute to SOC prediction (Jacques et al., 2014). There have been many studies on SOC prediction using remote sensing data, most notably optical data (e.g., Sentinel-2, Landsat, and MODIS satellite data); their bands range from visible to short-wave infrared, giving additional information. Additionally, soil surface characteristics (for example, plant cover, debris, soil wetness, and soil roughness) influence SOC prediction. The benefits of synthetic aperture radar (SAR) technology include the ability to image in all weather conditions and at all hours of the day or night. As a result, the spectral range of remote sensing utilized to forecast SOC is not restricted to visible and infrared; radar has also been employed (Dadhich et al., 2019; Mason et al., 2014). Recent research has shown that multi-temporal SAR data may be used to capture the soil–plant connection and forecast soil chemical properties, but this capability is contingent on collecting vegetation features from remote sensing pictures that signal soil property change. Fluctuating connections between the backscattering coefficients of Sentinel-1 multi-temporal pictures and SOC were observed (Yang & Guo, 2019). SAR data have a wide range of application prospects for predicting soil properties, but their application to predicting SOC is still limited and rarely reported in the literature due to their complexity, diversity, and scarcity (e.g., SAR data pre-processing is more complicated and contains less information than optical data). Sentinel-1 and Sentinel-2, developed by the European Space Agency (ESA), have the advantages of a short revisit time, high spatial resolution (5–20 m for Sentinel-1 and 10–60 m for Sentinel-2), free download, and high-quality physical calibration data. They have the potential to offer a significant quantity of free optical and SAR remote sensing data with a high spatial resolution for SOC prediction (Wang et al., 2021b).

2.2 SOIL CARBON MODELS

A model is a simplified representation of reality useful for achieving various goals. A model may be a physical miniature of a real object, such as a clay sculpture, a hypothetical description of a complicated entity or process, or a schematic/mathematical depiction of a real-world system, theory, occurrence, or situation. SOC models simulate the turnover or breakdown of SOC and its transformation within the soil–plant–atmosphere system. To achieve a variety of different objectives, SOC models have been developed.

2.3 APPLICATION OF SOIL ORGANIC CARBON (SOC) MODELS

There are numerous applications for models of SOC turnover, which integrate soil, agronomical, and environmental studies (Balesdent et al., 2018;

Jones et al., 2005; Sulman et al., 2018). These models aid us in understanding soil carbon turnover and the underlying soil carbon stabilization processes. Simultaneously, models are increasingly being utilized to extend our knowledge of SOC dynamics temporally (future predictions), geographically, and globally. SOC models can also be used in agronomy, where they are integrated into decision support systems to enhance agricultural efficiency and environmental quality, such as APSIM, DSSAT-CSM, and SUNDIAL-FRS. Beyond the domain of experimental work, under various management and environmental conditions, SOC models can be used to study the dynamics of SOC. They may be used to interpolate and extrapolate experimental data both spatially and temporally, as well as to predict the fate of SOC under present and future environmental and management circumstances. Policymakers are increasingly relying on SOC models across the length and breadth of the globe.

Categories of soil organic carbon models:

i. Process-oriented models;
ii. Organism-oriented/food-web model;
iii. Cohort model.

2.3.1 *PROCESS-ORIENTED MODELS*

These models focus on how matter or energy moves and changes over time (Yagasaki & Shirato, 2014). Process-based models may be classified into the following categories:

- Single compartmental model, in which SOC is seen as a single homogenous compartment;
- Models with two compartments in which SOC is divided into two parts; and
- Multi-compartmental model includes SOCs such as Roth, CENTURY, and others.

Generally, the first two types of SOC models are static, with constant environmental factors, while the models of type (c) are predominantly dynamic, with varying environmental attributes. The majority of SOC models are multi-compartmental in nature, representing SOC as a limited number of compartments or SOC pools, each of which is homogenous. Decay rates are often described using first-order kinetics in terms of the pool's concentration (C) as:

$$dC/dt = kC$$

where; 't' is the time; and 'k' is the first-order rate constant. With a lower decay rate constant, the organic pool's stability increases with a longer half-life and turnover time. The compartments interact via material exchanges and interactions with the surrounding environment. As carbon moves from plant and animal wastes to microbial biomass and then on to soil organic pools of increasing stability, most models show that this occurs in a series of successions. It is common practice to split the output of an organic carbon pool among microbial biomass pools, another carbon pool, and the atmosphere's CO_2. Biomass and humus pools receive decomposed carbon based on microbial efficiency and stabilization/humification. CO_2 emissions must be accounted for by a ratio of efficiency and humification factors < 1. The majority of models used nowadays are multi-compartmental process-based models.

2.3.2 ORGANISM-ORIENTED/FOOD-WEB MODEL

Organism-based models are concerned with the movement of energy and matter (C and N) via soil organisms' food webs. These are dynamic and explicitly account for soil biota's various trophic levels or functional groupings. A comprehensive evaluation of the biology of the system being modeled is required for the development of food-web models, which are frequently parameterized for use in specific locations (Batlle-Aguilar et al., 2011; Senapati et al., 2014a).

2.3.3 COHORT MODEL

This model treats each new contribution of plant detritus as a distinct cohort that decays in a continuous manner (Heinemeyer et al., 2010; Negash & Kanninen, 2015). These models characterize decomposition as a continuous process. Most of these models are dynamic in nature and consider SOC as a separate pool that decomposes through a feedback loop back into itself. Every component of the SOC pool is characterized by its degradability and impact on the physiology of decomposers, which means that the pool can contain an infinite number of elements. Q-SOIL is an example of this kind of model (Powlson et al., 2011; Vogel et al., 2015).

A single rate equation describes the dynamics of each SOC component in this model, which is reliant on quality. Rate equations have their equations solved analytically. The most common of the three kinds of models described are process-based multi-compartmental dynamic models applied as they are – (i) simple to use and transfer; (ii) well integrated with GIS; (iii) the ease with which their various internal parameters can be estimated and calibrated for specific purposes; and (iv) their scalability. Diverse compartments streamline the complicated physicochemical and biological properties of SOC, which aid in the comprehension and visualization of the underlying turnover process.

2.4 ROTHC MODEL

Soil carbon dynamics can be simulated using the Rothamsted carbon model (Roth) in land-use systems and practices (Coleman & Jenkinson, 1996; Francaviglia et al., 2012; Zimmermann et al., 2007). Using this method, it is possible to predict SOC dynamics fairly accurately. The model considers the average air temperature, rainfall, and evapotranspiration (ET) to simulate the landscape C pool every month. Crop production data was used to estimate the annual volume of crop residue and roots. Various sorts of manure have their C: N ratios determined. Harvested biomass is returned to the soil as above-ground biomass (Farina et al., 2013; Francaviglia et al., 2012). Soil carbon sequestration potential and SOC change, an indicator of soil quality, can be used to assess the cropping study's validity. The balance between C input and output after decomposition can be seen in SOC change.

Soil SOM turnover in upland soils can be modeled using the Roth C model (Figure 2.1). Calibration of the model requires long-term experimentation in various climates, soils, and management practices. Over the course of 401 different experiments, the model was put to the test and found to be an accurate representation of SOC dynamics under various treatment conditions. To simulate changes in SOC under various soil and climate conditions and cropping systems (Bhattacharyya et al., 2011). With the help of advanced soil and climate databases, the RothC model simulates SOC density in major cereal-growing systems around the world (Wang et al., 2017) (Table 2.1).

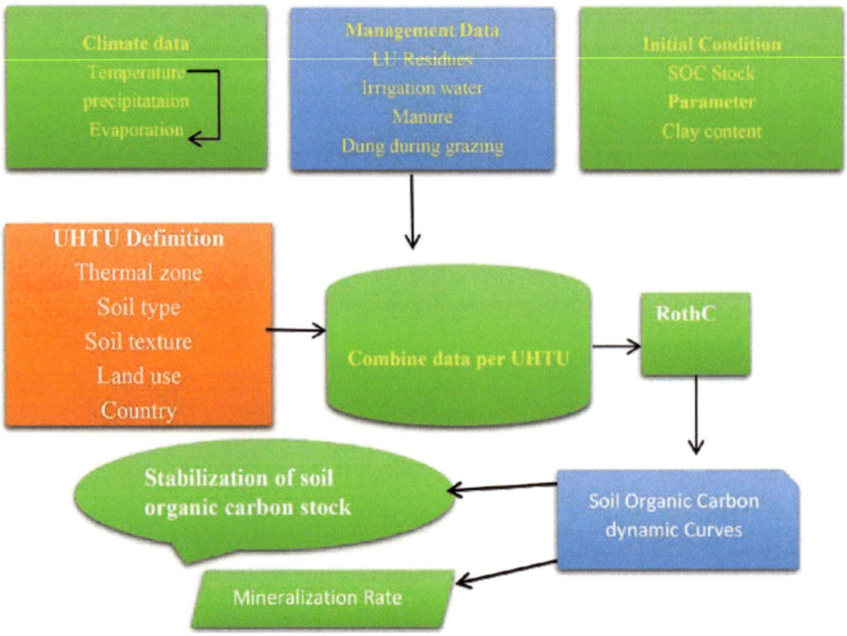

FIGURE 2.1 Workflow for data preparation in the RothC model.

Abbreviations: LU: Land use; SOC: Soil organic carbon; and UHTU: Unique homogeneous territorial unit.

TABLE 2.1 Soil Carbon Models

Model	Type	References
DAISY	Process-oriented multi-compartment.	Hansen et al. (1991)
NCSOIL	Process-oriented multi-compartment.	Molina et al. (1997)
CANDY (carbon and nitrogen dynamics)	Process-oriented multi-compartment.	Franko et al. (1995)
RothC (Rothamsted model)	Process-oriented multi-compartment.	Skjemstad et al. (2004)
CENTURY	Process-based	Parton (1996)
DAYCENT	Process-based	Parton (1996)
DNDC	Process-based	Li (1996)
SOCRATES (soil organic carbon reserves and transformations in ecosystems)	Process-based	Grace & Ladd (1995)
SOMM	Process-based	Chertov & Komarov (1997)

2.5 DAYCENT MODEL

Decent is a century biogeochemical model with a daily time step. C and N fluxes in the atmosphere, plant, and soil can be simulated using DAYCENT (Del Grosso et al., 2005, 2009; Necpálová et al., 2015). Plant production and NPP allocation, litter decay, nutrient mineralization, and N gas emissions from nitric and denitrifying bacteria are among the most critical sub-models. C and N flow between the pools is influenced by factors such as the C/N ratio, the lignin content of the material, and the abiotic water/temperature factors (Del Grosso et al., 2008; Oliveira et al., 2017; Zhang et al., 2018). Climate, water, and sunlight affect plant growth and yield. Water/nutrient stress and vegetation type determine how NPP is apportioned to plant components (e.g., roots or shoots) (Zhang et al., 2020). Variations in plant nutrient concentrations can be found depending on the type of vegetation and the number of nutrients available in the soil. It affects how quickly litter and SOM decompose and how well plants can use nutrients because of how much lignin is in the substrate, how much C/N is in the substrate, and how much water and temperature are available. It has been demonstrated that DAYCENT can accurately predict NPP and SOC, as well as N2O emissions and NO3– leaching, using data from native and managed systems. Various inputs are required to run a DayCent model, such as daily maximum, minimum temperature, precipitation, soil texture, pH, and organic carbon. Apart from these inputs, field management techniques include planting time and method, the cultivar being grown, fertilizer, and manure (Gurung et al., 2020).

2.6 SOCRATES (SOIL ORGANIC CARBON RESERVES AND TRANSFORMATIONS IN ECOSYSTEMS)

It's a process-based model that uses a small amount of soil, meteorological, and biological data to simulate effects on SOC. Relying on generic concepts of carbon cycling and biogeochemistry whilst being convenient to use and generally applicable, SOCRATES was developed (Gertsis & Vasilikiotis, 2018).

Additionally, it would not necessitate using fractionated carbon pools as an input. Annual precipitation (mm), mean annual temperature (°C), soil clay content (percent) or CEC (mmol kg–1), and initial soil organic C (percent) are the minimum driving variables for SOCRATES, which operates on a weekly time step (g cm–3). If bulk density data isn't available, a value of 1.3 g cm–3 is assumed. If needed, monthly climate data can be entered. First-order decomposition occurs due to temperature and relative humidity in the carbon model (Figure 2.2), which has five compartments (Grace & Robertson, 2021). Based on the conceptual fragments, all plant material can be divided into decomposable (DPM) and resistant (RPM) components.

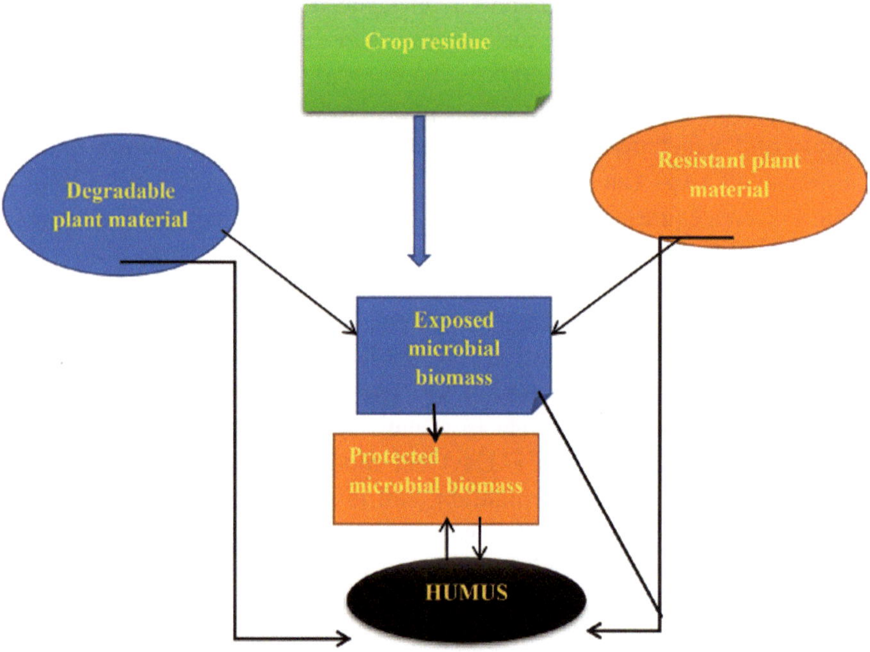

FIGURE 2.2 SOCRATES model components.

2.7 CENTURY MODEL

To simulate soil carbon, nitrogen, phosphorus, and sulfur dynamics, the CENTURY model was developed as a site-specific complex model (Bandaranayake et al., 2003; Berardi et al., 2020). First established for grassland, agricultural, and forest industries have now been added to the scope of the program. Some sub-models of the model include the SOM, water budget, and crop production model. Data for this model is based on average monthly rainfall and temperature for each month. Utilizing specific to a given site knowledge and pre-existing conditions of soil parameters (such as textural characteristics, depth, bulk density, total carbon, and nitrogen content, as well as crop growth parameters (Sowing date, fertilizer requirement, harvesting time, and varieties of crops) (Nicoloso et al., 2020). While simulating steady-state SOM, the model can also provide data on SOM turnover rates across various climates and management regimes. As per this model, better soil management increases the average soil carbon density. As per studies, soil C loss is influenced by various factors, viz., physical characteristics of the soil, management practices, and climate (Nicoloso et al., 2020). The rate constants for different decomposing compartments or SOC pools follow first-order kinetics, which is constant for a given set of biotic and abiotic circumstances. For less-than-ideal environmental conditions, the maximum value of the rate constant (k) is adjusted by multiplying it by a reduction factor (between 0 and 1). The temperature of the soil, water, oxygen, pH, nitrogen, clay content, CEC, salinity, vegetation type, plant cover, and tillage are all included in SOC models when k is modified (García-Palacios et al., 2021).

2.8 INITIALIZATION OF SOC MODELS

In simulation studies, SOC initialization across model pools is the most critical stage because soil C stock predictions can be affected (Dunstone et al., 2020; García-Palacios et al., 2021). Faulty initialization may result in an incorrect evaluation of intranasal variability and may also result in erroneous output trends when state variables drift back toward the model ideal. A problem with initialization may arise if SOC pools are partitioned because they are based on qualitative notions and do not always correspond to quantifiable fractions (Joensuu et al., 2021). It is thus possible to see how the model's goals and initialization method are reflected in the model. The most common way to deal with the initialization issue is to run a spin-up run of the model, that is, a run of the model for a few 100 to a few 1,000 years to figure out equilibrium/steady state SOC, presuming that the preliminary SOC and its distribution among the model pools are in equilibrium/steady state with current land use as well as weather factors. This is the most common way to deal with the initialization challenge (Duarte et al., 2022). However, the reported soils may be out of balance as a result of disturbances such as fire, erosion, changes in land use, and management. Models are run for so many years to accomplish a quasi-steady state distribution under prevailing land use and management conditions if they are initiated with default or steady state pool distribution. The concern, however, is that there isn't always enough information to adequately initialize the model for a large area. Getting the majority of pools to a steady state and altering SOC content in the slowest pool to fit the observed value is a third alternative (Premrov et al., 2021). The system's current state is represented by rapid turnover pools, whereas slow turnover pools represent the system's historical state. The pool distribution and observed C dynamics are combined as a final method using a statistical model-data fusion methodology. Using measured SOC pools can also be used to initialize a SOC model. Compared to other initialization methods, evaluated SOC pools have the advantage of not requiring knowledge of previous land use and management history or relying on model assumptions (Cécillon et al., 2018). Any process that is disregarded in the model but potentially affects SOC dynamics is also considered during SOC partitioning. Once the SOC pools have been quantified using a valid fractionation technique, model initialization problems and model performance may be addressed (Laub et al., 2019).

2.9 SOC FRACTION MEASUREMENT AND CONCEPTUALLY MODELED POOLS

Chemically and spatially, soil organic matter (SOM) is a complex mixture of organic elements which can be broken down over time (Lehmann & Kleber, 2015). Rather than being divided into separate pools, SOC appears to exist as a single, massive pool with an almost limitless number of possible turnover periods, ranging from minutes to millennia (Amthor, 1995). As a means of capturing SOC dynamics, process-based SOC models divide SOC into various conceptual pools based on their sizes and turnover periods, which are governed by first-order rate constants and influenced by various climate and soil variables (Waring et al., 2020). All of the material in a pool is identical in terms of chemical composition and accessibility to plants and soil microbes. It's a pool when a portion is non-composite and distinctive. On the other hand, several other studies have emphasized the methods and processes of SOC stability in measurable fractions/pools and the significance of modeling SOC dynamics. Using a simplified, predetermined, multi-compartment simulation largely based on empirical relationships, the SOC turnover in models was described Process-based multi-compartmental SOC models have a major drawback in that the distinct model compartments/pools seem to be conceptual and often do not correlate to measurable fractions (Krull et al.,

2003). Only microbial biomass and total SOC are considered in this experiment. As a result, it is often impossible to compare the dynamics of simulated SOC pools to those of observed SOC pools. However, measurable pools in models benefit from comprehensive model assessment and initialization. Several chemical techniques for separating SOC into distinct pools have been suggested during the past several decades. There is no consideration given to the impact of soil protection on the turnover rate of SOC pools by wet chemical fractionation approaches (von Lützow et al., 2007). Soil matrix and structure or aggregate of the soil significantly impact the methodology of SOC stabilization/decomposition. For soil C to be utilized by microbes, they need direct access to the carbon-rich soil. When the structure of the system doesn't quite affect the accessibility of SOC to microbiota, soil carbon dynamics are impacted by the soil's chemical makeup (Doetterl et al., 2016). While soil microorganisms eat the majority of organic substrate that falls or is deposited on the soil, organic carbon accumulates in the soil. Physical fractionation techniques have surpassed chemical fractionation methods in recent decades. Even though SOC fractionation is a physical process, it does not consider chemical or biochemical stabilization mechanisms, i.e., the chemical or biochemical nature of SOC on its stabilization cannot be accounted for by physical fractionation methods. Thus, in order to capture all of the SOC stabilizing processes, different techniques in SOC fractionation schemes have subsequently merged physical and chemical methodologies (Matilainen et al., 2011; Plante et al., 2006). Before chemical analysis, density or size fractionation was the most often used technique. Using density partitioning or particle size fractionation (POC), the light fraction (LF) or the sand-sized fraction (particulate organic matter: POC), which consists mainly of fresh plant matter, is separated from the rest fractions employing acid hydrolysis, base extraction, or oxidation with H2 O2/NaOCl/Na2S2O8. Certain physicochemical techniques have shown considerable potential for correlating measured SOC fractions to SOC pools. It was suggested that SOC be divided into POC, charcoal carbon (photo-oxidation), and humus (Schmidt et al., 2002). A similar approach for SOC partitioning into free LF/ POC, organic carbon (DOC), stable aggregate and sand-associated carbon (S + A, S + C), silt and clay-associated carbon (s + c), and an inert organic carbon (rSOC) was oxidized with NaOCl was suggested (Zimmermann et al., 2007). The RPM pool is represented by POC, the IOM pool by charcoal C, and the humus pool (HUM) is calculated as total SOC–(POC + Charcoal carbon) (Broos et al., 2011). The IOM has a direct impact on the C in the rSOC fraction. Under equilibria, the DPM: RPM ratio of the Roth model divides the total carbon in the POC and DOC fractions into the degradable plant matter pool (DPM) and the residual plant matter pool (RPM). S+A fraction and silt/clay fractions (s + c – rSOC) are separated into biomass pool (BIO) and hydrogen pool (H) using the same technique (HUM) (Finke et al., 2019). The SOC model relies heavily on the inert organic matter pool (IOM). According to the model, IOM is a non-dynamic inert pool that does not disintegrate. Despite the fact that many chemical techniques can produce a SOC fraction that is chemically resistant and stable, they are not completely inert. An extra study may be necessary to determine why the IOM proportion of SOC is so high. SOC stabilizing processes can be improved in multi-compartment models if the SOC fractions in each pool/compartment are based on distinctive, non-composite SOC fractions.

2.10 BLACK CARBON AND MODELING

Graphitic organic molecules undergo incomplete combustion, resulting in the inorganic carbon compound known as black carbon (C) (Vignati et al., 2010; Wang et al., 2018). Due to the resistance of black carbon to oxidation and microbial activity, it remains durable across geological timescales. Due to its refractory character, it is believed that black carbon is a component of the passive organic carbon pool (Kahnert & Kanngießer, 2020). This pool is anticipated to account for a greater percentage of total SOC in fire-prone areas. It was shown that black carbon was critical in the simulation of SOC. As a result, in responding to a 30°C warming over 100 years, approximated CO_2 emissions in two Australian savannah areas were lowered by 18.3 and 24.4% by including realistic black carbon stocks in prediction models (Lehmann et al., 2008). As a result, quantifying the quantity of black carbon is critical in SOC modeling. Historically, black carbon in soils has been determined by decomposing organic materials with nitric acid, heating, and then observing the results manually or under a microscope, sifting through char particles (Wagner et al., 2017). The use of benzene polycarboxylic acids (BPDA) as a marker for black carbon in nitric acid digestions, chemical digestion, or a combination of chemical oxidation followed by solid-state 13°C nuclear magnetic resonance spectroscopy are recent analytical techniques for evaluating char materials in soils. Among the types of oxidations were thermal, chemical oxidation, and molecular marker method (Schmidt & Noack, 2000). The resultant black C values differed by two orders of magnitude for each sample, suggesting a significant difference across individual techniques.

2.11 SOC MODELS EVALUATION

SOC model evaluation is a necessary but insufficient condition for model application and helps evaluate model performance. Model is evaluated due to follows: (i) To put it another way, it tells you how accurate your model will be in a given situation; (ii) analysis of model behavior; (iii) identifies the most important model components that affect the result; and (iv) its graphical analysis provides a quick approximation while also helping to improve the system's understanding, while quantitative analysis determines the simulation's statistical correctness (Cerri et al., 2007; Falloon & Smith, 2002; He et al., 2011). Certain statistics quantify association viz., correlation coefficient, while others quantify coincidence (RMSE, relative error, mean difference, modeling efficiency, and lack of fit, among others) between measured and simulated data. Greater correlation implies that the measured and simulated values follow similar patterns, while a higher coincidence suggests that the simulated values are near the corresponding observed values. A sensitivity analysis assesses the model's behavior, while an uncertainty analysis quantifies the amount of uncertainty created in each component's output (Senapati et al., 2014a). Generally, model performance is evaluated using long-term experimental data collection. Different levels of models can be evaluated, such as the individual processes, subset processes, or model outputs (for example, changes in total SOM over time (Porter et al., 2010).

2.12 SOC MODELS EVALUATION UNDER CHANGING LAND USE

Land use change is a worldwide issue that has a detrimental impact on the climate through greenhouse gas (GHG) emissions (Ripple et al., 2014). Almost all soils have lost between 40 and 90 Pg C to the atmosphere due to agriculture and disturbance, and changes in land use alone are estimated to have released 1.60.8 Pg C yr–1 to the sky during the 1990s. A comparison of several hypothetical model performances with varying degrees of correlation and connection between measured and simulated data:

- Strong association but weak correlation leads to poor model performance;
- Strong correlation but weak association leads to poor model performance;
- Strong association and coincidence lead to excellent model performance;

- Strong association and coincidence in time scale lead to excellent model performance.

The amount of SOC changes after LUC varies significantly across sites due to variations in environmental circumstances like the type of soil, texture, mineralogy, etc.), climate, ecosystem productivity, plant species, and management methods (Duarte-Guardia et al., 2019; Nayak et al., 2019; Singh et al., 2020). Predicting SOC changes as a result of fast land use changes is a critical challenge for the future, and SOC dynamic models are one of the most promising methods in this regard (Borrelli et al., 2017; Stockmann et al., 2013; Thuiller et al., 2008). Generally, model performance is assessed using a chronosequence long-term field experimental data set. Chronosequence data sets may also be used to evaluate model performance in simulating changes in SOC under LUC (Boysen et al., 2021; Poeplau & Don, 2013; Senapati et al., 2014a). However, there are just a few instances in which chronosequence data have been obtained before and after a LUC. In this regard, paired sites give a chance to investigate the impact of LUC on SOC and may serve as a foundation for evaluating SOM models. The paired-site concept comprises two locations, one where LUC happened many years ago and another nearby remnant, a native-vegetation site, both of which have a similar soil type, landscape position, and slope angle (de Oliveira Bordonal et al., 2017; Durigan et al., 2017; Pal, 2021).

2.13 CLIMATE CHANGE AND SOIL ORGANIC CARBON (SOC)

Various studies have examined the possibility of soil carbon sequestration to reduce climate change, and soil carbon stock changes will be a major driver in the 21st century. Increment in average air temperature is projected to decrease SOC breakdown and loss in the yet to come if soil moisture would not be a constraint (Black et al., 2017; Davidson & Janssens, 2006; Fan et al., 2013). On the other hand, any gain in soil carbon intake due to improved ecosystem productivity due to CO_2 fertilization or climate change, as well as technology and management advancements, may compensate for the loss (Calzadilla et al., 2013; Pant, 2011; Stevanović et al., 2016). Increased severe climatic events and decreased rainfall may lower ecosystem production, thus canceling any beneficial effects of climate change and CO_2 fertilization on soil C inputs (Gampe et al., 2021; Kalra et al., 2007; Xiao et al., 2018). Ultimately, the rate of SOC breakdown and the amount of soil C added to the soil will decide the destiny of the soil C reserve during future climatic shifts. Thus, it is critical to simulate the destiny of SOC in the face of climate change. Recent studies have examined the projection of SOC under climate change in arable, woodland, and grassland ecosystems. With a land area of approximately 40% and a carbon storage capacity of 39%, grasslands are among the most considerable and comprehensive terrestrial ecosystems on earth (580 Gt C) (Bardgett et al., 2021; Bartlett et al., 2020; Dlamini et al., 2016). Carbon cycles around the world will suffer if grasslands' SOC storage capacity is reduced in any way. As a result, it's critical to monitor how climate change affects grassland soil carbon stores. Over 90 years (1990–2080), a loss of 6–10% of European grassland SOC stocks was predicted (Smith et al., 2020; White et al., 2020). Similarly, a 10–11% decline in grassland SOC in Australia as a direct result of climate change was predicted (Senapati et al., 2014b). Systems incorporating agroforestry can help moderate climate change by slowing these increases (Lorenz & Lal, 2014). Agroforestry is growing crops, raising livestock, and harvesting wood on the same plot of land. Agroforestry practices can specifically increase the SOC pool, which is the only terrestrial pool capable of storing carbon (C) for centuries (Navarro-Pedreño et al., 2021). Land management, changes in land use, and climate change are the main factors that influence increases or decreases in SOC. These drivers of SOC change have been the subject of thousands of original studies, which have now been compiled in an increasing number of meta-analyzes (Beillouin et al., 2022; Smith et al., 2008).

2.14 LIMITATIONS OF SOC MODELS

Models of SOC turnover often treat soil microbial biomass as an active pool, so it's impossible to know exactly how it works in practice. Soil microbial biomass input is often not modeled in models because of changes in various management approaches. Soil microflora dynamics and feedback may improve the performance of the model, even though there is little indication that soil microflora accumulation restricts SOC turnover. Soil aggregates protect SOC from being mineralized. SOC is lost when soil aggregates are broken down, whereas SOC retention is enhanced when soil aggregates are formed. In contrast, most models fail to consider the entire process of soil aggregate production or obliteration and thus fail to consider the dynamics of soil structure and its effect on SOC stability or decomposition. The model could benefit from being enhanced by including aggregate soil dynamics. The feedback between soil structure and microbial activity in SOC turnover models must be addressed, despite some attempts to incorporate soil aggregation. Diffusion is critical for the movement and transport of solutes and gaseous substances, but SOC models often fail to explicitly model the diffusive process. The role of diffusive processes in SOC dynamics may also require model research. Many macronutrient dynamics (N, P, K) and micronutrient dynamics in the soil and SOC dynamics are overestimated in models. Decomposition in acidic soils occurs up to two-thirds slower than in alkaline soils, making SOC models overestimate SOC turnover. SOC turnover is a complex process; some models incorporate soil pH into the equation, while others don't. The SOC turnover model's overall performance may be improved if these variables, along with soil C, are properly modeled. It is also important that SOC models are better integrated into the landscape and whole-ecosystem processes.

2.15 CONCLUSION

SOC models are crucial for understanding how SOC behaves under present and future management approaches, land use change, and climatic conditions. Agronomic decision support systems use them to increase agricultural productivity while protecting the environment. These models can aid data exploration, extrapolation, interpolation, and explanation. SOC models also help us better understand soil carbon turnover and stabilization mechanisms. Process-based multi-compartment SOC turnover models are widely used. This simplifies understanding SOC's complex physical, biological, and chemical features. However, in multi-compartmental models, the pools are based on conceptual rather than quantitative ideas. To accurately represent a pool in the model, SOC fractions must be distinct and non-composite. As promising as it may be to link observed SOC fractions to model pools, more research is required. Using non-composite SOC fractions and various SOC stabilization methods can improve multiple pools/compartments in process-oriented multi-compartmental models. Once measured, SOC pools can be used to explicitly initialize the model. Compared to other initialization methods, measured SOC pools do not rely on model assumptions. However, details of previous land management are not necessary when using fractions because they better represent true site-specific conditions. To use a model, it must first be thoroughly evaluated. The model assessment ensures a model's trustworthiness and dependability. Before being used in a new context, SOC models must be thoroughly tested on as many reliable datasets as possible.

An evaluation chronosequence of long-term field experiments is usually used. A lack of long-term chronological data does not preclude the use of pair-site data sets. Climate change has altered soil C dynamics significantly over the last century. Increases in temperature are predicted to hasten the breakdown of SOC and its subsequent loss of SOC. Climate change can increase ecosystem productivity, while decreasing rainfall can reduce it. As a result, predicting SOC dynamics in the face of climate change is critical. SOC dynamics modeling relies on accurate black carbon measurements in wildfire-prone areas. Many SOC models misrepresent macro- and micro-nutrient dynamics in relation to SOC dynamics. As a result, they frequently overlook pH, soil aggregation, and biota dynamics, which is a major issue. Improved model representation of complex actual situations can be achieved by including all of these procedures, factors, and parameters.

ACKNOWLEDGMENTS

The author expresses gratitude to Tajamul Mansoor for his significant contribution in preparing this chapter.

KEYWORDS

- **black carbon**
- **CENTURY model**
- **climate change**
- **modeling**
- **RothC model**
- **SOC stocks**
- **soil carbon**

REFERENCES

Amthor, J. S., (1995). Terrestrial higher-plant response to increasing atmospheric [CO2] in relation to the global carbon cycle. Global Change Biology, 1, 243–274.

Bandaranayake, W., Qian, Y., Parton, W., Ojima, D., & Follett, R., (2003). Estimation of soil organic carbon changes in turfgrass systems using the century model. Agronomy Journal, 95, 558–563.

Bardgett, R. D., Bullock, J. M., Lavorel, S., Manning, P., Schaffner, U., Ostle, N., Chomel, M., et al., (2021). Combatting global grassland degradation. Nature Reviews Earth & Environment, 2, 720–735.

Bartlett, J., Rusch, G. M., Kyrkjeeide, M. O., Sandvik, H., & Nordén, J., (2020). Carbon Storage in Norwegian Ecosystems (Revised Edition).

Beillouin, D., Cardinael, R., Berre, D., Boyer, A., Corbeels, M., Fallot, A., Feder, F., & Demenois, J., (2022). A global overview of studies about land management, land-use change, and climate change effects on soil organic carbon. Global Change Biology, 28, 1690–1702.

Berardi, D., Brzostek, E., Blanc-Betes, E., Davison, B., DeLucia, E. H., Hartman, M. D., Kent, J., et al., (2020). 21st-century biogeochemical modeling: Challenges for Century-based models and where do we go from here? GCB Bioenergy, 12, 774–788.

Bhattacharyya, T., Pal, D., Deshmukh, A., Deshmukh, R., Ray, S., Chandran, P., Mandal, C., et al., (2011). Evaluation of RothC model using four long term fertilizer experiments in black soils, India. Agriculture, Ecosystems & Environment, 144, 222–234.

Black, C. K., Davis, S. C., Hudiburg, T. W., Bernacchi, C. J., & DeLucia, E. H., (2017). Elevated CO2 and temperature increase soil C losses from a soybean–maize ecosystem. Global Change Biology, 23, 435–445.

Borrelli, P., Robinson, D. A., Fleischer, L. R., Lugato, E., Ballabio, C., Alewell, C., Meusburger, K., et al., (2017). An assessment of the global impact of 21st century land use change on soil erosion. Nature Communications, 8, 1–13.

Boysen, L. R., Brovkin, V., Wårlind, D., Peano, D., Lansø, A. S., Delire, C., Burke, E., et al., (2021). Evaluation of soil carbon dynamics after forest cover change in CMIP6 land models using chrono sequences. Environmental Research Letters, 16, 074030.

Broos, K., Feys, H. B., De Meyer, S. F., Vanhoorelbeke, K., & Deckmyn, H., (2011). Platelets at work in primary hemostasis. Blood Reviews, 25, 155–167.

Calzadilla, A., Rehdanz, K., Betts, R., Falloon, P., Wiltshire, A., & Tol, R. S., (2013). Climate change impacts on global agriculture. Climatic Change, 120, 357–374.

Cécillon, L., Baudin, F., Chenu, C., Houot, S., Jolivet, R., Kätterer, T., Lutfalla, S., et al., (2018). A model based on rock-eval thermal analysis to quantify the size of the centennially persistent organic carbon pool in temperate soils. Biogeosciences, 15, 2835–2849.

Cerri, C. E., Easter, M., Paustian, K., Killian, K., Coleman, K., Bernoux, M., Falloon, P., et al., (2007). Simulating SOC changes in 11 land use change chronosequences from the Brazilian Amazon with RothC and century models. Agriculture, Ecosystems & Environment, 122, 46–57.

Chertov, O., & Komarov, A., (1997). SOMM: A model of soil organic matter dynamics. Ecological Modeling, 94, 177–189.

Davidson, E. A., & Janssens, I. A., (2006). Temperature sensitivity of soil carbon decomposition and feedbacks to climate change. Nature, 440, 165–173.

De Oliveira, B. R., Lal, R., Ronquim, C. C., De Figueiredo, E. B., Carvalho, J. L. N., Maldonado, Jr. W., Milori, D. M. B. P., & La Scala, Jr. N., (2017). Changes in quantity and quality of soil carbon due to the land-use conversion to sugarcane (Saccharum officinarum) plantation in southern Brazil. Agriculture, Ecosystems & Environment, 240, 54–65.

Del Grosso, S. J., Halvorson, A., & Parton, W., (2008). Testing DAYCENT model simulations of corn yields and nitrous oxide emissions in irrigated tillage systems in Colorado. Journal of Environmental Quality, 37, 1383–1389.

Del Grosso, S. J., Mosier, A. R., Parton, W. J., & Ojima, D. S., (2005). DAYCENT model analysis of past and contemporary soil N2O and net greenhouse gas flux for major crops in the USA. Soil and Tillage Research, 83, 9–24.

Del Grosso, S. J., Ojima, D. S., Parton, W. J., Stehfest, E., Heistemann, M., DeAngelo, B., & Rose, S., (2009). Global scale DAYCENT model analysis of greenhouse gas emissions and mitigation strategies for cropped soils. Global and Planetary Change, 67, 44–50.

Dlamini, P., Chivenge, P., & Chaplot, V., (2016). Overgrazing decreases soil organic carbon stocks the most under dry climates and low soil pH: A meta-analysis shows. Agriculture, Ecosystems & Environment, 221, 258–269.

Doetterl, S., Berhe, A. A., Nadeu, E., Wang, Z., Sommer, M., & Fiener, P., (2016). Erosion, deposition, and soil carbon: A review of process-level controls, experimental tools, and models to address C cycling in dynamic landscapes. Earth-Science Reviews, 154, 102–122.

Duarte, A. C., Ferreira, C., & Vitali, G., (2022). Use of simulation models to aid soil and water conservation actions for sustainable agro-forested systems. Natural Resources Conservation and Advances for Sustainability (pp. 389–412). Elsevier.

Duarte-Guardia, S., Peri, P. L., Amelung, W., Sheil, D., Laffan, S. W., Borchard, N., Bird, M. I., et al., (2019). Better estimates of soil carbon from geographical data: A revised global approach. Mitigation and Adaptation Strategies for Global Change, 24, 355–372.

Dunstone, N., Smith, D., Yeager, S., Danabasoglu, G., Monerie, P. A., Hermanson, L., Eade, R., Ineson, S., Robson, J., & Scaife, A., (2020). Skilful interannual climate prediction from two large initialized model ensembles. Environmental Research Letters, 15, 094083.

Durigan, M. R., Cherubin, M. R., De Camargo, P. B., Nunes, F. J., Berenguer, E., Gardner, T. A., Barlow, J., et al., (2017). Soil organic matter responses to anthropogenic forest disturbance and land use change in the Eastern Brazilian Amazon. Sustainability, 9, 379.

Falloon, P., & Smith, P., (2002). Simulating SOC changes in long-term experiments with RothC and century: Model evaluation for a regional scale application. Soil Use and Management, 18, 101–111.

Fan, Z., David McGuire, A., Turetsky, M. R., Harden, J. W., Michael, W. J., & Kane, E. S., (2013). The response of soil organic carbon of a rich fen peatland in interior Alaska to projected climate change. Global Change Biology, 19, 604–620.

Finke, P., Opolot, E., Balesdent, J., Berhe, A. A., Boeckx, P., Cornu, S., Harden, J., et al., (2019). Can SOC modeling be improved by accounting for pedogenesis? Geoderma, 338, 513–524.

Franko, U., Oelschlägel, B., & Schenk, S., (1995). Simulation of temperature-, water-and nitrogen dynamics using the model CANDY. Ecological Modeling, 81, 213–222.

Gampe, D., Zscheischler, J., Reichstein, M., O'Sullivan, M., Smith, W. K., Sitch, S., & Buermann, W., (2021). Increasing impact of warm droughts on northern ecosystem productivity over recent decades. Nature Climate Change, 11, 772–779.

García-Palacios, P., Crowther, T. W., Dacal, M., Hartley, I. P., Reinsch, S., Rinnan, R., Rousk, J., et al., (2021). Evidence for large microbial-mediated losses of soil carbon under anthropogenic warming. Nature Reviews Earth & Environment, 2, 507–517.

Gertsis, A. C., & Vasilikiotis, C., (2018). The LISA and SOCRATEES© approach for sustainable crop and soil management. Eco-friendly Agro-biological Techniques for Enhancing Crop Productivity (pp. 89–110). Springer.

Grace, P., & Ladd, J., (1995). SOCRATES (Soil Organic Carbon Reserves and Transformations in Agro-Ecosystems): A Decision Support System for Sustainable Farming Systems in Southern Australia. Cooperative Research Centre for Soil and Land Management, Adelaide.

Grace, P., & Robertson, G. P., (2021). Soil carbon sequestration potential and the identification of hotspots in the eastern corn belt of the United States. Soil Science Society of America Journal, 85, 1410–1424.

Gurung, R. B., Ogle, S. M., Breidt, F. J., Williams, S. A., & Parton, W. J., (2020). Bayesian calibration of the DayCent ecosystem model to simulate soil organic carbon dynamics and reduce model uncertainty. Geoderma, 376, 114529.

Hansen, S., Jensen, H., Nielsen, N., & Svendsen, H., (1991). Simulation of nitrogen dynamics and biomass production in winter wheat using the Danish simulation model DAISY. Fertilizer Research, 27, 245–259.

He, H., Xiong, R., & Fan, J., (2011). Evaluation of lithium-ion battery equivalent circuit models for state of charge estimation by an experimental approach. Energies, 4, 582–598.

Joensuu, K., Rimhanen, K., Heusala, H., Saarinen, M., Usva, K., Leinonen, I., & Palosuo, T., (2021). Challenges in using soil carbon modeling in LCA of agricultural products—The devil is in the detail. The International Journal of Life Cycle Assessment, 26, 1764–1778.

Kahnert, M., & Kanngießer, F., (2020). Modeling optical properties of atmospheric black carbon aerosols. Journal of Quantitative Spectroscopy and Radiative Transfer, 244, 106849.

Kalra, N., Chander, S., Pathak, H., Aggarwal, P., Gupta, N., Sehgal, M., & Chakraborty, D., (2007). Impacts of climate change on agriculture. Outlook on Agriculture, 36, 109–118.

Krull, E. S., Baldock, J. A., & Skjemstad, J. O., (2003). Importance of mechanisms and processes of the stabilization of soil organic matter for modeling carbon turnover. Functional Plant Biology, 30, 207–222.

Laub, M., Demyan, M. S., Nkwain, Y. F., Blagodatsky, S., Kätterer, T., Piepho, H. P., & Cadisch, G., (2019). DRIFTS peaks as measured pool size proxy to reduce parameter uncertainty of soil organic matter models. Biogeosci. Discuss, 1–29.

Lehmann, J., & Kleber, M., (2015). The contentious nature of soil organic matter. Nature, 528, 60–68.

Lehmann, J., Skjemstad, J., Sohi, S., Carter, J., Barson, M., Falloon, P., Coleman, K., et al., (2008). Australian climate–carbon cycle feedback reduced by soil black carbon. Nature Geoscience, 1, 832–835.

Li, C., (1996). The DNDC model. Evaluation of Soil Organic Matter Models (pp. 263–267). Springer.

Lorenz, K., & Lal, R., (2014). Soil organic carbon sequestration in agroforestry systems. A review. Agronomy for Sustainable Development, 34, 443–454.

Matilainen, A., Gjessing, E. T., Lahtinen, T., Hed, L., Bhatnagar, A., & Sillanpää, M., (2011). An overview of the methods used in the characterization of natural organic matter (NOM) in relation to drinking water treatment. Chemosphere, 83, 1431–1442.

Molina, J., Crocker, G., Grace, P., Klir, J., Körschens, M., Poulton, P., & Richter, D., (1997). Simulating trends in soil organic carbon in long-term experiments using the NCSOIL and NCSWAP models. Geoderma, 81, 91–107.

Navarro-Pedreño, J., Almendro-Candel, M. B., & Zorpas, A. A., (2021). The increase of soil organic matter reduces global warming, myth or reality? Sci., 3, 18.

Nayak, A. K., Rahman, M. M., Naidu, R., Dhal, B., Swain, C. K., Nayak, A. D., Tripathi, R., et al., (2019). Current and emerging methodologies for estimating carbon sequestration in agricultural soils: A review. Science of the Total Environment, 665, 890–912.

Necpálová, M., Anex, R. P., Fienen, M. N., Del Grosso, S. J., Castellano, M. J., Sawyer, J. E., Iqbal, J., et al., (2015). Understanding the DayCent model: Calibration, sensitivity, and identifiability through inverse modeling. Environmental Modeling & Software, 66, 110–130.

Nicoloso, R. S., Amado, T. J., & Rice, C. W., (2020). Assessing strategies to enhance soil carbon sequestration with the DSSAT-century model. European Journal of Soil Science, 71, 1034–1049.

Oliveira, D. M., Williams, S., Cerri, C. E., & Paustian, K., (2017). Predicting soil C changes over sugarcane expansion in Brazil using the DayCent model. Gcb Bioenergy, 9, 1436–1446.

Pal, S., (2021). From Groundwater to the Atmosphere: Using RELAMPAGO Observations and Modeling to Understand Changes in the Hydrologic Cycle of Southeastern South America. In: University of Illinois at Urbana-Champaign.

Pant, K. P., (2011). Economics of climate change for smallholder farmers in Nepal: A review. Journal of Agriculture and Environment, 12, 113–126.

Parton, W., (1996). The century model. Evaluation of Soil Organic Matter Models (pp. 283–291). Springer.

Plante, A. F., Conant, R. T., Stewart, C. E., Paustian, K., & Six, J., (2006). Impact of soil texture on the distribution of soil organic matter in physical and chemical fractions. Soil Science Society of America Journal, 70, 287–296.

Poeplau, C., & Don, A., (2013). Sensitivity of soil organic carbon stocks and fractions to different land-use changes across Europe. Geoderma, 192, 189–201.

Porter, C. H., Jones, J. W., Adiku, S., Gijsman, A. J., Gargiulo, O., & Naab, J., (2010). Modeling organic carbon and carbon-mediated soil processes in DSSAT v4. 5. Operational Research, 10, 247–278.

Premrov, A., Wilson, D., Saunders, M., Yeluripati, J., & Renou-Wilson, F., (2021). CO2 fluxes from drained and rewetted peatlands using a new ECOSSE model water table simulation approach. Science of the Total Environment, 754, 142433.

Ripple, W. J., Smith, P., Haberl, H., Montzka, S. A., McAlpine, C., & Boucher, D. H., (2014). Ruminants, climate change and climate policy. Nature Climate Change, 4, 2–5.

Schmidt, M. W., & Noack, A. G., (2000). Black carbon in soils and sediments: Analysis, distribution, implications, and current challenges. Global Biogeochemical Cycles, 14, 777–793.

Schmidt, M. W., Skjemstad, J. O., & Jäger, C., (2002). Carbon isotope geochemistry and nanomorphology of soil black carbon: Black chernozemic soils in central Europe originate from ancient biomass burning. Global Biogeochemical Cycles, 16, 70-71–70-78.

Senapati, N., Ghosh, S., Daniel, H., & Rakshit, A., (2014a). A basic concept on modeling soil organic carbon. Modeling and Simulation of Diffusive Processes (pp. 293–313). Springer.

Senapati, N., Hulugalle, N. R., Smith, P., Wilson, B. R., Yeluripati, J. B., Daniel, H., Ghosh, S., & Lockwood, P., (2014b). Modeling soil organic carbon storage with RothC in irrigated Vertisols under cotton cropping systems in the sub-tropics. Soil and Tillage Research, 143, 38–49.

Singh, A. K., Jiang, X. J., Yang, B., Wu, J., Rai, A., Chen, C., Ahirwal, J., et al., (2020). Biological indicators affected by land use change, soil resource availability and seasonality in dry tropics. Ecological Indicators, 115, 106369.

Skjemstad, J., Spouncer, L., Cowie, B., & Swift, R., (2004). Calibration of the Rothamsted organic carbon turnover model (RothC ver. 26.3), using measurable soil organic carbon pools. Soil Research, 42, 79–88.

Smith, P., Fang, C., Dawson, J. J., & Moncrieff, J. B., (2008). Impact of global warming on soil organic carbon. Advances in Agronomy, 97, 1–43.

Smith, P., Soussana, J. F., Angers, D., Schipper, L., Chenu, C., Rasse, D. P., Batjes, N. H., et al., (2020). How to measure, report and verify soil carbon change to realize the potential of soil carbon sequestration for atmospheric greenhouse gas removal. Global Change Biology, 26, 219–241.

Stevanović, M., Popp, A., Lotze-Campen, H., Dietrich, J. P., Müller, C., Bonsch, M., Schmitz, C., et al., (2016). The impact of high-end climate change on agricultural welfare. Science Advances, 2, e1501452.

Stockmann, U., Adams, M. A., Crawford, J. W., Field, D. J., Henakaarchchi, N., Jenkins, M., Minasny, B., et al., (2013). The knowns, known unknowns and unknowns of sequestration of soil organic carbon. Agriculture, Ecosystems & Environment, 164, 80–99.

Thuiller, W., Albert, C., Araújo, M. B., Berry, P. M., Cabeza, M., Guisan, A., Hickler, T., et al., (2008). Predicting global change impacts on plant species' distributions: Future challenges. Perspectives in Plant Ecology, Evolution and Systematics, 9, 137–152.

Vignati, E., Karl, M., Krol, M., Wilson, J., Stier, P., & Cavalli, F., (2010). Sources of uncertainties in modeling black carbon at the global scale. Atmospheric Chemistry and Physics, 10, 2595–2611.

Von, L. M., Kögel-Knabner, I., Ekschmitt, K., Flessa, H., Guggenberger, G., Matzner, E., & Marschner, B., (2007). SOM fractionation methods: Relevance to functional pools and to stabilization mechanisms. Soil Biology and Biochemistry, 39, 2183–2207.

Wagner, S., Ding, Y., & Jaffé, R., (2017). A new perspective on the apparent solubility of dissolved black carbon. Frontiers in Earth Science, 5, 75.

Wang, G., Zhang, W., Sun, W., Li, T., & Han, P., (2017). Modeling soil organic carbon dynamics and their driving factors in the main global cereal cropping systems. Atmospheric Chemistry and Physics, 17, 11849–11859.

Wang, Z., Huang, X., & Ding, A., (2018). Dome effect of black carbon and its key influencing factors: A one-dimensional modeling study. Atmospheric Chemistry and Physics, 18, 2821–2834.

Waring, B. G., Sulman, B. N., Reed, S., Smith, A. P., Averill, C., Creamer, C. A., Cusack, D. F., et al., (2020). From pools to flow: The PROMISE framework for new insights on soil carbon cycling in a changing world. Global Change Biology, 26, 6631–6643.

White, K. E., Brennan, E. B., Cavigelli, M. A., & Smith, R. F., (2020). Winter cover crops increase readily decomposable soil carbon, but compost drives total soil carbon during eight years of intensive, organic vegetable production in California. PLoS One, 15, e0228677.

Xiao, W., Chen, X., Jing, X., & Zhu, B., (2018). A meta-analysis of soil extracellular enzyme activities in response to global change. Soil Biology and Biochemistry, 123, 21–32.

Zhang, Y., Gurung, R., Marx, E., Williams, S., Ogle, S. M., & Paustian, K., (2020). DayCent model predictions of NPP and grain yields for agricultural lands in the contiguous US. Journal of Geophysical Research: Biogeosciences, 125, e2020JG005750.

Zhang, Y., Hansen, N., Trout, T., Nielsen, D., & Paustian, K., (2018). Modeling deficit irrigation of maize with the DayCent model. Agronomy Journal, 110, 1754–1764.

Zimmermann, M., Leifeld, J., Schmidt, M., Smith, P., & Fuhrer, J., (2007). Measured soil organic matter fractions can be related to pools in the RothC model. European Journal of Soil Science, 58, 658–667.

CHAPTER 3

Modeling for Soil Processes, Soil Quality, and Soil Biodiversity from an Agricultural and Ecosystem Perspective

NASIR BASHIR NAIKOO,[1] OWAIS BASHIR,[1] SHABIR AHMAD BANGROO,[1] NAZIR AHMAD GANIE,[2] ABDUL ROUF MALIK,[1] REHANA RASOOL,[1] BURGISH BHAT,[3] SHAFEEQ UR REHMAN,[4] and KHALID REHMAN HAKEEM[5]

[1]*Division of Soil Science, Sher-e-Kashmir University of Agricultural Sciences and Technology of Kashmir, Kashmir, India*

[2]*Division of Animal Genetics and Breeding, Sher-e-Kashmir University of Agricultural Sciences and Technology of Kashmir, Kashmir, India*

[3]*Division of Environmental Science, Sher-e-Kashmir University of Agricultural Sciences and Technology of Kashmir, Kashmir, India*

[4]*School of Environment and Civil Engineering, Dongguan University of Technology, Dongguan, Guangdong, China*

[5]*Department of Biological Sciences, Faculty of Science, King Abdulaziz University, Jeddah, Saudi Arabia:*

ABSTRACT

Soil is becoming scarce for food, energy, and raw materials production. Thus, understanding concepts have become increasingly important. One must first assess its functions to make wise decisions for healthy and long-lasting soil. Several soil science fields have evolved advanced techniques that examine soil processes, including physical, chemical, and biological ones. You'll find

Soil, Agriculture, and Ecosystem Modelling: Smart Technologies for Sustainable Solutions.
Owais Bashir, Khalid Rehman Hakeem, Surinder S. Kukal, and Razeef Mohd (Eds.)

a new systemic modeling framework allowing regular coupling of recognizable reductionist models. Various soil functional attributes are linked by a web of mutually reinforcing process networks. Despite significant progress, soil models are still fragmented across disciplines and fields of study. Better knowledge and experience exchange are required to understand soil characteristics. One of the most important indicators of soil erosion is soil organic carbon (SOC). Soil SOC stock reductions negatively affect habitats, biological diversity, and soil nutrient retention, resulting in degradation and negative effects. Soils are vital in agricultural landscapes for many reasons. Improved land use and farm management information can help achieve this goal. The soil biological function is evaluated as a proof of concept for an agricultural decision support tool. No action has been taken yet, but the evaluation of the soil's biological function has progressed. Agriculture production, hydrological cycle regulation, quality of water, as well as soil quality are all negatively impacted by unsustainable land use. Land management practice's long-term effects on soil quality can be better understood when considered in the context of current soil quality measurements. Improved land use and management plans will be created due to this information.

3.1 INTRODUCTION

Soil is an amazing resource because of its diversity and changing properties (Bardgett, 2005; Burns et al., 2013). Wild and domesticated plants benefit from the diversity of habitats provided by the forest and its resilience to environmental stressors such as extreme weather (Goettsch et al., 2021; Sharma & Gupta, 2020). It is impossible to get a complete picture of the soil by sampling it at the same place and time every time. We frequently have to make inferences between the points of sampling. For instance, we may have to interpolate in space if we want to accurately map soil properties. To make comparisons of various scenarios or management practices, we may have to forecast. Prediction necessitates the development and use of realistic models. Soil surveyors use conceptual models of soil variation to create a map of the soils they study. Others use more explicit soil variation models. For example, computer simulation models predict water flow and nutrient leaching in soil (Vereecken et al., 2016). Geostatistical models soil properties as random fields. Soil scientists use models to predict how soil will behave in space and time (Goovaerts, 2010; Sharififar, Sarmadian, & Minasny, 2019). These models may differ greatly, but they all include some ambiguity. The soil varies greatly at all scales, and a deterministic model cannot capture this variation fully. This chapter focuses on soil models that recognize the associated uncertainty.

3.2 HISTORY OF SOIL MODELING

Since the beginning of soil science, researchers have sought to quantify the soil's physical, chemical, and biological interactions at various levels of refinement (Yuxin Ma, Minasny, Malone, & Mcbratney, 2019; Wadoux, Román-Dobarco, & McBratney, 2021). To begin, soil science has relied on numerical and analytical models that date back to the early 20th century. These models focused on simulating water flow, heat flow and solute transport processes, SOC dynamics, and nutrients dynamics (Arnold et al., 2015; Bashir et al., 2016, 2021; He, Jansson, & Gärdenäs, 2021; Mohi-Ud-Din et al., 2017). For soils and porous media with well-defined properties, using analog and digital computers to solve single partial differential equations constituted the majority of these models. In the early days of digital computing, only physical and chemical processes were taken into account in the models, and soil structure processes were not taken into account. The earliest models consider soil structure when predicting solute transport and leaching (Addiscott, 1977). A number of studies in the 1990s focused on the role of soil structure in water flow and solute transport. Soil microbes and soil fauna were examined in relation to food webs and nutrient dynamics in these studies (Fry et al., 2019; Larsen, Jaramillo-López, Nájera-Rincon, & González-Esquivel, 2015). For modeling soil biodiversity and its deterioration, soil fauna and microbes' roles in nutrient transfer processes have recently been developed in soil ecosystem models. Soil models can now be developed and validated using data from X-ray tomography, soil neutron tomography, and molecular technology that enables quantitative assessment of molecular-driven soil biochemical functions have been developed as well as soil microbial composition (Helliwell et al., 2013; Tracy et al., 2015). It has been possible to design dynamic soil models integrating physical, biological, and chemical methodologies from pore scales to global scales; advancements in computers and developments in analytical and improved numerical algorithms, notably at the end of the 1980s. Even though initial modeling efforts made much progress, basic soil processes and their interactions are still lacking and insufficient, making it hard to predict and measure essential soil features. Soil processes are hard to figure out and measure because there isn't a lot of communication and coordination among research organizations and disciplines (Baveye et al., 2018; Vereecken et al., 2016). In this chapter, we examine soil modeling from an environmental and agricultural perspective, looking at soil characteristics, soil quality, and soil as habitat.

3.3 NOVEL APPROACH FOR MODELING SOIL FUNCTIONS

Soil dynamics cannot be simulated with first principles; therefore, importance among the most relevant and noticeable soil characteristics that would provide critical information on soil functioning becomes evident (Fredlund, Rahardjo, & Fredlund, 2012; Tate III, 2020). First, we propose categorizing observable soil attributes into three groups based on their change timescale. In terms of "soil characteristics properties," these include mineral composition, texture, layering, and depth, but soil management does not affect these things. Due to external stimuli, many soil characteristics (such as water content and temperature) can change dramatically (e.g., redox potential, microbiological activity) (e.g., soil state variables). "Functional soil characteristics" fall somewhere in the middle of these two extremes, able to change quickly in response to external forces but changing more slowly (days to months) due to internal activities and interplay. The most vital information on soil characteristics can be found in the classification of functional soil characteristics. Finding a set of functional characteristics that can be used to derive meaningful indicators is difficult. There seems to be agreement on which physical and chemical properties are required (Fierer et al., 2012; Wösten, Pachepsky, & Rawls, 2001). Continued study of bio-attributes is underway. The community's diversity and key (stone) species driving specific methods, such as bioturbation, should precede taxa community characteristics in understanding soil functions. The evaluation of obtained indicators is highly dependent on intrinsic soil parameters. The impacts of various forcing on functional properties could depend on the state variables (e.g., traffic compaction depending on the water content) (e.g., the relevance of macropores depends on soil texture). State variables (such as soil moisture, redox potential, and respiration), related fluxes (such as evaporation, leaching of nutrients, and greenhouse gas (GHG) emissions), and transformations (such as nutrient uptake and decomposition) are the focus of this study (e.g., mineral N, biomass production, nutrient transfer). It is common to treat a functional soil characteristic as if it were a permanent soil characteristic (e.g., water retention curve, water capacity, hydraulic conductivity, porosity, pH, cation exchange capacity, biological composition or group abundances) (Chethan Kumar, Dutta, Chaudhary, & Meena, 2021; Coggan, Hayward, & Gibb, 2018; Creamer, Barel, Bongiorno, & Zwetsloot, 2022). Soil management affects soil features in complex ways that it is difficult to oversimplify. Consider that soil's functional characteristics are dynamic and subject to external factors in this particular instance. This suggests that studying functional soil characteristics and how they interact in real-time is a daunting scientific challenge that should be taken on by researchers specializing in soil processes (Arnold, 2018; Johnson, Boyd, O'Connor, & Smith, 2022; Rosenzweig, Carolan, & Schipanski, 2020). There are two extremes in complexity: small-scale complexity and extremely simplified static indicators, which this notion seeks to bridge, and it is predicated on the comprehension that all these functional properties emerge from various processing interactions at smaller scales. However, state variables should be included in the model concept, as soil management and global warming impact the functional properties of soil because these state variables change over time. The bulk density of soil is affected by its moisture, which should be considered. A long-held belief is that soils are self-organizing systems that are complex and dynamic. However, there is currently no corresponding model approach. External perturbation has a relatively small effect on soil behavior, but when perturbation exceeds a critical threshold, the soil state can shift into a different mode or configuration. Soils exhibit numerous important components of complex systems in responding to environmental perturbation. When SOC levels fall below a threshold that can't be compensated for by internal structure-forming processes, the soil is degraded (Bestelmeyer, Duniway, James, Burkett, & Havstad, 2013; Borgogno, D'Odorico, Laio, & Ridolfi, 2009). This leads to a positive feedback loop that reduces OM production as well as the ability of soil organisms to perform their functions. This integrates the complex process interactions that underlie them to provide a systematic perspective. A century of quantifiable pedology shows that the state of these functional soil parameters, especially their combination, is not arbitrary but instead continues to follow typical patterns based on soil types defined by pedogenic factors. Although this is true for abiotic factors like soil type, it is less so for biological species distributional patterns (Amundson, 2021; Yuxin Ma et al., 2019). Abiotic functional characteristics have been explicitly incorporated into the United States Soil Taxonomy and the World Reference Base for Soil Resources. Each location's unique microclimate influences the introduction of soil functional characteristics, such as parent material climatic factors, topography, vegetation, and land use. This serves as the foundation for quantifiable pedology, for example, in the SCORPION methods. For a deeper understanding of soil, it is necessary to reconsider the traditional explanation of soil types as mixtures of functional soil characteristics. Many variables can be considered in terms of complex systems, and in the complex and multifaceted state space of functional soil characteristics, an active type of soil is an attractor (Thompson, 2010). This also implies that these attractors are reasonably constant inside emerging pressure. The consequence of this idea would be that a set of functional properties is more than just a collection of distinct traits; rather, the set members are interdependent. This is demonstrated by the fact that they all arise from the same interplay of soil processes. As a direct result of the physical, chemical, and biological processes they are a part of, knowledge of these processes can be gained through observational science and comprehension. Soil's response to instabilities from the outside world is influenced by "effective" relationships between the functional characteristics of the soil that produce the attractor. When we look at these relationships, we can see them clearly, but we cannot see or measure the fundamental processes that lead to them (Vogel et al., 2017). They will be highly non-linear and enclosed, as are most biological ecosystems. Examples of associations that are not expected to be linear include those between SOM and aggregate stability, tunneling macrofauna, and soil bulk density, both of which have clearly defined minimum or maximum values.

3.4 SOIL MODELING AND ENVIRONMENTAL SERVICES

Ecosystem services (ES) are useful for analyzing soil model-based problems and providing diverse approaches. In addition to regulating and supplying ecological functions, soil also degrades and aids in the process, all of which are associated with social and demography concerns and are at the core of modern science as to how the world tends to work and how it will change in the future (Greiner, Keller, Grêt-Regamey, & Papritz, 2017; Paul, Kuhn, Steinhoff-Knopp, Weißhuhn, & Helming, 2021; Schwilch et al., 2016). It helps us to think about and perform soil modeling if we follow it. This can now better describe and predict soil processes and properties. The dominant framework examines the impact of natural and anthropogenic factors on soil processes and ES. It also examines how soil natural capital and procedures have been impacted. Subcomponents like processes, soil capital, and ecological processes can be used to meet basic needs. Modifications in land use and agricultural practices can also affect the above consequences (Paul et al., 2021; Pavan & Ometto, 2018). Natural capital includes the mass and energy of soil and the organization of these resources. Something to do with the concept of inherent and manageable properties in soil. Several authors discuss a wide range of activities to quantify and forecast soil degradation and provide services that can be regulated or regulated through soil modeling. For soils to function properly and accumulate and preserve their natural capital, they must have a basic process known as "supporting processes" (Paustian et al., 2019; Vereecken et al., 2016). Soil microbial interaction and biodiversity are involved in the formation and structure of soil, the cycling of nutrients, and the primary production of food. In addition to erosion and surface sealant, soil compaction, salinity, and acidification are all factors contributing to soil degradation. Services provided by the government ensure that people have a safe, healthy, and sustainable environment (Creamer et al., 2016; Zhang, Maltais-Landry, & Liao, 2021). In addition to climate control and water quality, they also prevent erosion. One definition of the term is regulating climate-altering energy, water, and matter fluxes by soils. In terms of water regulation, soil services that aid in water storage and retention are considered. Irrigation and drainage are among the water management practices affected by this. Soils can store and release different chemicals that impact soil, water, crops, and the atmosphere. To provide these services, soils must be used because they are critical to a wide selection of products from the ecosystems (including food, fibers, wood, freshwater, infrastructure, and genetic resources). These advancements are based on fundamental biological, physical, and geochemical procedures (Sadatshojaei, Wood, & Rahimpour, 2021). A few other fundamental aspects have been the focus of soil model-based research, largely ignoring the many other basic mechanisms. For this review, we will focus on the most important aspects of modern soil modeling techniques. We will look at just how soil modeling is used to assess soil processes and associated ES, as well as the components and challenges associated with the potential implementation and effectiveness of current soil models. Research in soil physics has grown substantially in recent years, with the focus shifting from the status and challenges of soil water properties to a wide range of other topics, including the impact of plants on transportation, the identification diversity of microbial population, and the role of soil ecology in providing ES. According to studies on ecological systems, soil functions are rarely described in terms of their mechanisms. ES can be quantified using one-dimensional proxy data, which is not validated models based on probable mixtures of independent variables, validated equations calibrated on primary and secondary data sets, representative datasets collected within the area used to evaluate ES or assumed modeling of the ecosystem service amount within a financial value (Van Looy et al., 2017; Vereecken et al., 2016). The supply side of ecological systems has been quantified using various models, including the poly-scale landscape model, the ARIES, and SOIVES). However, there are unresolved issues regarding the testing and validation of these models and approaches regarding precision and accuracy. In order to assess particular ecosystem functions, quite complex mechanistic models could be necessary in order to evaluate ambiguity in ES, following approaches similar to those used by the IPCC (Troy, Pavao-Zuckerman, & Evans, 2015). In order to perform these types of analyzes, the model and comparison platform we'll be presenting later on in this chapter might be an excellent option. Models of ES can be compared using mechanistic soil representations as a basis for comparison. These models could be simplified to make it easier for these models to incorporate complex modeling techniques. The lack of available data is a major reason to use simplified models instead of more complex models for assessing and quantifying soil characteristics. For example, soil scientists use the pedotransfer function, remote soil sensing, and proximal soil sensors to resolve the lack of available data in their research (Padarian, Minasny, & McBratney, 2020; Wadoux & McBratney, 2021). ES have yet to be quantified using these techniques. Soil science research has been at the heart of dealing with issues of spatial heterogeneity of key ecosystem properties, and this has become increasingly apparent in ES investigation as well (Feng, Wang, Bai, & Reading, 2019). Geographical information systems (GIS) and spatial maps of ES help to quantify larger areas and identify the spatial patterns of ES quite clearly. Using a GIS framework such as this, ecosystem service quantification can consider both vertical and lateral control, such as overland flow and routing, erosion, and sediment transport (e.g., playscape model). Additional steps include gauging the quality and accuracy of the forecasts, as well as verifying those results against actual data. When it comes to ecosystem service demand, economists have focused on estimating how many these characteristics and drivers are worth (Martínez Pastur, Peri, Lencinas, García-Llorente, & Martín-López, 2016).

3.5 WATER CYCLING

As part of the soil water cycle, evapotranspiration (ET) or leaching is used to restore this water vapor into the atmosphere and groundwater (Han et al.,

2020; Ripl & Eiseltová, 2009). In order to accurately describe and predict soil-based ES, measurements of rainfall, interception, soil infiltration, soil moisture redistribution, and root water uptake are essential. Rainfall is one of the most difficult and unpredictable of these processes to study because of its wide spatial and temporal variability. A rise in the spatial and temporal variability and strength of infiltration and overland flow processes is a consequence of climate change (e.g., strong convective rainfall events) (Jiang, Gan, Xie, & Wang, 2014). Soil flow of water models typically employs the Richards equation, which also integrates both Darcy and continuity equations, to better account for root-extracted moisture in the soil.

$$\frac{d\theta}{dt} = \nabla.(k\nabla H) - S \tag{1}$$

where; 'θ' is the volumetric water content ($L^3\ L^{-3}$); 't' is the time; 'k' is the unsaturated soil hydraulic conductivity tensor ($L\ T^{-1}$); 'H' is the hydraulic head (L); and 'S' is the sink term accounting for root water uptake ($L^3\ L^{-3}\ T^{-1}$).

SWAP model is an agro-hydrological model with a physical basis (Figure 3.1). During the simulation, a basic crop model is used (Noory et al., 2011; Rallo, Agnese, Minacapilli, & Provenzano, 2012). Linear or fixed-length crop cycles are used to model wheat development. In order to calculate the root water extraction rate, Sp (z) (d–1), we need to know how dense roots are at a particular depth, root (z) ($cm^3\ cm^{-3}$), concerning the total root length density. Water stress can be represented by the function in the SWAP model (Jiang et al., 2011).

$$\frac{Sp(Z) = \pi root(z)}{\int^{0}_{-Droot} \pi root(z)\ \partial Z * T_p} \tag{2}$$

where; D_{root} is the root layer thickness in cm.

3.6 SOIL QUALITY

3.6.1 SOIL ORGANIC CARBON (SOC) MODELING

Soil organic carbon (SOC) modeling is widely accepted as a reliable indicator of soil quality. It improves the soil's structure, fertility, and ability to store water, all of which contribute to the ecosystem's overall health (Lehmann, Bossio, Kögel-Knabner, & Rillig, 2020). Land cover, vegetation, soil parent material, weather patterns, and landforms affect SOC content, as does human activity. Soil quality has been analyzed in natural and artificial ecosystems using a model.

3.6.1.1 TRENDS IN SOIL ORGANIC CARBON (SOC) MODELING

To determine soil quality, we can look at the amount of SOC in it. Long-term agricultural sustainability depends critically on this soil microbial variable (Li, Cui, Chang, & Zhang, 2019; Su et al., 2015). There are several ways in which this nutrient can influence soil fertility, C cycling, soil-atmosphere interactions, and even the sequestration of carbon dioxide (CO_2). A soil's SOC content can be used to assess its quality in both natural and artificial ecosystems. SOC is a critical indicator of land use and management changes. SOC content varies depending on the type of land use.

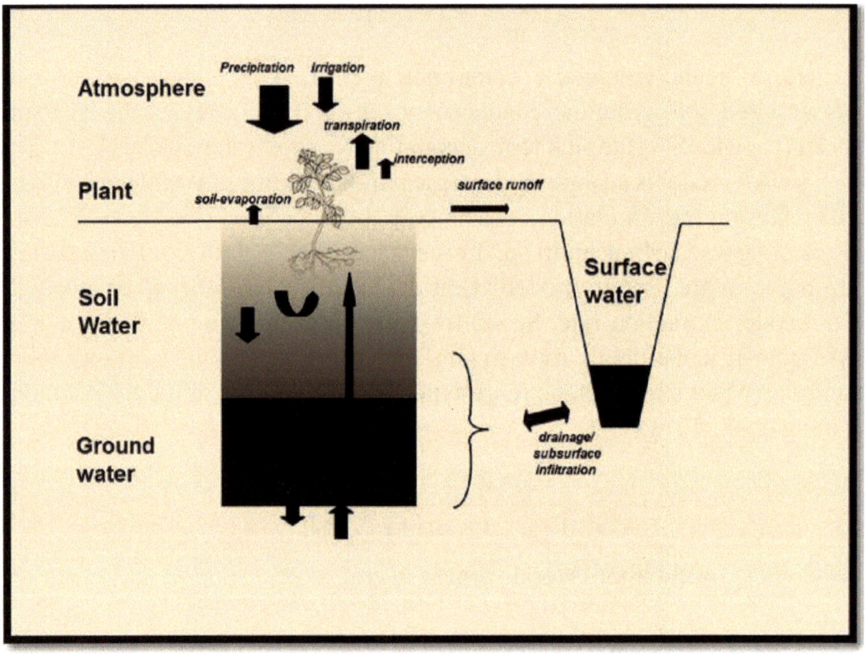

FIGURE 3.1 SWAP model.

Process-based soil C models require numerous model inputs, but they are difficult to obtain on a large scale. These models can be used to study soil C dynamics at large spatial and temporal scales under various environmental and management conditions and provide useful information. A breakdown of the soil C models in use are discussed in subsections.

3.6.1.2 SOIL CARBON DYNAMICS MODELING

Like other models of ecosystem functioning, models of soil organic matter (SOM) turnover are needed to assess processes, draw inferences or interpolate experimental findings, and explore situations and suppositions outside the scope of experimentation (Sulman et al., 2018). Because of this, there are many different SOM models, each with a different formulation and purpose. Multi-compartment models, process-based multi-component models, and models that consider the carbon and nitrogen transfers through different trophic levels of a soil ecosystem are all examples of strategies for simulating SOM turnover. SOM models, highlighting their strengths and weaknesses as well as future developments. SOM models like RothC and CENTURY are frequently used.

3.6.1.3 SOIL ORGANIC MATTER MODELS (SOMS)

For SOM models, there are multiple sources of metadata as well as information (Laub et al., 2020; Wutzler & Reichstein, 2013). The global change and terrestrial ecosystems SOM network (GCTESOMNET) includes more than 30 operational SOM models. Several authors have thoroughly examined SOM models in the past. Organic matter (OM) models can be divided into one, two, or four compartments based on the amount of OM in the soil. However, some models are generally static, while some are dynamic. As a result, a SOM model may be used to predict SOM degradation, soil nitrogen, and other essential nutrients. The CENTURY model incorporates a variety of sub-models for plant management and growth, soil moisture and temperature, and other aspects of SOM modeling (Han, Sun, Gan, Fu, & Zhu, 2020).

3.6.1.4 SOM MODELS; MULTI-COMPARTMENT

Most models seem to be process-based, which indicates they specialize in the movement and transformation of energy and matter (Saiti, Macaš, Lhotská, Štechová, & Pithová, 2020). 1st rate kinetics is the most common assumption. The SOM was initially viewed as a single, homogenous compartment in early versions. The GCTE-SOMNET database contains 33 SOM models, which are multi-compartment process models. A model's compartments and SOM pools are defined by their location within the model's structure and the rate at which they degenerate. The pool's concentration (C) is commonly used to measure the decay rate. First-order kinetics is used to show this.

$$\frac{dC}{dt} = -kC \tag{3}$$

where; 't' is the time required to reduce the concentration to ½ of the pool when there is no input; and 'k' is the rate constant of first-order kinetics.

The organic pool's half-life, turnover time, and stability increase with decreasing k values. Carbon in plant and animal waste is typically transferred into soil organic pools via microbial biomass (Gunina, Smith, Kuzyakov, & Jones, 2017). They use feedback loops to think about things such as the evolution of organisms over time. It is not uncommon for an organic pool's output flow to be divided into two distinct streams. If it is in an open area, it will end up in a microbial biomass pool, an organic pool, and a CO_2 pool. This split is like a microbial population that is eating and growing on the same thing at the same time. Two things are needed to measure the split flow. Their efficiency and stability (humification) factors are often used to describe them. These factors influence the amount of carbon that enters the biomass and humus pools, made up of carbon that microorganisms have degraded. For environmental reasons, the sum of the efficiency and humification factors must all be less than one (Zhang et al., 2019).

3.6.1.5 APPLICATION OF SOIL ORGANIC MATTER (SOM) MODEL

As a research tool, the different representations of soil carbon and nitrogen dynamics are frequently used to differentiate among competing hypotheses or to improve general knowledge of SOM turnover mechanisms in soil (Zhang et al., 2021). Several SOM models, such as SUNDIAL-FRS, DSSAT, and APSIM, are now used in decision support systems to improve agronomic performance and environmental efficiency in an increasingly wide range of agricultural applications. When attempting to extrapolate our knowledge of SOM dynamics over long periods and across large geographic areas, SOM models are more critical than before. In the central United States, the CENTURY models have been used to simulate the effects of various management practices and policies on agricultural ecologies. For countries' UNFCCC and Kyoto Protocol findings on terrestrial carbon storage, models can also be used. SOM models need to be open, recorded, and evaluated thoroughly. Our existing SOM models are still based on many assertions and understandings. Without site-specific calibration, new SOM model advancements in the future should make models more predictive. SOM model prediction estimates will be improved, and ambiguity will be reduced due to these developments (Mehta et al., 2017).

3.7 FOOD WEB MODELS

Soil biodiversity at various trophic levels and in different functional groups are explicitly modeled in these models that simulate carbon and nitrogen transfer through a network of soil organisms. Soil biota descriptions and process-based approaches have been combined in some models. Food web models are parameterized for use at specific locations and necessitate an in-depth understanding of the biology of the system being simulated (Libralato, Christensen, & Pauly, 2006).

3.7.1 JENA SOIL MODEL

JSM is a biogeochemical soil model derived from the vertically explicit SOC model COMMISSION, which uses C as the only input parameter (Yu, Ahrens, Wutzler, Schrumpf, & Zaehle, 2020). As part of COMISSION's evolution from the conventional model, soil texture was used to determine the soil's maximum capacity to absorb organic carbon and the microbial residues of dissolved organic C. Temperature and moisture rate modifiers were used for microbiome procedures. In a study conducted, we will look into how mineral-associated organic C's highest adsorption capacities affect the trends in SOC content and radiocarbon age that we see. An integrated soil model and one incorporating vegetation and surface processes are possible with QUINCY and TBM modeling. JSM was used as a standalone model in this research. JSM cannot simulate litter production as it only describes processes at the atmosphere-soil interface or the depth of the soil profile in terms of energy and water. With the increase in soil depth, the SOM layer's thickness grows exponentially, as shown in the JSM model.

There are different layers for each biogeochemical process and each pool of C, N, and P. As a result of percolation and bioturbation, biogeochemical pools move from one layer to the next. Sediment organic matter (SOM) is depicted as pools of dissolving organic matter (DOM), minerals-associated DOM, as well as microbial biomass and mineral-associated microbial residue. As with CO_2, some procedures connect the inorganic nutrient cycles to OM turnover. These processes include mineralization and plant uptake. According to the model's assumption, MB is supposed to have an established stoichiometry. DOM is a mixture of inorganic and organic forms of carbon, nitrogen, and phosphorus (C, N, and P) from dissolved mineral pools. As a general rule, microbes are assumed to be striving for rapid growth by using as much C as possible. However, when there are insufficient nutrients for growth, microbes lower their C use productivity and enhance nutrient mineralization. Based on C: N: P ratios of influx and efflux, all other SOM pools can be calculated. As long as respiration is a component that produces these fluxes, they will retain their stoichiometry relative to their source pools (Gerphagnon et al., 2015).

3.8 MODELLING NUTRIENT ACQUISITION AND ENZYME DISTRIBUTION IN SOIL

While the major soil stocks, as well as biogeochemical fluxes in the SEAM-off scenario, were similar to those in the Base Scenario (Ma et al., 2018), although its composition, as well as radiocarbon pattern, have both been modified, it is feasible that entire amount of SOC was not impacted by either scenario. At three different depths, the uptake of inorganic PO_4 was shown to compete with phosphate-adsorption, and seasonal and vertical uptakes of the two inorganic ions also were observed, including microbes and plants. Regarding inorganic P uptake in JSM, the simulations revealed that microbes outcompeted roots at all depths. A reduction in plant P uptake, but a rise in competitive root nature for phosphate uptake, was observed as soil depth continued to increase. However, as soil depth increased, so did the phosphate adsorption rate; consequently, the plant and microbial uptake in the deeper soil layers increased (Postma, Dathe, & Lynch, 2014). Due to higher microbial and plant activity in the summer months, the phosphate adsorption capacity of the topsoil decreased significantly.

On the other hand, microbes were outcompeted by plants throughout the year and throughout the soil profile, especially in the summer when DOM is the primary source of N for microbes. The model's nutrient acquisition competition feature has been disabled. In the ECA-off scenario, topsoil microbial biomass and Po-to-Pi ratios have been considerably lower. Due to reduced microbe competition, plants seem to take up more inorganic P in the soil surface layer than in the Base Scenario. Consider the fact that fluxes under the ECA-off scenario varied far less seasonally. While the decrease in P flow rate with a soil depth is less prominent in the scenario without ECA, it is more pronounced in the scenario where net N mineralization is reduced with soil depth. Adsorption and absorption processes, which are more significant in the soil P cycle than in the soil N cycle and exhibit seasonal and vertical patterns quite distinct from biological pathways, such as mineralization, are to blame for this discrepancy (Oroo, 2021).

3.9 MODELING SOIL STRUCTURE

Solid particles were made to extend into the entire space by degrees using the RPF model, which proposes that the pore phase has the fractal property. In contrast to the regularity and structure of real soil, its apparent symmetry conflicts. The steps outlined in Figure 3.2 must be completed to create an RPF model. Determine how many white square units (depiction of pore phase) to divide (m*m square units) using the Initial dimension (m). The iteration units are chosen randomly from among the pore units following PGP, Ga. A uniform distribution function within the system will randomly assign a number to each pore unit (0, 1). Random numbers larger than Ga will serve as iteration units for this procedure. On the other hand, the remaining pore units retain their pore phase characteristics. Iteration graphs based on PDIU and Dm are generated (Cao, Li, & Li, 2019) (Table 3.1).

TABLE 3.1 Soil Processes Models

Soil Processes	Model	References
Soil organic carbon dynamics	Single compartmental model. Two compartmental models RothC and CENTURY	Coleman & Jenkinson (1996); Parton (1996)
	Cohort model	Bosatta & Ågren (1995)
	Jena soil model	Yu et al. (2020)
Formation of soil	SoilGen2	Keyvanshokouhi, Cornu, Samouëlian, & Finke (2016)
	Soil-landscape model	Gobin, Campling, & Feyen (2001)
	Orthod model	Minasny, McBratney, Mendonça-Santos, Odeh, & Guyon (2006)
	MILESD	Temme & Vanwalleghem (2016)
Nutrient cycling	HP1	Jacques & Simunek (2005)
	SWAP	Kroes, Wesselling, & Van Dam (2000)
Soil water	HYDRUS 1D	Ying Ma, Feng, Su, Gao, & Huo (2010)
	WaSim-ETH	Cullmann, Mishra, & Peters (2006)
	Community land model (CLM)	Drewniak, Song, Prell, Kotamarthi, & Jacob (2013)
	SiSPAT-isotope	Braud, Bariac, Gaudet, & Vauclin (2005)

It is necessary to divide each iteration unit into m2 subunits first (*see* Step 1). Every time an iteration unit generates new m2 sub-units, the sub-units will turn black if the random new number they receive exceeds Dm (representing solid phase). Take the graphs of two 4×4 squares as a starting point. Iteration units that retain their original properties were substituted for the many disordered iteration graphs produced by Step 2. Repeat Steps 2, 3, as well as four till the number of iterations, N, satisfies the setting's requirements.

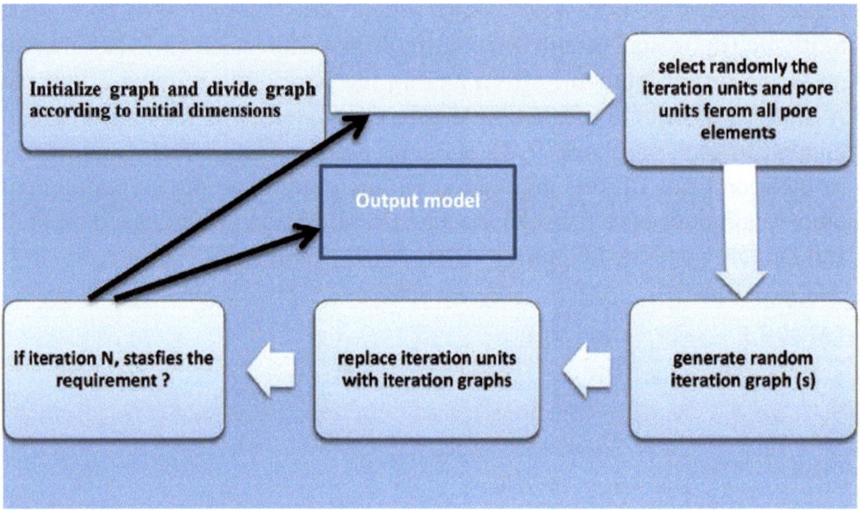

FIGURE 3.2 Flow chart modeling.

3.9.1 SOILGEN2 MODEL

The Richards' equation for unsaturated water flow and the Convection-Dispersion Equation for solute transport regulates the SoilGen transport model (Yu, Finke, Wu, & Guo, 2013). Heat flow is also used to evaluate constant chemical values, OM mineralization, and water flow to determine the temperature of frozen soils. If water fluxes are relatively large, the 0.05-day time intervals are reduced to a minimum of 10–7 days, but this is rarely less than 10–4 days. Whether such a procedure formulation seems complex to predict over several 1,000 years is an open question. Even so, there have been no model simplification efforts, even though flow rates play a role in clay migration. As seen in Figure 3.3, the SoilGen2 chemical system includes Al and H in the exchange phase, Gibbsite precipitation and Al solution, and calcium, magnesium, sodium, potassium, and aluminum in the un-weathered phase following the first stage of soil development. Calculate chemical equilibrium every day and every 80-time step (Finke & Hutson, 2008). SoilGen2 utilizes the four vegetation types (grass/scrub; conifers; deciduous wood; agriculture; barley) to estimate root allocation, nutrient absorption, and C-mineralization. The above sub-model was derived from the RothC26.3 model. SoilGen1's cation uptake forcing function already contains AI and follows the comparative abundances, but even then, it is identical to SoilGen1. For example, bioturbation affects the physical weathering and new mechanisms for clay migration that were added to SoilGen2 but were absent from SoilGen1. If a depth-dependent input mass fraction is fed into a bioturbation pool, this pool is simulated as an insufficient mixing process (vertical mixing). Then it is re-distributed to each soil compartment based on the mass fractions of the input material (horizontal mixing). During this procedure, we deal with a wide range of different substances, including dissolved solids, soil water, cations exchanged, and recrystallized salts (Opolot & Finke, 2015).

3.10 WEATHERING OF PRIMARY MINERALS

Mineral weathering is the prime source of cations Ca, Mg, K, and Na in natural unfertilized soil. It is impossible to deny leaching in a millennium-scale environment because it quickly depletes initial soil solution, exchange complexes, and OM quantities. Water weathering of primary silicate minerals to soil cations Ca, Mg, K, and Na. The equation to understand this phenomenon is given below (Van Grinsven et al., 1988).

$$FX = \rho^* \, T^* \, Kx^* \, cH^{\alpha(X)} \quad (4)$$

where; 'FX' is the weathering flux ($mol_c\ ha^{-1}\ y^{-1}$); 'ρ' is the dry soil bulk density ($kg\ m^{-3}$); 'T' is the thickness soil compartment (m); 'Kx' is the weathering rate constant ($m^3\ mol_c\ ha^{-1}\ y^{-1}$) for cation X; 'cH' is the H concentration ($mol_c\ m^{-3}$); and α(X) is the parameter.

Al concentrations are brought to pH equilibrium with gibbsite in the soil solution. According to the LEACHC-code, the simulated chemical system can be seen in Figure 3.3.

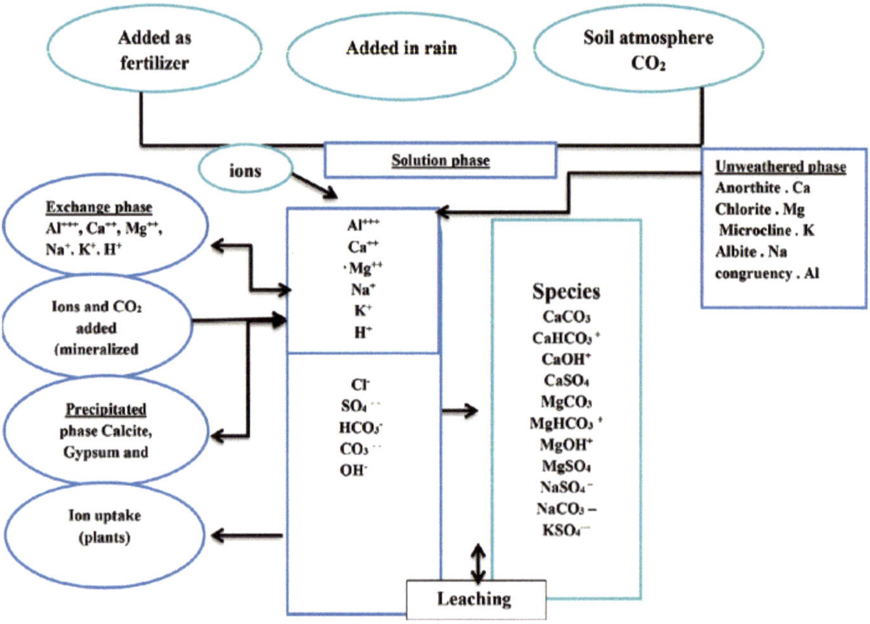

FIGURE 3.3 Chemical system simulated by SoilGen2.

3.11 SOIL EROSION MODELING

One of the biggest environmental dangers that lead to the loss of fertility is soil erosion and loss. It also causes other problems, such as landslides in hilly locations (Chalise, Kumar, & Kristiansen, 2019; Karki & Ojha, 2021). Climate change and land use changes connected to soil erosion are both factors to consider. Increased soil erosion causes nutrient loss, lowering agricultural yield. Soil is a critical resource for human food sources, and its depletion has both direct and indirect consequences for sustainability. Off-site erosion effects can be severe in terms of pollution and siltation of reservoirs and other structures (Rashmi et al., 2022). One of the world's most pressing environmental concerns is soil erosion. Agricultural land loss and runoff rates can be estimated using soil erosion models. These models can also be used to help with land use planning and guide government strategy and action plans on soil and water conservation. At least 3.2 billion people are negatively affected by land degradation caused by humans, and the planet is being driven to extinction, with one-sixth of all species suffering as a result (Fernández-Palacios et al., 2021). It will take a significant response from the government to slow down the rate of soil erosion in degraded environments.

Because of climate change and intensive land use, soil systems risk eroding. As of 2012, an estimated 10.1 million km^2 of land had been affected by water and wind erosion, which seriously affected 7.5 million km^2. People who cultivate face a real and persistent problem with soil erosion and sedimentation. Sheet erosion, rill erosion, gully erosion, and bank erosion are the most common types of soil erosion. The disintegration of sheets and rills is the primary cause of soil loss (Lal, 2017; Niu et al., 2020). Models are organized depending on the input parameters and the standard with which they are associated. It is critical for determining the source of silt and is legitimately only inside the confinement points. The erosion rate is influenced by geography, climate, soil type, and land use. Topographic modeling in erosion investigations can vary in complexity, complexity, and data necessary for model use and calibration, which might be empirical, physical, or conceptual.

Accelerated soil erosion is a major environmental issue in many agricultural areas worldwide (Devia, Ganasri, & Dwarakish, 2015). The problem is most severe in humid tropical mountain environments (HTMEs), such as the Himalayas, Andes, and Amazon rainforests. Soil loss tolerance is substantially lower in HTMEs than in temperate lowland environments. Because of their simplicity and availability of information, empirical models are the most widely utilized for predicting erosion. Distance from the origin to the point on the slope (L) is known as slope length, at which each slope gradient lowers sufficiently to allow the deposition to commence or when the flow begins to concentrate in a defined channel (Blaikie & Brookfield, 2015). At agricultural field scales, soil erosion models are rarely examined. Unmanned aerial vehicles (UAVs) were used to measure soil erosion in farm fields.

As determined by UAV, the erosion rate seemed to be 18.83 t ha^{-1} $year^{-1}$. Each year, 75 billion tons of soil loss from agricultural land is due to soil erosion. Soil erosion has an annual economic cost of 400 billion dollars. The rising demand for agricultural commodities is expected to exacerbate these trends. Soil loss in agricultural systems is difficult to estimate accurately (Karamage et al., 2016). Soil erosion is a widespread problem that has jeopardized most Mediterranean countries. Land management strategies are routinely used at the watershed scale to reduce runoff and sediment burdens. Their effectiveness, on the other hand, is difficult to quantify. As a result, models have been developed to assess their impact on runoff dynamics and soil erosion, as well as to determine the most effective conservation measure for given land use. Soil erosion can cause hillslopes to deteriorate over time, especially in steep-slope agricultural areas. Erosion processes can cause a steady depletion of soil resources, which can have a negative influence on ES and agricultural production. Various methods are used to forecast soil erosion-affected areas. Different studies employ different models to estimate soil erosion. The accuracy of different models and the rate of erosion may vary depending on land characteristics, slope, and soil type in different places. Agriculture has the potential to disrupt long-term landscape archaeological alignment by being a primary driver of soil erosion (Homburg & Sandor, 2011).

3.12 CONCLUSION

Climate change and land management policymakers increasingly rely on SOM models. Short-term soil carbon dynamics datasets, such as microbial biomass or CO_2 flux, have not been rigorously tested by SOM models. Model SOM pools and subsoil processes are two specific research areas that need improvement. Deep soils met the water needs of plants with good hydrodynamic properties. Soils with a coarse texture and a shallow depth had the greatest groundwater recharge, but their ability to buffer water quality was the lowest. As a whole, the spatial distribution of SES and the interactions between SES must be studied in the future. Improved PSF models are better at predicting soil hydraulic properties than the previous model. Image data analysis shows that the change point does not change significantly with adding biochar. Using GMER and GSDER as the standard for judging bias is preferable. The DA parameter negatively correlates with soil conductivity accuracy rate, and the improved model is better suited for soils with inadequate anisotropy. It was shown that a new SOC model, JSM, is based on the organo-organo-mineral association enabled SOC model that is mathematically formulated in this way. The SWAP model accurately depicts water and salt transport during spring wheat growth. A decrease in irrigation water quantity and an increase in water salinity led to a decrease in soil water content and an increase in salt concentration in the soil. Alternative uses of saline water for crops necessitate more investigation. The results of this study show that it is possible to develop a reliable model for predicting the stability of soil aggregates using readily available soil input data. The models can guide soil conservation and evaluation. One of the most critical issues today is figuring out how to measure and model the myriad processes in soils accurately. Soils are involved in everything from ES to climate change to food security. Soil process modeling faces four major challenges that we believe will shape the future of the field. A community modeling platform could be set up to facilitate the exchange of knowledge about modeling soil processes. Integrated models and benchmarks might be developed to test both existing and new models and approaches for efficiently and effectively coupling soil processes.

ACKNOWLEDGMENTS

The author expresses gratitude to M. Auyoub Bhat and Aamir Hassan Mir for their significant contribution in preparing this chapter.

KEYWORDS

- **food web models**
- **Jena soil model**
- **nutrient acquisition**
- **soil organic matter models**
- **soil processes**
- **soil quality**
- **SoilGen2 model**

REFERENCES

Addiscott, T., (1977). A simple computer model for leaching in structured soils. *Journal of Soil Science, 28*(4), 554–563.

Amundson, R., (2021). Factors of soil formation in the 21st century. *Geoderma, 391*, 114960.

Arnold, C. A. T., (2018). *Adaptive law Research Handbook on Climate Disaster Law*. Edward Elgar Publishing.

Arnold, J. G., Youssef, M. A., Yen, H., White, M. J., Sheshukov, A. Y., Sadeghi, A. M., & Skaggs, R. W., (2015). Hydrological processes and model representation: Impact of soft data on calibration. *Transactions of the ASABE, 58*(6), 1637–1660.

Bardgett, R., (2005). *The Biology of Soil: A Community and Ecosystem Approach*. Oxford University Press.

Bashir, O., Ali, T., Baba, Z. A., Rather, G., Bangroo, S., Mukhtar, S. D., & Bhat, R. A., (2021). Soil organic matter and its impact on soil properties and nutrient status *Microbiota and Biofertilizers* (Vol. 2, pp. 129–159). Springer.

Bashir, O., Khan, K., Hakeem, K. R., Mir, N. A., Rather, G. H., & Mohiuddin, R., (2016). Soil microbe diversity and root exudates as important aspects of rhizosphere ecosystem *Plant, Soil and Microbes* (pp. 337–357). Springer.

Baveye, P. C., Otten, W., Kravchenko, A., Balseiro-Romero, M., Beckers, É., Chalhoub, M., & Hapca, S., (2018). Emergent properties of microbial activity in heterogeneous soil microenvironments: Different research approaches are slowly converging, yet major challenges remain. *Frontiers in Microbiology*, 1929.

Bestelmeyer, B. T., Duniway, M. C., James, D. K., Burkett, L. M., & Havstad, K. M., (2013). A test of critical thresholds and their indicators in a desertification-prone ecosystem: More resilience than we thought. *Ecology Letters, 16*(3), 339–345.

Blaikie, P., & Brookfield, H., (2015). *Land Degradation and Society*. Routledge.

Borgogno, F., D'Odorico, P., Laio, F., & Ridolfi, L., (2009). Mathematical models of vegetation pattern formation in ecohydrology. *Reviews of Geophysics, 47*(1).

Bosatta, E., & Ågren, G., (1995). Theoretical analyses of interactions between inorganic nitrogen and soil organic matter. *European Journal of Soil Science, 46*(1), 109–114.

Braud, I., Bariac, T., Gaudet, J. P., & Vauclin, M., (2005). SiSPAT-Isotope, a coupled heat, water and stable isotope (HDO and $H_2{}^{18}O$) transport model for bare soil. Part I. Model description and first verifications. *Journal of Hydrology, 309*(1–4), 277–300.

Burns, R. G., DeForest, J. L., Marxsen, J., Sinsabaugh, R. L., Stromberger, M. E., Wallenstein, M. D., & Zoppini, A., (2013). Soil enzymes in a changing environment: Current knowledge and future directions. *Soil Biology and Biochemistry, 58*, 216–234.

Cao, X., Li, Z., & Li, R., (2019). *A Method to Model Soil Structure Based on the Randomness and Fractality of Soil Pore*. Paper presented at the 2019 11th Asia-Pacific International Conference on Lightning (APL).

Chalise, D., Kumar, L., & Kristiansen, P., (2019). Land degradation by soil erosion in Nepal: A review. *Soil Systems, 3*(1), 12.

Chethan, K. G., Dutta, D., Chaudhary, J., & Meena, A. L., (2021). Functional diversity management through microbial integrity for sustainability. *Soil Science: Fundamentals to Recent Advances* (pp. 361–387). Springer.

Coggan, N. V., Hayward, M. W., & Gibb, H., (2018). A global database and "state of the field" review of research into ecosystem engineering by land animals. *Journal of Animal Ecology, 87*(4), 974–994.

Coleman, K., & Jenkinson, D., (1996). RothC-26.3-A model for the turnover of carbon in soil. *Evaluation of Soil Organic Matter Models* (pp. 237–246). Springer.

Creamer, R., Barel, J., Bongiorno, G., & Zwetsloot, M., (2022). The life of soils: Integrating the who and how of multifunctionality. *Soil Biology and Biochemistry, 166*, 108561.

Creamer, R., Hannula, S., Van, L. J., Stone, D., Rutgers, M., Schmelz, R., & Bouffaud, M. L., (2016). Ecological network analysis reveals the inter-connection between soil biodiversity and ecosystem function as affected by land use across Europe. *Applied Soil Ecology, 97*, 112–124.

Cullmann, J., Mishra, V., & Peters, R., (2006). Flow analysis with WaSiM-ETH–model parameter sensitivity at different scales. *Advances in Geosciences, 9*, 73–77.

Devia, G. K., Ganasri, B. P., & Dwarakish, G. S., (2015). A review on hydrological models. *Aquatic Procedia, 4*, 1001–1007.

Drewniak, B., Song, J., Prell, J., Kotamarthi, V., & Jacob, R., (2013). Modeling agriculture in the community land model. *Geoscientific Model Development, 6*(2), 495–515.

Feng, Y., Wang, J., Bai, Z., & Reading, L., (2019). Effects of surface coal mining and land reclamation on soil properties: A review. *Earth-Science Reviews, 191*, 12–25.

Fernández-Palacios, J. M., Kreft, H., Irl, S. D., Norder, S., Ah-Peng, C., Borges, P. A., & Montes, E., (2021). Scientists' warning–The outstanding biodiversity of islands is in peril. *Global Ecology and Conservation, 31*, e01847.

Fierer, N., Leff, J. W., Adams, B. J., Nielsen, U. N., Bates, S. T., Lauber, C. L., & Caporaso, J. G., (2012). Cross-biome metagenomic analyses of soil microbial communities and their functional attributes. *Proceedings of the National Academy of Sciences, 109*(52), 21390–21395.

Finke, P. A., & Hutson, J. L., (2008). Modeling soil genesis in calcareous loess. *Geoderma, 145*(3, 4), 462–479.

Fredlund, D. G., Rahardjo, H., & Fredlund, M. D., (2012). *Unsaturated Soil Mechanics in Engineering Practice*. John Wiley & Sons.

Fry, E. L., De Long, J. R., Álvarez, G. L., Alvarez, N., Carrillo, Y., Castañeda-Gómez, L., & Hasegawa, S., (2019). Using plant, microbe, and soil fauna traits to improve the predictive power of biogeochemical models. *Methods in Ecology and Evolution, 10*(1), 146–157.

Gerphagnon, M., Macarthur, D. J., Latour, D., Gachon, C. M., Van, O. F., Gleason, F. H., & Sime-Ngando, T., (2015). Microbial players involved in the decline of filamentous and colonial cyanobacterial blooms with a focus on fungal parasitism. *Environmental Microbiology, 17*(8), 2573–2587.

Gobin, A., Campling, P., & Feyen, J., (2001). Soil-landscape modeling to quantify spatial variability of soil texture. *Physics and Chemistry of the Earth, Part B: Hydrology, Oceans and Atmosphere, 26*(1), 41–45.

Goettsch, B., Urquiza-Haas, T., Koleff, P., Acevedo, G. F., Aguilar-Meléndez, A., Alavez, V., & Carr, J. A., (2021). Extinction risk of Mesoamerican crop wild relatives. *Plants, People, Planet, 3*(6), 775–795.

Goovaerts, P., (2010). Combining areal and point data in geostatistical interpolation: Applications to soil science and medical geography. *Mathematical Geosciences, 42*(5), 535–554.

Greiner, L., Keller, A., Grêt-Regamey, A., & Papritz, A., (2017). Soil function assessment: Review of methods for quantifying the contributions of soils to ecosystem services. *Land Use Policy, 69*, 224–237.

Gunina, A., Smith, A. R., Kuzyakov, Y., & Jones, D. L., (2017). Microbial uptake and utilization of low molecular weight organic substrates in soil depend on carbon oxidation state. *Biogeochemistry, 133*(1), 89–100.

Han, M., Sun, L., Gan, D., Fu, L., & Zhu, B., (2020). Root functional traits are key determinants of the rhizosphere effect on soil organic matter decomposition across 14 temperate hardwood species. *Soil Biology and Biochemistry, 151*, 108019.

Han, Z., Huang, S., Huang, Q., Bai, Q., Leng, G., Wang, H., & Zheng, X., (2020). Effects of vegetation restoration on groundwater drought in the Loess Plateau, China. *Journal of Hydrology, 591*, 125566.

He, H., Jansson, P. E., & Gärdenäs, A. I., (2021). CoupModel (v6. 0): An ecosystem model for coupled phosphorus, nitrogen, and carbon dynamics–evaluated against empirical data from a climatic and fertility gradient in Sweden. *Geoscientific Model Development, 14*(2), 735–761.

Helliwell, J. R., Sturrock, C. J., Grayling, K. M., Tracy, S. R., Flavel, R., Young, I., & Mooney, S. J., (2013). Applications of X-ray computed tomography for examining biophysical interactions and structural development in soil systems: A review. *European Journal of Soil Science, 64*(3), 279–297.

Homburg, J. A., & Sandor, J. A., (2011). Anthropogenic effects on soil quality of ancient agricultural systems of the American Southwest. *Catena, 85*(2), 144–154.

Jacques, D., & Simunek, J., (2005). *User Manual of the Multicomponent Variably-Saturated Flow and Transport Model HP1*.

Jiang, R., Gan, T. Y., Xie, J., & Wang, N., (2014). Spatiotemporal variability of Alberta's seasonal precipitation, their teleconnection with large-scale climate anomalies and sea surface temperature. *International Journal of Climatology, 34*(9), 2899–2917.

Johnson, D., Boyd, C., O'Connor, R. C., & Smith, D., (2022). *Ratcheting up Resilience in the Northern Great Basin.* Rangelands.

Karamage, F., Zhang, C., Ndayisaba, F., Shao, H., Kayiranga, A., Fang, X., & Tian, G., (2016). Extent of cropland and related soil erosion risk in Rwanda. *Sustainability, 8*(7), 609.

Karki, K. B., & Ojha, R. B., (2021). *Land Degradation the Soils of Nepal* (pp. 127–142). Springer.

Keyvanshokouhi, S., Cornu, S., Samouëlian, A., & Finke, P., (2016). Evaluating SoilGen2 as a tool for projecting soil evolution induced by global change. *Science of the Total Environment, 571*, 110–123.

Kroes, J., Wesseling, J., & Van, D. J., (2000). Integrated modeling of the soil–water–atmosphere–plant system using the model SWAP 2·0 an overview of theory and an application. *Hydrological Processes, 14*(11, 12), 1993–2002.

Lal, R., (2017). Soil erosion and global warming in India. *Journal of Soil and Water Conservation, 16*(4), 297–305.

Larsen, J., Jaramillo-López, P., Nájera-Rincon, M., & González-Esquivel, C., (2015). Biotic interactions in the rhizosphere in relation to plant and soil nutrient dynamics. *Journal of Soil Science and Plant Nutrition, 15*(2), 449–463.

Laub, M., Demyan, M. S., Nkwain, Y. F., Blagodatsky, S., Kätterer, T., Piepho, H. P., & Cadisch, G., (2020). DRIFTS band areas as measured pool size proxy to reduce parameter uncertainty in soil organic matter models. *Biogeosciences, 17*(6), 1393–1413.

Lehmann, J., Bossio, D. A., Kögel-Knabner, I., & Rillig, M. C., (2020). The concept and future prospects of soil health. *Nature Reviews Earth & Environment, 1*(10), 544–553.

Li, Y., Cui, S., Chang, S. X., & Zhang, Q., (2019). Liming effects on soil pH and crop yield depend on lime material type, application method and rate, and crop species: A global meta-analysis. *Journal of Soils and Sediments, 19*(3), 1393–1406.

Libralato, S., Christensen, V., & Pauly, D., (2006). A method for identifying keystone species in food web models. *Ecological Modeling, 195*(3, 4), 153–171.

Ma, X., Zarebanadkouki, M., Kuzyakov, Y., Blagodatskaya, E., Pausch, J., & Razavi, B. S., (2018). Spatial patterns of enzyme activities in the rhizosphere: Effects of root hairs and root radius. *Soil Biology and Biochemistry, 118*, 69–78.

Ma, Y., Feng, S., Su, D., Gao, G., & Huo, Z., (2010). Modeling water infiltration in a large layered soil column with a modified green–Ampt model and HYDRUS-1D. *Computers and Electronics in Agriculture, 71*, S40–S47.

Ma, Y., Minasny, B., Malone, B. P., & Mcbratney, A. B., (2019). Pedology and digital soil mapping (DSM). *European Journal of Soil Science, 70*(2), 216–235.

Martínez, P. G., Peri, P. L., Lencinas, M. V., García-Llorente, M., & Martín-López, B., (2016). Spatial patterns of cultural ecosystem services provision in Southern Patagonia. *Landscape Ecology, 31*(2), 383–399.

Mehta, D., Rhodin, H., Casas, D., Fua, P., Sotnychenko, O., Xu, W., & Theobalt, C., (2017). *Monocular 3d Human Pose Estimation in the Wild Using Improved CNN Supervision.* Paper presented at the 2017 international conference on 3D vision (3DV).

Minasny, B., McBratney, A. B., Mendonça-Santos, M., Odeh, I., & Guyon, B., (2006). Prediction and digital mapping of soil carbon storage in the lower Namoi valley. *Soil Research, 44*(3), 233–244.

Mohi-ud-Din, R., Dar, N., Lal, S., Rukhsana, J., Khan, K., Owais, B., & Asmat, A., (2017). Agronomic practices influencing yield attributes of sweet corn (*Zea mays saccharata* L.). *Vegetos, 30*(Special Issue 2), 459–462.

Niu, B., Tan, L., Zhang, X. J., Qu, J., An, Z., Wang, J., & Li, K., (2020). Targeted control of sand hazards for a railway in extremely arid regions using fingerprinting approaches. *Geomorphology, 361*, 107189.

Noory, H., Van, D. Z. S., Liaghat, A. M., Parsinejad, M., & Van, D. J., (2011). Distributed agro-hydrological modeling with SWAP to improve water and salt management of the Voshmgir irrigation and drainage network in Northern Iran. *Agricultural Water Management, 98*(6), 1062–1070.

Opolot, E., & Finke, P., (2015). Evaluating sensitivity of silicate mineral dissolution rates to physical weathering using a soil evolution model (SoilGen2. 25). *Biogeosciences, 12*(22), 6791–6808.

Oroo, P. A., (2021). *Determination of the Effects of Crop Rotation, Water.* Egerton University.

Padarian, J., Minasny, B., & McBratney, A. B., (2020). Machine learning and soil sciences: A review aided by machine learning tools. *Soil, 6*(1), 35–52.

Parton, W., (1996). The century model. *Evaluation of Soil Organic Matter Models* (pp. 283–291). Springer.

Paul, C., Kuhn, K., Steinhoff-Knopp, B., Weißhuhn, P., & Helming, K., (2021). Towards a standardization of soil-related ecosystem service assessments. *European Journal of Soil Science, 72*(4), 1543–1558.

Paustian, K., Collier, S., Baldock, J., Burgess, R., Creque, J., DeLonge, M., & Goddard, T., (2019). Quantifying carbon for agricultural soil management: From the current status toward a global soil information system. *Carbon Management, 10*(6), 567–587.

Pavan, A. L. R., & Ometto, A. R., (2018). Ecosystem services in life cycle assessment: A novel conceptual framework for soil. *Science of the Total Environment, 643*, 1337–1347.

Postma, J. A., Dathe, A., & Lynch, J. P., (2014). The optimal lateral root branching density for maize depends on nitrogen and phosphorus availability. *Plant Physiology, 166*(2), 590–602.

Rallo, G., Agnese, C., Minacapilli, M., & Provenzano, G., (2012). Comparison of SWAP and FAO agro-hydrological models to schedule irrigation of wine grapes. *Journal of Irrigation and Drainage Engineering, 138*(7), 581–591. https://ascelibrary.org/doi/abs/10.1061/(ASCE)IR.1943-4774.0000435 (accessed on 30 November 2023).

Rashmi, I., Karthika, K., Roy, T., Shinoji, K., Kumawat, A., Kala, S., & Pal, R., (2022). Soil erosion and sediments: A source of contamination and impact on agriculture productivity. *Agrochemicals in Soil and Environment* (pp. 313–345). Springer.

Ripl, W., & Eiseltová, M., (2009). Sustainable land management by restoration of short water cycles and prevention of irreversible matter losses from top soils. *Plant, Soil and Environment, 55*(9), 404–410.

Rosenzweig, S. T., Carolan, M. S., & Schipanski, M. E., (2020). A dryland cropping revolution? Linking an emerging soil health paradigm with shifting social fields among wheat growers of the high plains. *Rural Sociology, 85*(2), 545–574.

Sadatshojaei, E., Wood, D. A., & Rahimpour, M. R., (2021). Potential and challenges of carbon sequestration in soils. *Applied Soil Chemistry*, 1–21.

Saiti, K., Macaš, M., Lhotská, L., Štechová, K., & Pithová, P., (2020). Ensemble methods in combination with compartment models for blood glucose level prediction in type 1 diabetes mellitus. *Computer Methods and Programs in Biomedicine, 196*, 105628.

Schwilch, G., Bernet, L., Fleskens, L., Giannakis, E., Leventon, J., Maranon, T., & Van, D. H., (2016). Operationalizing ecosystem services for the mitigation of soil threats: A proposed framework. *Ecological Indicators, 67*, 586–597.

Sharififar, A., Sarmadian, F., & Minasny, B., (2019). Mapping imbalanced soil classes using Markov chain random fields models treated with data resampling technique. *Computers and Electronics in Agriculture, 159*, 110–118.

Sharma, M., & Gupta, B. B., (2020). Role of wild relatives for development of climate-resilient varieties. *Rediscovery of Genetic and Genomic Resources for Future Food Security* (pp. 303–314). Springer.

Su, J. Q., Ding, L. J., Xue, K., Yao, H. Y., Quensen, J., Bai, S. J., & Tiedje, J. M., (2015). Long-term balanced fertilization increases the soil microbial functional diversity in a phosphorus-limited paddy soil. *Molecular Ecology, 24*(1), 136–150.

Sulman, B. N., Moore, J. A., Abramoff, R., Averill, C., Kivlin, S., Georgiou, K., & Wieder, W. R., (2018). Multiple models and experiments underscore large uncertainty in soil carbon dynamics. *Biogeochemistry, 141*(2), 109–123.

Tate, III. R. L., (2020). *Soil Microbiology*. John Wiley & Sons.

Temme, A. J., & Vanwalleghem, T., (2016). LORICA–a new model for linking landscape and soil profile evolution: Development and sensitivity analysis. *Computers & Geosciences, 90*, 131–143.

Thompson, E., (2010). *Mind in Life: Biology, Phenomenology, and the Sciences of Mind*: Harvard University Press.

Tracy, S. R., Daly, K. R., Sturrock, C. J., Crout, N. M., Mooney, S. J., & Roose, T., (2015). Three-dimensional quantification of soil hydraulic properties using X-ray computed tomography and image-based modeling. *Water Resources Research, 51*(2), 1006–1022.

Troy, T. J., Pavao-Zuckerman, M., & Evans, T. P., (2015). Debates—Perspectives on socio-hydrology: Socio-hydrologic modeling: Tradeoffs, hypothesis testing, and validation. *Water Resources Research, 51*(6), 4806–4814.

Van, G. J., Booltink, H., Dirksen, C., Van, B. N., Bongers, N., & Waringa, N., (1988). Automated in situ measurement of unsaturated soil water flux. *Soil Science Society of America Journal, 52*(5), 1215–1218.

Van, L. K., Bouma, J., Herbst, M., Koestel, J., Minasny, B., Mishra, U., & Padarian, J., (2017). Pedotransfer functions in Earth system science: Challenges and perspectives. *Reviews of Geophysics, 55*(4), 1199–1256.

Vereecken, H., Schnepf, A., Hopmans, J. W., Javaux, M., Or, D., Roose, T., & Aitkenhead, M., (2016). Modeling soil processes: Review, key challenges, and new perspectives. *Vadose Zone Journal, 15*(5).

Vogel, H., Bartke, S., Daedlow, K., Helming, K., Kögel, I., & Lang, B., (2017). Sustainable soil management requires a systemic approach. *Soil Discuss.*

Wadoux, A. M. C., & McBratney, A. B., (2021). Digital soil science and beyond. *Soil Science Society of America Journal, 85*(5), 1313–1331.

Wadoux, A. M. C., Román-Dobarco, M., & McBratney, A. B., (2021). Perspectives on data-driven soil research. *European Journal of Soil Science, 72*(4), 1675–1689.

Wösten, J., Pachepsky, Y. A., & Rawls, W., (2001). Pedotransfer functions: Bridging the gap between available basic soil data and missing soil hydraulic characteristics. *Journal of Hydrology, 251*(3, 4), 123–150.

Wutzler, T., & Reichstein, M., (2013). Priming and substrate quality interactions in soil organic matter models. *Biogeosciences, 10*(3), 2089–2103.

Yu, L., Ahrens, B., Wutzler, T., Schrumpf, M., & Zaehle, S., (2020). Jena soil model (JSM v1. 0; revision 1934): A microbial soil organic carbon model integrated with nitrogen and phosphorus processes. *Geoscientific Model Development, 13*(2), 783–803.

Yu, Y., Finke, P., Wu, H., & Guo, Z., (2013). Sensitivity analysis and calibration of a soil carbon model (SoilGen2) in two contrasting loess forest soils. *Geoscientific Model Development, 6*(1), 29–44.

Zhang, K., Maltais-Landry, G., & Liao, H. L., (2021). How soil biota regulate C cycling and soil C pools in diversified crop rotations. *Soil Biology and Biochemistry, 156*, 108219.

Zhang, Y., Lavallee, J. M., Robertson, A. D., Even, R., Ogle, S. M., Paustian, K., & Cotrufo, M. F., (2021). Simulating measurable ecosystem carbon and nitrogen dynamics with the mechanistically defined MEMS 2.0 model. *Biogeosciences, 18*(10), 3147–3171.

Zhang, Z., Zhao, Y., Yang, T., Wei, Z., Li, Y., Wei, Y., & Wang, L., (2019). Effects of exogenous protein-like precursors on humification process during lignocellulose-like biomass composting: Amino acids as the key linker to promote humification process. *Bioresource Technology, 291*, 121882.

CHAPTER 4

Holistic Approach and New Assessment Techniques for Modeling Soil Erosion Under Changing Climatic Conditions

NASIR BASHIR NAIKOO,[1] OWAIS BASHIR,[1] SHABIR AHMAD BANGROO,[1] AAMIR HASSAN MIR,[1] M. AUYOUB BHAT,[1] SHAKEEL AHMAD MIR,[1] OWAIS BHAT,[2] SHAFEEQ UR REHMAN,[3] and KHALID REHMAN HAKEEM[4]

[1]*Division of Soil Science, Sher-e-Kashmir University of Agricultural Sciences and Technology of Kashmir, Kashmir, India*

[2]*Division of Agriculture Engineering, Sher-e-Kashmir University of Agricultural Sciences and Technology of Kashmir, Kashmir, India*

[3]*School of Environment and Civil Engineering, Dongguan University of Technology, Dongguan, Guangdong, China*

[4]*Department of Biological Sciences, Faculty of Science, King Abdulaziz University, Jeddah, Saudi Arabia*

ABSTRACT

Natural and man-made factors contribute to soil erosion, which is a worldwide issue. Numerous soil erosion models have been employed. Around 75 billion tons of soil is lost from agricultural land every year because of erosion. Soil conservation and quantification of soil loss are critical in today's world. Soil loss from cultivating land may result in diminished crop potential, poorer surface water quality, and impaired drainage networks. We can employ cutting-edge technology models to accurately monitor soil erosion and deposition. Soil erosion and sediment flux models are possible

Soil, Agriculture, and Ecosystem Modelling: Smart Technologies for Sustainable Solutions.
Owais Bashir, Khalid Rehman Hakeem, Surinder S. Kukal, and Razeef Mohd (Eds.)

instruments for predicting soil erosion and sediment flux and assessing the efficacy of conservation management approaches. Various models are designed for "flat agricultural areas" in temperate climates. They can be used to assess the impact of climate change and soil erosion on a large scale. We scoured academic journals, bulletins, textbooks, and research reports to learn more about these models and how they are used. According to the three most commonly used climate change models, soil erosion in Mediterranean catchments is projected to increase or decrease. In this review, commonly used soil erosion models are examined.

4.1 INTRODUCTION

One of the biggest environmental dangers that lead to declined fertility is soil erosion and loss (Eswaran, Lal, & Reich, 2019). It also causes other problems, such as landslides in hilly locations. Climate change and land use changes are critical factors for soil erosion. Increased soil erosion causes nutrient loss, lowering agricultural yield. Soil exhaustion has direct and indirect effects on the sustainability of human food sources (Borrelli et al., 2020). Off-site erosion effects can be severe in terms of pollution and siltation of reservoirs and other structures. Soil erosion has emerged as one of the severe environmental issues across the globe. Planning land use and government policy on soil and water conservation can be helped by using soil erosion models when estimating soil loss and runoff rates from farmland (Singh & Singh, 2017). Land degradation caused by humans affects at least 3.2 billion people, and the planet as a whole is being driven to extinction, with one-sixth of all species suffering the consequences. A significant governmental response is necessary to counteract the effects of soil erosion in degraded areas. Because of climate change and intensive land use, soil systems risk deterioration and degradation due to erosion. In 2012, an estimated 10.1 million km^2 of land was affected by water and wind erosion, with 7.5 million km^2 of that area severely damaged. Sheet, rill, gully, and bank erosion are the most common forms of soil erosion. The disintegration of soil sheets and rills is the most serious problem (Nieder, Benbi, & Reichl, 2018). Models are organized depending on the input parameters and the standard with which they are associated. There are three general models: empirical, physical, and conceptual (Supratak, Dong, Wu, & Guo, 2017). It is critical for determining the source of silt and is legitimately only inside the confinement points. Geographic location, climate, soil type, and land use all affect the erosion rate. Topographic modeling in erosion investigations can vary in complexity,

complexity, and data necessary for model use and calibration, which might be empirical, physical, or conceptual (Devia, Ganasri, & Dwarakish, 2015). Many agricultural areas worldwide face a major environmental problem due to rapid soil erosion. The problem is most severe in HTMEs, such as the Himalayas, Andes, and Amazon rainforests. Soil loss tolerance is substantially lower in HTMEs than in temperate lowland environments. Because of their simplicity and availability of information, empirical models are the most widely utilized for predicting erosion (Blaikie & Brookfield, 2015). The slope length (L) is the distance from the origin to the point at which each slope gradient lowers sufficiently to allow the deposition to commence or when the flow begins to concentrate in a defined channel. At agricultural field scales, soil erosion models are rarely examined.

A unique UAV methodology was applied. The yearly erosion rate, as determined by the UAV, was 18.83 t ha^{-1} yr^{-1}. Soil erosion has an annual economic cost of $400 billion dollars (Eisenbeiss, 2004). The rising demand for agricultural commodities is expected to exacerbate these trends. Some have referred to agricultural systems as stochastic due to the fact that even seemingly identical farming systems can have vastly different hydrologic responses to rainfall. The majority of Mediterranean countries are at risk due to soil erosion. At the watershed scale, land management strategies are routinely used to reduce runoff and sediment burdens. Their effectiveness, on the other hand, is difficult to quantify. As a result, models have been developed to assess their impact on runoff dynamics and soil erosion, as well as to determine the most effective conservation measure for the given land use (Fales et al., 2016). Soil erosion can cause hillslopes to deteriorate over time, especially in steep-slope agricultural areas. According to academics at the University of Siena in Italy, assessing the danger of soil erosion should be an important part of vineyard design and management. Erosion processes can cause a steady depletion of soil resources, which can have a negative influence on ecosystem services (ES) and agricultural production (Biasi, Brunori, Smiraglia, & Salvati, 2015). Several techniques are employed to predict the effects of soil erosion. Soil erosion is estimated using a variety of models in various studies. In Spain, soil erosion and rehabilitation were examined using RUSLE and Morgan-Morgan and Finney (MMF) models. Erosion rates and model accuracy can vary widely based on land characteristics, slope, soil type, and climate. It may be possible to identify the best model for the area by comparing three commonly utilized models (USLE, MMF, and RUSLE) (Zema, Nunes, & Lucas-Borja, 2020). The erosion rates of all models are compared using a geographic information system (GIS). Once a model has been validated, it is selected as the best fit for the study area. Agriculture has the potential to disrupt long-term landscape archaeological alignment by being a primary driver of soil erosion (Barthel, Isendahl, Vis, Drescher, Evans, & Van, 2019).

4.2 SOIL EROSION MODELS

4.2.1 *RUSLE*

Field or farm-sized units subject to various management practices are subject to RUSLE's broadly used method for simulating long-term erosion rates (Ghosal & Das, 2020). It indicates the position of rain gauge stations that have an impact on the basin. To retrieve the study area from the MMF dataset, we use a car to DEM and a satellite image (from the Indian Remote Sensing satellite, the linear imaging self-scan sensor-3 IRS LISS-3). According to RUSLE's theory, detachment and deposition are influenced by the flow's sediment content. The flow's carrying capacity, not the origin of the eroded material, sets a constraint on erosion. Detachment is impossible when the sediment load exceeds the flow's carrying capacity. During the hydrograph's receding phase, sedimentation must also take place (Kaffas, Hrissanthou, & Sevastas, 2018). A few factors in the RUSLE equation have been modified, but the basic formula has remained the same. For predicting erosion at the regional or local level, RUSLE is the most practical model. Slope and aspect, for example, are easy to incorporate into RUSLE when derived from DEM and LULC from satellite images. Using RUSLE is restricted to a small area due to its inability to route sediment via channels. Consequently, the model is not used for very large watersheds. The land area is depicted as a grid of square cells, and soil erosion is calculated for each cell using the RUSLE model. On field slopes, RUSLE estimates annual erosion by using the equation below:

$$A = R*K*L*S*C*P \tag{1}$$

where; A, R, K, L, S, C, and P are soil loss ton per hectare per year (t ha^{-1} yr^{-1}), Rainfall erosivity factor (MJ mm ha^{-1} hr^{-1} yr^{-1}, soil erodibility factor (ha^{-1} MJ mm^{-1}), slope length factor, slope steepness factor, cover management factor and Crop management practice, respectively.

4.2.2 *MMF MODEL*

There must be two phases for the model to work: water and sediment. Exponential relationships between rainfall, topography, soil water storage, and routing are used to determine runoff volume during the wet phase (Mondal, Khare, & Kundu, 2018). Runoff volume, slope gradient, and crop management are used to evaluate runoff transport capacity in the sediment phase. Rainfall energy and interception are used to estimate rainfall splash. When comparing soil particle detachment and transport capacity, the lower the value, the lower the annual sediment yield. As a result of these and other changes, the MMF model has undergone three revisions: integration with GIS, flow detachment, and the effects of particle-size specificity and vegetation cover.

It was found that there is a wide variation in soil erosion across the United States and that the most vulnerable areas are located in the southwest and southeast (Wang & Dai, 2020). Across the three models, the average annual soil loss is < 5 t/ha/yr, with the MMF model losing the least amount of soil while the RUSLE model loses the most, averaging 61.87%. RUSLE's erosion rate is more severe than the other two models, but the extent of this erosion in RUSLE is greater. Those areas of RUSLE that experience the greatest erosion (> 80 t ha^{-1} yr^{-1}) suffer the most. Moderate (5 to 10 t ha^{-1} yr^{-1}) and high (11 to 20 t ha^{-1} yr^{-1}) rates of soil erosion reveal considerably more damaged areas in MMF and USLE than RUSLE, and validation shows that RUSLE has higher accuracy and precision in its measurements than MMF and USLE (Mohammed et al., 2020). According to this model, deposition, sediment yields, channel eroding, and gully erosion cannot be accurately predicted. RULSE scores for the C factor outperform USLE scores by a wide margin. Based on land use, both the RUSLE and USLE models have different C factors that yield better results because of the actual variation in land surface character in RUSLE and a generalized value in USLE, respectively. Slopes with more than 20° are more susceptible to erosion than those with fewer than 20°, depending on the soil type and land use (Biswas & Pani, 2015) (Figure 4.1).

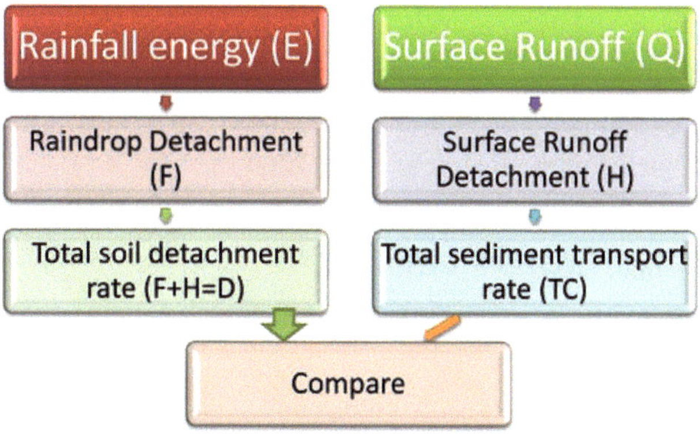

FIGURE 4.1 Flowchart depicting methodology used for MMF modeling.

More clay-rich regions resist erosion, while areas with sandy loam soil are the most vulnerable. Soil erosion seems more likely in open fallow areas than in forest ones since the soil is naked and loose following harvest. As a result, settlements have less soil exposure, which suggests low erosion (Tully, Sullivan, Weil, & Sanchez, 2015). Locations with high LS factors were found to be the most susceptible to erosion in all three models. Soil nutrient protection must receive more attention as erosion rates rise on agricultural lands. We need to pay particular attention to erosion rates in our research area's northern, northeastern, and southwest boundaries because of their undulating topography, which makes these areas particularly vulnerable.

4.2.3 *EROSION POTENTIAL MODEL*

Empirical soil erosion models like the Gavrilovic method, also recognized as the EPM, incorporate data from multiple sources, including climatic, geological, topographic, land-use, and erosion severity (Elhag, Kojchevska, & Boteva, 2019). Detached soil volume is calculated using the following formula:

$$w = T.h.\pi.\sqrt{z3} * F \tag{2}$$

where; W, T, h, z, and F are average annual erosion (m^3 $year^{-1}$), temperature coefficient, mean annual rainfall (mm), erosion coefficient and catchment area (km^2), respectively.

4.2.4 *ZHENG'S MODEL*

Using a regression coefficient and runoff depth, calculated the mean sediment yield per unit runoff, or the mean sediment concentration for all events considered (Li et al., 2017). At 17 gauging stations and two runoff experimental plots in the Dalihe watershed, collected data on 717 flow events. The event means sediment concentration was found to follow a power function in drainage basins with an area of up to 700 km^2.

4.3 PHYSICAL-BASED MODEL

4.3.1 *WERS*

Based on cell availability and coherence, the model divides the catchment into square components (matrix cells) (Adu & Kumarasamy, 2018).

Three kinds of disintegration are the raindrop effect, overland stream, and particle transport. Four information records are required for WEPP and ANSWERS show Documents on topography, climate, soil, and land use patterns (Walling, He, & Whelan, 2003). The model's dirt disintegration in water is calculated using two procedures. The first segment depicts the disintegration of dirt due to the impact of precipitation's kinetic vitality. The shear and lift powers of the spillover are used to evacuate soil molecules in the second part. The relationship is used to calculate soil erosion (Figure 4.2).

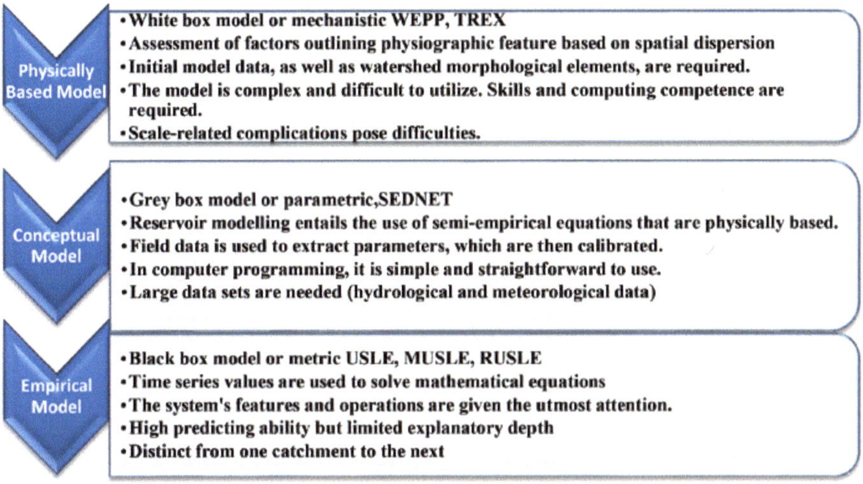

FIGURE 4.2 Models of soil erosion and their characteristics.

4.3.2 LUMPED PARAMETER MODELS (LPMS)

Lumped parameter models (LPMs) use basin-averaged precipitation for runoff calculations to incorporate the effects of non-uniform spatial processes (Ghosh, 2022). For example, RUSLE and SLEMA were developed to evaluate long-term average annual soil erosion at the field level. Field management practices and soil erosion made water quality and soil productivity a top priority in the late 1960s and early 1970s. As a result, other LPMs were created, such as methods for analyzing the influence of agricultural practices on the loss of nutrients, groundwater pollution, and crop yield. EPIC, CREAMS, and GLEAMS are some of these models (Nwokocha, 2021). The models of LPM are described in subsections.

4.3.2.1 SOIL LOSS ESTIMATION MODEL FOR SOUTHERN AFRICA (SLEMSA)

Soils, climate, terrain, vegetation, and human influence all contribute to the erosion process, according to the technique (Smith, 1999). The model can be used to make better land use decisions for increased production. Several methods have been employed in the past to estimate overall soil losses. Geomorphologists and conservationists have developed reliable yield estimations on the basin exits while researching erosion processes in drainage basin studies (Walling, 2017). This strategy does not account for the spatial distribution of erosion in the watershed. It is assumed that the amount of sediment trapped and stored in the basin during the measuring period is insignificant. The approach disregards the bedload in favor of focusing on the suspended load. It's also difficult to determine the source of silt in the catchment. Within the catchment, it could be due to landslides, gullies, or highways. This would provide a clearer picture of the sediment's nature and source within the basin (Joshi & Xu, 2017).

4.3.2.2 CREAMS (CHEMICALS, RUNOFF, EROSION, AGRICULTURAL MANAGEMENT SYSTEMS)

The CREAMS model is software that predicts how much runoff, soil erosion, plant nutrients, and pesticides will be delivered in the field due to agricultural operations (Silburn & Loch, 1989). USDA created this model, which was constructed for personal computers (PCs) rather than mainframes and incorporated irrigation as an optional feature for the PC version. CREAMS is well suited to predicting hydrological processes in small watersheds or fields with specific land use, reasonably homogenous soils, spatially even rainfall, and a distinctive management approach (Yuan, Sinshaw, & Forshay, 2020). This model is appropriate for less than 100 acres (40 hectares).

4.3.2.3 GLEAMS (GROUNDWATER LOADING EFFECTS OF AGRICULTURAL MANAGEMENT SYSTEMS)

With GLEAMS, a mathematical model, large-scale climate-soil management relationships can be predicted for field areas (Knisel & Turtola, 2000). It was developed to evaluate water, erosion material, and agricultural chemical loadings at the edge of the field and the bottom of the root system. There had to be an addition made to the model for the soil's nitrogen and phosphorus cycles to be accurately depicted. It is not the model's primary goal to predict how much water, soil erosion, or chemicals would be lost in the field through these methods but to evaluate management approaches. Numerous GLEAMS applications are available for assessing management alternatives. For the GLEAMS model, watershed management, erosion/sediment yield, pesticides, and soil nutrients work together. GLEAMS' erosion model is based on the USLE's storm-by-storm adaptation. A prominent overland flow element of the field with a non-uniform slope is used to compute rill and inter-rill erosion. Storm-by-storm rainfall energy is used to replace the erosivity factor (R) of the USLE. There are no changes to the user's management factors, such as the soil loss ratio (C) and the practice factors (P). In addition to depicting overland flow, the field can also show a concentrated or channel flow. The influence of raindrops separates soil particles and organic matter (OM), and the delivery process in the field directs this separated material. By using the hydrology component's simulation of storm runoff at the field outlet, unusual discharge is used to determine the particle movement in the soil (Acharya, Cochrane, Davies, & Bowman, 2011). Each computational segment's particle transport capacity and deposition can be calculated using the characteristic discharge for the topmost element. Eroding soil surface area to the initial surface area is calculated at the end of each flow element and field outlet.

4.3.2.4 EPIC (EROSION-PRODUCTIVITY IMPACT CALCULATOR)

EPIC was created to help researchers understand the relationship between soil erosion and productivity (Williams, Renard, & Dyke, 1983). It utilizes a regular time step and freely accessible inputs to predict the procedures in real-time. The model was used to predict hundreds of years of erosion if required because it is a long-term process. EPIC can be used to estimate the impact of management changes on outputs, a general-purpose programming language. When it comes to calculating erosion costs and determining the best management practices, physically-based elements for simulating erosion and plant growth and economically-based components are included in the EPIC model. Hydrology, weather simulations, erosion, and sedimentation, nutrient cycling, and plant growth are some of EPIC's physical components. EPIC can simulate the physical processes simultaneously and practically since it uses inputs (Gassman et al., 2010). Common input data (weather, crop, tillage, and soil factors) can be accessed through a computer filing system specifically designed for EPIC. Even though erosion is slow, EPIC can approximate centuries of erosion if needed. Easy to use, efficiency and the ability to calculate the impact of management changes are all advantages that EPIC provides. Growth of plants, soil temperature, plowing, and economics can all be classified as EPIC components for discussion. It is assumed that soils and management are homogeneous across the region, so EPIC uses a small drainage area (less than 1 hectare). When used with a vertically variable soil profile, the model can handle any variability of soil properties up to 10 layers deep. By interpolation, the thickness of the second layer is lowered to compensate for any eroded material, and all of the top layer characteristics are remedied according to movement in the next layer. When the thickness of the second layer is zero, the upper layer starts to slide down into the third layer (Ribolzi et al., 2011).

4.3.2.5 CAESAR MODEL

For CAESAR LEM, erosion and deposition induced by fluvial and gradient procedures are being used as a basis for adjusting heights in a regular grid of cells (Lowry, Coulthard, Hancock, & Jones, 2011). The catchment and reach modes of CAESAR allow the system to take in sediment and water from outside the catchment area. Various factors or initial conditions need to be inputted into the model for both modes (catchment and flow modes), such as elevation, particle sizes, rainfall, and flow input. In the CAESAR model, topography influences fluvial and hillslope processes that govern the distribution of erosion and sedimentation. As a result of the shifting topography, the reference point for the next time step also shifts. For example, elevation and sediment distributions over time and space and release and flux at the exit are outputs of the model's simulation (Coulthard, Hancock, & Lowry, 2012). Three basic components of CAESAR are the hydrological model, flow model, erosion and deposition of sediments, and slope processes. The catchment model in CAESAR uses rainfall as an input to produce runoff over the catchment. A modified TOPMODEL calculates the surface runoff that will occur if a certain threshold value is exceeded (Beven, Kirkby, Schofield, & Tagg, 1984).

4.3.3 DISTRIBUTED PARAMETER MODELS (DPMS)

Distributed parameter models (DPMs), as opposed to LPMs, take runoff into consideration and natural variation in parameter values. Reflecting the physical process's spatial and temporal changes is the goal of this project

(Liu, Du, Qi, & Zhao, 2018). A few examples of such models are WEPP and EUROSEM. Other models in this field entail GAMES, ANSWERS, and AGNPS, which could predict runoff loadings of nutrients and agricultural chemicals. While LPMs can be studied in various settings, several efforts have been made to investigate DPMs outside of their original contexts, such as EUROSEM in Mexico, Costa Rica, and Nicaragua and LISEM in South Africa. Because of this, it is difficult to confirm the usefulness of HTMEs. Compared to RUSLE, WEPP, EROSEM, and other DPMs, WEPP has significant advantages. They can predict the allocation of net soil loss and deposition over any period for an entire hillside. WEPP, in particular, uses a stochastic weather generator to generate its climate and weather data owing to its unique process-based sub-models for modeling the physical processes driving erosion (Raj, Kumar, & Sooryamol, 2022).

4.3.1.1 KYERMO MODEL

As a research tool, KYERMO was designed to isolate relevant sub-processes within the overall erosion process (Hirschi & Barfield, 1988). The aims are to provide a framework for further studies, to model erosion and to assess sediment collection and alteration in the middle sections of rivers, on geometric surfaces properly monitored on accentuated slopes, as a first step toward modeling the erosion process over a stochastic surface. Segments of slopes and furrow basins are divided into separate furrows, and the primary component uses a conceptual model for erosion to evaluate transport and sediment removal rates. Sediment production, flow targeting, and sediment targeting are the four key components of the main programmed and supporting routines. You can run it on other machines by making small adjustments to the code written in Microsoft FORTRAN, a subset of the standard ANSIFORTRAN language. ANSWERS simulate sediment transport in watersheds. Using 1 ha cells, it can cover basins up to 100 km^2, but the insightful version has a practical limit of 17 km^2 due to the limited number of 1,700 cells that the model can support. Subsurface drainage, laminar and rill flow orientation, increased rainfall simulation, and sediment transport simulation—are all included in this model. There is a disparity between rainfall and the sum of its capture, retention, and infiltration. With the continuity and Manning equations working together, the surface flow is guided from one cell to the next using the continuity equation. In contrast, the concentrated flow is estimated using both the continuity and Manning equations, with all streams presumed to have a rectangular cross-section (Endreny, 2005).

4.3.1.2 EUROSEM (EUROPEAN SOIL EROSION MODEL)

Soil stewardship approaches have been suggested using the European Soil Erosion Model (EUROSEM) (Morgan et al., 1998). Evaluating erosion risks at plot and watershed levels, determining solute distribution, providing credible estimates of erosion and solute concentration for comparing with acceptable standards, operating on a one-time event basis and being suitable as a design tool to select protective soil measures are its main objectives. Based on current and forecasted precipitation and runoff data, it uses plant cover, the impact of a raindrop and runoff concentration to remove soil particles and transport capacity to remove other soil particles (Pierson, Bates, Svejcar, & Hardegree, 2007). It is possible to regulate erosion by removing sediment or the flow carrying capacity by measuring soil loss and comparing it to the sediment supply readily accessible for transportation during the removal process. Deposition occurs whenever the transport capacity is less than the sediment removal capacity. Starting with knowing how much rain falls in a given period, a model's ability to capture rain is one way to estimate its volume and intensity. The quantity of precipitation that reaches the soil's surface is defined depending on the vegetation cover and soil protection. Using the ratio of straight line distance between two points to the actual distance measured overall micro-topographic irregularities, the roughness of the soil surface is defined. Measured data could be used to gather information on runoff generation for the model using a hydrological model. A depression in the model's surface separates the flow into two halves, one flowing into it and the other on the remaining surface. Particle breaking speed, a function of particle movement above a threshold value, describes how much soil raindrops remove at the saturation point (Morgan, 2009). Estimates of soil loss, liquid erosion, and deposition can be made using this technique.

4.3.1.3 WEPP (WATER EROSION PREDICTION PROJECT)

There are several updates for different conditions and abilities in the WEPP, which is a process-based, geographically diverse model for runoff and erosion modeling at a field or small catchment scale (Flanagan, Gilley, & Franti, 2007). There is only a chance of runoff if the amount of annual precipitation estimated by using the method exceeds the capacity of the depression. Therefore, soil infiltration influences water balance and crop progress, which affects runoff routing and erosion parameters. Predicting rill and inter-rill dynamics is done using the steady-state sediment continuity equation. Flow-induced shear stress can lead to drill erosion if it exceeds a certain threshold value. The term "inter-rill erosion" refers to delivering sediment to the rills. The sediment continuity equation models sediment detachment, transportation, and deposition in channels (Sun, Guo, Liu, Wu, Weckler, & Yang, 2021).

4.4 EROSION 2D/3D MODEL

Using a process-based approach, the EROSION 2D/3D model can calculate rainfall-induced soil erosion and deposition in 2D and small watersheds in 3D (Beitlerová et al., 2021). The model was created to provide a simple tool for modeling erosion to assist in conservation planning and to assess soil and water. The model needed to fulfill the specified user requirements.

It should:

- Have a few input parameters as possible to make it easy to use;
- Produce precise results without the need for individual calibration;
- Work on an as-needed basis;
- Work with currently-existing GIS systems.

Erosion 2D/3D is primarily related to physical concepts. The amount of material that can be separated from the soil surface and the flow's ability to move it are important factors in limiting erosion. Detachment occurs when the erosive force of the overland flow and falling droplets exceeds the soil's shear strength. As the particles settle for deposition, the vertical flow component must work to move the single particles through the flow. According to the model's primary presumption, the erosive impact of streamflow and droplets is inversely proportional to their corresponding momentum fluxes (Alewell, Borrelli, Meusburger, & Panagos, 2019). The soil's erosional resilience can be considered a critical momentum flux. EROSION 2D/3D's runoff subroutine utilizes the modified infiltration equation to determine rainfall surpluses. The Green and Ampt approach doesn't account for macro-pore flow and the effects of surface crusting in the infiltration model; a simple empirical adjustment strategy has been used to improve the model's ability to account for these factors.

4.5 SWAT (SOIL AND WATER ANALYSIS TOOL)

The SWAT model is highly effective in the watershed-to-stream bowl scale models to simulate the quality and quantity of surface and groundwater and to foresee the natural effects of land uses, administration practices, and environmental shifts (Douglas-Mankin, Srinivasan, & Arnold, 2010). SWAT is commonly used to assess soil disintegration avoidance and control, non-point source pollution control, and territorial oversight. Water, dregs, and rural concoction yields in large, complicated watersheds with various soils and long periods under various management circumstances can all be assessed using SWAT, a waterway bowl scale display. It is devised to evaluate the influence of land management practices on these variables. Climatic condition, surface spillover, return stream, permeability, evapotranspiration (ET), transmission blunders, lake and repository, modified advancement and water system, underground water stream, reach directing, supplement and pesticide stacking, and water flow in catchments are the major components of SWAT. MUSCLE condition was used to evaluate soil disintegration and sedimentation in this model. The hydrology display provides approximations of spillover volume and peak overflow rate, which are used to calculate the spillover erosive vitality variable in conjunction with the sub-watershed region (Dixit, 2021).

4.6 GAMES

Agricultural management techniques are evaluated for their impact on erosion and sedimentation using the Guelph Model (Guo et al., 2019). It was designed as a screening instrument for watershed management. For the model to work, it relies on the USLE to evaluate soil erosion, as well as a micro-delivery ratio function which takes surface roughness, the slope of land hydrologic condition, and the length of the overland flow path into consideration when calculating field soil loss percentages. The model can use a variety of spatial units of varying sizes and shapes. Soil erodibility, rainfall characteristics, and land management practices are all affected by seasonal changes, but the period over which these changes occur separates them. Each field in the watershed affected by soil erosion has its own set of soil degradation and erosion risks, as well as sediment load and yield rates, which are all calculated for each field in the watershed. Soil loss and sediment delivery are the two main components of GAMES. An essential component of soil loss is based on the equation USLE. For each season, the equation proposed characterizes the yearly sheet and rill erosion (Kalambukattu & Kumar, 2017).

4.7 LISEM (LIMBURG SOIL EROSION MODEL)

It is possible to use the Limburg soil Erosion Model for scheduling and conservation purposes because it is based on physical hydrology and soil erosion (LISEM) (Kværnø & Stolte, 2012). For the first time, a raster GIS uses a physical model in conjunction with the LISEM model. This integration makes it easier to implement in larger catchments by eliminating conversion steps, increasing user-friendliness, and the inclusion of remotely sensed data. Annual precipitation, interception, surface storage in micro-depressions, penetration, and vertical water movement in the soil, overland flow, channel flow, detachment by rainfall and overland flow are all considered in the model. Surface sealing, small roads, and tractor wheeling all received special attention. Vertical water movement in the soil can be simulated using Richard's equation. The Holtan or Green-Ampt infiltration models are available to the user. It is one of the first models to integrate a physically-based model into a RAW GIS; the LISEM model is an important milestone. No conversion techniques are utilized, and the model is fully described in the GIS command structure when integrated (Zhang, Hou, Wang, Pan, & Yan, 2020).

4.7.1 LISEM INPUT AND OUTPUT

1. **Rainfall File and Rain Gauge Map:** A time series file is used to store data from one or additional rain gauges. Each pixel on a map is assigned a rain gauge identification number. As a result, the model accommodates rainfall variability in both space and time.
2. **Tables for the Soil Water Model:** A modified version of the SWAT soil water model is included in LISEM, which mimics the vertical movement of water in the soil. Soil profiles are defined within the catchment. Subdividing a soil profile into a user-defined number of levels simulates vertical soil water movement (e.g., 12). The measured K-8-h relations are read from the horizon-specific tables for each unique soil horizon.
3. **Maps of Relevant Topographical, Soil, and Land-Use Variables:**
   - Maps in PC Raster format is required to run LISEM.
   - Maps that depict the morphology of the catchment:
   - An 'area map' that specifies the major catchment;
   - An 'id. Map' that specifies the spatial rainfall pattern;
   - Map indicating the location of the main and sub-catchment outlets;
   - Map indicating the 'local drain direction,' which is related to aspect;
   - Map with a steep slope;

- Chart of overland flow with Manning's n;
- Map depicting the main channels' slope gradient;
- Channel flow map based on Manning's n;
- Two maps depicting the morphology of the canal;
- Map showing the location of highways and their widths.

Because it contains a lot of information about the distribution of land attributes, LISEM is extremely useful in raster-based GIS. Many factors make LISEM an excellent tool for urban planners, including its use of physically based mathematical relationships, the ease with which new relationships can be added, the inclusion of data on spatial heterogeneity in land characteristics, and the model's detailed spatial display (Pandey, Himanshu, Mishra, & Singh, 2016).

4.8 WATER/SEDEM

A basic erosion and sediment delivery model, WATEM/SEDEM, is based on RUSLE (Krasa, Dostal, Jachymova, Bauer, & Devaty, 2019). The RUSLE model is incapable of simulating erosional deposition of eroded material, but the WATEM/SEDEM model does, in addition to soil erosion, using additional input parameters. The model has been put through its paces in a wide range of settings across Europe and other countries. In the Czech Republic, sediment accumulation in water reservoirs was used as a calibration tool. Similarly, the model was validated in additional sample areas. Sediment flows into reservoirs from a vast watershed, where grape cultivation accounted for a portion of the total arable land, was quantified using this model. An improved version of the RUSLE is used to estimate the yearly average rate of soil erosion. In general, the model consists of three major parts: (i) determination of the mean annual soil erosion rate for each grid cell; (ii) determination for each grid cell of the mean annual transport capacity; and (iii) sediment routing algorithm which redistributes the produced sediment across the catchment while taking into account the catchment's topology and the spatial information of the transport capacity. It can quantify the erosivity of rainfall (R factor) as a function of the total kinetic energy of a storm at its peak 30-minute intensity by using the formula (I_{max30}).

$$R = a \times (I_{max30})b$$

where; 'R' denotes rainfall erosivity factor; I_{max30} represents maximum 30-minute rainfall intensity; and 'a' and 'b' are region-specific empirical coefficients.

4.9 SHETRAN

River basin-scale model-based system SHETRAN is based on the physical, geographical diverse flow of water, sediment transport, and pollutant migration models (De Hipt et al., 2019). Incorporating land-phase hydrology provides a comprehensive picture of water flow in a river basin, both above and below the surface. Land degradation by raindrops, leaf drip, overland flow, and channel flow are all estimated to account for in the sediment transport component of the equation. Vertical distribution of catchment features, rainfall input, and hydrological response is achieved using an orthogonal grid network and a section of horizontal layers at each grid square. Evaluated rainfall data designed to simulate overland and channel flow data are used to model sediment erosion and transport. SHETRAN requires data such as spatially distributed topography, soil, vegetation, and sediment property to characterize a location.

4.10 DYRIM MODEL

This model consists of four layers, the data and model layers, the application layer, and the post-processing layers (Cai, Zhou, Liu, Wang, Jia, & Xu, 2019). Digital topography files, remote sensing images and field measurements can all be incorporated into the model thanks to the information available in the "data layer," which is the model's foundation. Before being accessed by the model and application layers, GIS-processed data can be edited or changed. A variety of parameters required to calculate different physical procedures can be acquired and altered by the model. To make large-scale river basin modeling a lot easier, the database allows it to easily share and transmit data between different users. A wide range of DYRIM-supporting techniques include databases, GIS, remote sensing, virtual reality (VR), and parallel computations. Digital elevation model (DEM), drainage network codification and segmentation, parameter acquisition and parallel computing are some of the DYRIM's distinguishing features from other information systems. In terms of digital terrain data, the DEM serves as the foundation of the DYRIM. Data from simulation models such as channel density, hillslope gradient, and a variety of channel segment properties are all extracted using DEM software. Upstream discharge routing and sediment transport simulation must be carried out first on hillslope-channel units. In terms of drainage network affluxion order, the Horton-Strahler order is a well-known one. A new drainage network codification method is proposed in the DYRIM to improve topological algorithms. When a dendritic river is analyzed, it is represented as a binary tree, with each channel segment having short and long lengths (BS Length, BS Value). The BS length is the level number of the binary tree segment that signifies the logical distance to the watershed's outlet. Spatial dispersion of the DYRIM's parameters is evident. The DEM provides the geometrical parameters, whereas spatial data photos in raster data provide the underlying surface parameters. These include land use, natural vegetation, soil type, potential evaporation, and more (Wang, Fu, Shi, & Li, 2015).

4.11 ANN MODELING

Artificial neural networks (ANNs) are one of the emerging modeling approaches for soil erosion modeling (El Bilali, Taleb, El Idrissi, Brouziyne, & Mazigh, 2020). Soil measurements are not possible throughout the entire research region or for a large number of sites. ANN can also estimate hydrological parameters using data from the field or observations. Thus, using an ANN to evaluate soil erosion saves money and study time, and ANN can provide estimates of soil degradation at any time and location. As a primary advantage of neural networks, they seem to be data-driven and do not make rigid assertions about the form of the basic model. Because they understand the evaluated data and do not need reprogramed, neural networks are known as "black box models." ANNs must be trained, optimized, and assessed during the modeling process. A neural network can only be trained if a set of candidate models is narrowed down to a single one. When parameters are tweaked in an effort to achieve the best possible set of parameters while staying within a set of predetermined parameters, the process is called optimization (Bai, Hu, He, He, & Zhao, 2022). Soil erosion and other features can be mapped with GIS as well. It is possible to use ANN to evaluate soil erosion with high precision and in a short time; however, its consequences (without geographic coordinates) will be challenging for other users. For ecological concerns and designed to simulate soil erosion, however, GIS is an effective tool. An ANN model and a GIS technique can predict soil erosion rates (Chakrabortty et al., 2020) (Table 4.1).

TABLE 4.1 Literature on Soil Erosion Models

Title	Model Applied	Result	Conclusion	Author(s)
Soil erosion assessment using RUSLE model and GIS in Huluka watershed, Central Ethiopia.	RUSLE	Estimates of the erosivity factor were done.	The spatial pattern of soil erosion over a large area can be accurately estimated and mapped using remote sensing and GIS in conjunction with the RUSLE model. Human-induced land-use and land cover changes have been shown to have significant effects on soil erosion in huge areas.	Atoma, Suryabhagavan, & Balakrishnan (2020)
Modeling the effect of vegetation cover and different tillage Practices on soil erosion in vineyards: a case study in Vráble (Slovakia) using WATEM/SEDEM.	WATEM/ SEDM	Among the year of planting the vineyard during the year of measurement, the exposure vineyard pole heights were measured to estimate an erosion and deposition rates. Deposition and erosion zones have different pole heights.	In the recent studies, WATEM/SEDEM model on the catchment level have been applied; this study uses the model on the fine field level. While the model does a good job of predicting soil erosion, it is less accurate in predicting soil deposition at this scale.	Lieskovský, & Kenderessy (2014)
SLEMSA model application for land use management in Semi-arid Kenya.	SLEMSA	Overgrazing in Kimutwa led to the highest soil loss estimate of 42.3 t ha^{-1}.	Due to its ability to make field-scale predictions of soil loss, the soil loss estimator model could be useful in this area. Climate, soils, vegetation, and topography data all matched up well when applied to the area in question.	Kinama, Stocking, & Maingi (2007)
Use of cesium-137 data to evaluate SHETRAN simulated long-term erosion patterns in arable lands.	SHETRAN	In other words, the net erosion rate is the difference between the gross erosion and the gross deposition of soil at a given location. Measured erosion rates at Dalicott Farm and Rufford Forest Farm are similar (6.05 t ha^{-1} $year^{-1}$ and 10.4 t ha^{-1} yr^{-1}).	The long-term erosion effects of potential land management techniques can be accurately predicted using physical and spatially distributed models. There has been a lack of appropriate data to demonstrate this capacity. Using Cs-137 in the field could be advantageous.	Norouzi, Bathurst, & Walling (2004)

TABLE 4.1 *(Continued)*

Title	Model Applied	Result	Conclusion	Author(s)
LISEM: a single-event physically-based hydrological and soil erosion model for drainage basins. I: theory, input, and output.	LISEM	In the early days of raster GIS, it was unusual to find a physically based model that was also fully integrated. Allowing for the use of remotely sensed data as well as avoiding the need for conversions, this feature makes it easier to implement in larger areas.	Watershed-scale hydrological and soil erosion processes are efficiently simulated by LISEM, a powerful model. Soil conservation scenarios can be explored, and the impacts of land use changes can be calculated with LISEM. Based on hypothetical storms with known recurrence probabilities, the cost-effectiveness of runoff and erosion mitigation measures can be calculated with LISEM.	De Roo, Wesseling, & Ritsema (1996)
Digital yellow river model	DYRIM	Stability analysis-based riverbank erosion is used to calculate gravitational erosion. Gravitic erosion is identified in this chapter using the attributes of unsaturated loess and the induction impact of slope flow. During the simulation, which takes place in the gully region, the soil body collapses or slides, and mechanical conditions are analyzed.	The DYRIM, a collection of physically-based models, provides distributed modeling at the river basin scale using information technology. In the Yellow River's Loess Plateau, the model was able to simulate the allocation of sediment yield and transportation in a large watershed.	Shi, Li, Wang, Zhang, Wang, & Fu (2016)
ANSWERS: A model for watershed planning	ANSWERS	A variety of simulations based on rainfall simulator plot data have been carried out on varying plot sizes. Data from these plots has helped us understand how different land use, conservation, and management practices affect runoff and soil loss in a particular area.	With the ANSWERS program, a comprehensive model is used to quantitatively evaluate non-point source pollution in an ungaged watershed and to determine the relative efficiency of alternative corrective measures.	Beasley, Huggins, & Monke (1980)

TABLE 4.1 *(Continued)*

Title	Model Applied	Result	Conclusion	Author(s)
GAMES – A screening model of soil erosion and fluvial sedimentation on agricultural watershed.	GAMES	The 1974–1975 cropping year data sets were used to develop soil erosion per hectare and cell delivery ratios for adjacent cells. The rate at which cells enter the stream. Delivery of sediment to a stream and sediment production ha^{-1}.	The games were created to provide seasonal mapping of soil loss and sediment yield in small agricultural watersheds using the most up-to-date soil erosion and sediment transport ideas.	Rudra, Dickinson, Clark, & Wall (1986)
Soil erosion prediction using Morgan-Morgan-Finney model in a GIS environment in Northern Ethiopia catchment.	MMF	Soil loss at catchment level has been especially in comparison to sediment yield evaluated at the reservoir's outlet. According to the MMF model, the mean soil loss (TC) at the catchment scale was 26 t ha^{-1} yr^{-1}, whilst the sediment yield from the reservoir survey was 20 t ha^{-1} yr^{-1}.	When it comes to integrating and managing spatially distributed hydro physical data, GIS has proven to be an effective tool. There was less soil detachment, as measured by the MMF model, than was expected for the soil transport capacity of overland flow (TC).	Tesfahunegn, Tamene, & Vlek (2014)

4.12 CONCLUSION

Increased soil erosion has recently been observed in the field, a development that should be taken seriously. It's important to consider rainfall erosivity, soil erodibility factor, topography, practice factor, and land use configuration when assessing. Around 11 soil erosion models were evaluated for the Loess Plateau in this study. Process-based models could be used to estimate soil erosion and sediment yield in great detail. Soil erosion on hillslopes and sediment production in small catchments can be modeled using WEPP, a powerful instrument. Analyzing and acquiring topographic parameters in a digital context has become a critical component of erosion model development. It should be noted that the accuracy of topographic surveys used to derive the elevation model is linked to the accuracy of these estimates. These methods can be used on complex geometries without requiring huge spatiotemporal databases like spatial and temporal databases. Predictive erosion models for temperate mid-latitude zones should be developed for humid tropical mountainous terrain. It is possible to use LPMs for watershed planning in humid tropical mountainous regions when coupled with GIS to conserve soil. Rainfall could be significantly reduced, as well as increased, as a result of climate change. Soil loss is greatly impacted by climate change. Predicted soil loss was found to be lower by RUSLE compared to MUSLE and MMF. There are many advantages to using dry-stone walls in vineyards, such as preserving soil and water. According to the report, the soil erosion models currently in use do not account for the full extent of the problem. A key component of effective soil erosion control is the involvement of residents and landowners, whose indigenous knowledge gives them an advantage in spotting rill erosion before it becomes serious.

ACKNOWLEDGMENTS

The author expresses gratitude to Tajamul Mansoor for his significant contribution in preparing this chapter.

KEYWORDS

- **distributed parameter models**
- **erosion**
- **erosion-productivity impact calculator**
- **European soil erosion model**
- **Limburg soil erosion model**
- **soil and water analysis tool**
- **water erosion prediction project**

REFERENCES

Acharya, G., Cochrane, T., Davies, T., & Bowman, E., (2011). Quantifying and modeling post-failure sediment yields from laboratory-scale soil erosion and shallow landslide experiments with silty loess. *Geomorphology, 129*(1, 2), 49–58.

Adu, J. T., & Kumarasamy, M. V., (2018). Assessing non-point source pollution models: A review. *Polish Journal of Environmental Studies, 27*(5).

Alewell, C., Borrelli, P., Meusburger, K., & Panagos, P., (2019). Using the USLE: Chances, challenges, and limitations of soil erosion modeling. *International Soil and Water Conservation Research, 7*(3), 203–225.

Atoma, H., Suryabhagavan, K., & Balakrishnan, M., (2020). Soil erosion assessment using RUSLE model and GIS in Huluka watershed, Central Ethiopia. *Sustainable Water Resources Management, 6*(1), 1–17.

Bai, W., Hu, H., He, L., He, Z., & Zhao, Y., (2022). Simulation and prediction of soil erosion in Daning River Basin. *IEEE Journal of Selected Topics in Applied Earth Observations and Remote Sensing.*

Barthel, S., Isendahl, C., Vis, B. N., Drescher, A., Evans, D. L., & Van, T. A., (2019). Global urbanization and food production in direct competition for land: Leverage places to mitigate impacts on SDG2 and on the earth system. *The Anthropocene Review, 6*(1, 2), 71–97.

Beasley, D., Huggins, L., & Monke, A., (1980). Answers: A model for watershed planning. *Transactions of the ASAE., 23*(4), 938–944.

Beitlerová, H., Lenz, J., Devátý, J., Mistr, M., Kapička, J., Buchholz, A., et al., (2021). Improved calibration of the Green–Ampt infiltration module in the EROSION-2D/3D model using a rainfall-runoff experiment database. *Soil, 7*(1), 241–253.

Beven, K. J., Kirkby, M., Schofield, N., & Tagg, A., (1984). Testing a physically-based flood forecasting model (TOPMODEL) for three UK catchments. *Journal of Hydrology, 69*(1–4), 119–143.

Biasi, R., Brunori, E., Smiraglia, D., & Salvati, L., (2015). Linking traditional tree-crop landscapes and agro-biodiversity in Central Italy using a database of typical and traditional products: A multiple risk assessment through a data mining analysis. *Biodiversity and Conservation, 24*(12), 3009–3031.

Biswas, S. S., & Pani, P., (2015). Estimation of soil erosion using RUSLE and GIS techniques: A case study of Barakar River basin, Jharkhand, India. *Modeling Earth Systems and Environment, 1*(4), 1–13.

Blaikie, P., & Brookfield, H., (2015). *Land Degradation and Society*. Routledge.

Borrelli, P., Robinson, D. A., Panagos, P., Lugato, E., Yang, J. E., Alewell, C., et al., (2020). Land use and climate change impacts on global soil erosion by water (2015–2070). *Proceedings of the National Academy of Sciences, 117*(36), 21994–22001.

Cai, J., Zhou, Z., Liu, J., Wang, H., Jia, Y., & Xu, C. Y., (2019). A three-process-based distributed soil erosion model at catchment scale on the Loess Plateau of China. *Journal of Hydrology, 578*, 124005.

Chakrabortty, R., Pal, S. C., Sahana, M., Mondal, A., Dou, J., Pham, B. T., et al., (2020). Soil erosion potential hotspot zone identification using machine learning and statistical approaches in eastern India. *Natural Hazards, 104*(2), 1259–1294.

Coulthard, T. J., Hancock, G. R., & Lowry, J. B., (2012). Modeling soil erosion with a downscaled landscape evolution model. *Earth Surface Processes and Landforms, 37*(10), 1046–1055.

De Hipt, F. O., Diekkrüger, B., Steup, G., Yira, Y., Hoffmann, T., Rode, M., et al., (2019). Modeling the effect of land use and climate change on water resources and soil erosion in a tropical West African catch-ment (Dano, Burkina Faso) using SHETRAN. *Science of the Total Environment, 653*, 431–445.

De Roo, A., Wesseling, C., & Ritsema, C., (1996). LISEM: A single-event physically based hydrological and soil erosion model for drainage basins. I: Theory, input, and output. *Hydrological Processes, 10*(8), 1107–1117.

Devia, G. K., Ganasri, B. P., & Dwarakish, G. S., (2015). A review on hydrological models. *Aquatic Procedia, 4*, 1001–1007.

Dixit, P., (2021). *Hydrological Modeling of Narmada River Using ARCSWAT* Software: Delhi Technological University.

Douglas-Mankin, K., Srinivasan, R., & Arnold, J., (2010). Soil and water assessment tool (SWAT) model: Current developments and applications. *Transactions of the ASABE, 53*(5), 1423–1431.

Eisenbeiss, H., (2004). A mini unmanned aerial vehicle (UAV): System overview and image acquisition. *International Archives of Photogrammetry Remote Sensing and Spatial Information Sciences, 36*(5/W1), 1–7.

El Bilali, A., Taleb, A., El Idrissi, B., Brouziyne, Y., & Mazigh, N., (2020). Comparison of a data-based model and a soil erosion model coupled with multiple linear regression for the prediction of reservoir sedimentation in a semi-arid environment. *Euro-Mediterranean Journal for Environmental Integration, 5*(3), 1–13.

Elhag, M., Kojchevska, T., & Boteva, S., (2019). EPM for soil loss estimation in different geomorphologic conditions and data conversion by using GIS. *IOP Conference Series: Earth and Environmental Science*. IOP Publishing.

Endreny, T. A., (2005). *126: Land Use and Land Cover Effects on Runoff Processes.* Urban and Suburban Development.

Eswaran, H., Lal, R., & Reich, P., (2019). Land degradation: An overview. *Response to Land Degradation*, 20–35.

Fales, M., Dell, R., Herbert, M. E., Sowa, S. P., Asher, J., O'Neil, G., et al., (2016). Making the leap from science to implementation: Strategic agricultural conservation in Michigan's Saginaw Bay watershed. *Journal of Great Lakes Research, 42*(6), 1372–1385.

Flanagan, D. C., Gilley, J. E., & Franti, T. G., (2007). Water erosion prediction project (WEPP): Development history, model capabilities, and future enhancements. *Transactions of the ASABE, 50*(5), 1603–1612.

Gassman, P., Williams, J., Wang, X., Saleh, A., Osei, E., Hauck, L., et al., (2010). Invited review article: The agricultural policy/environmental eXtender (APEX) model: An emerging tool for landscape and watershed environmental analyses. *Transactions of the ASABE, 53*(3), 711–740.

Ghosal, K., & Das, B. S., (2020). A review of RUSLE model. *Journal of the Indian Society of Remote Sensing, 48*(4), 689–707.

Ghosh, S., (2022). An assessment of RUSLE model and erosion vulnerability in the slopes of Dwarka–Brahmani lateritic interfluve, Eastern India. *Climate, Environment and Disaster in Developing Countries* (pp. 475–506). Springer.

Guo, Y., Peng, C., Zhu, Q., Wang, M., Wang, H., Peng, S., et al., (2019). Modeling the impacts of climate and land use changes on soil water erosion: Model applications, limitations, and future challenges. *Journal of Environmental Management, 250*, 109403.

Hirschi, M. C., & Barfield, B. J., (1988). KYERMO—A physically based research erosion model part II. Model sensitivity analysis and testing. *Transactions of the ASAE, 31*(3), 814–0826.

Joshi, S., & Xu, Y. J., (2017). Bedload and suspended load transport in the 140-km reach downstream of the Mississippi River avulsion to the Atchafalaya River. *Water, 9*(9), 716.

Kaffas, K., Hrissanthou, V., & Sevastas, S., (2018). Modeling hydromorphological processes in a mountainous basin using a composite mathematical model and ArcSWAT. *Catena, 162*, 108–129.

Kalambukattu, J., & Kumar, S., (2017). Modeling soil erosion risk in a mountainous watershed of Mid-Himalaya by integrating RUSLE model with GIS. *Eurasian Journal of Soil Science, 6*(2), 92–105.

Kinama, J., Stocking, M., & Maingi, P., (2007). SLEMSA model application for land use management in Semi-Arid Kenya. *Proceedings of the Soil Science Society of East Africa Embu, Kenya, East Africa*, 26–30.

Knisel, W. G., & Turtola, E., (2000). GLEAMS model application on a heavy clay soil in Finland. *Agricultural Water Management, 43*(3), 285–309.

Krasa, J., Dostal, T., Jachymova, B., Bauer, M., & Devaty, J., (2019). Soil erosion as a source of sediment and phosphorus in rivers and reservoirs–Watershed analyses using WaTEM/SEDEM. *Environmental Research, 171*, 470–483.

Kværnø, S. H., & Stolte, J., (2012). Effects of soil physical data sources on discharge and soil loss simulated by the LISEM model. *Catena, 97*, 137–149.

Li, P., Mu, X., Holden, J., Wu, Y., Irvine, B., Wang, F., et al., (2017). Comparison of soil erosion models used to study the Chinese loess plateau. *Earth-Science Reviews, 170*, 17–30.

Lieskovský, J., & Kenderessy, P., (2014). Modeling the effect of vegetation cover and different tillage practices on soil erosion in vineyards: A case study in Vráble (Slovakia) using WATEM/SEDEM. *Land Degradation & Development, 25*(3), 288–296.

Liu, Y., Du, D., Qi, N., & Zhao, J., (2018). A distributed parameter Maxwell-slip model for the hysteresis in piezoelectric actuators. *IEEE Transactions on Industrial Electronics, 66*(9), 7150–7158.

Lowry, J., Coulthard, T., Hancock, G., & Jones, D., (2011). Assessing soil erosion on a rehabilitated landform using the CAESAR landscape evolution model©. *Mine Closure 2011: Proceedings of the Sixth International Conference on Mine Closure; 2011*. Australian Centre for Geomechanics.

Mohammed, S., Alsafadi, K., Talukdar, S., Kiwan, S., Hennawi, S., Alshihabi, O., et al., (2020). Estimation of soil erosion risk in southern part of Syria by using RUSLE integrating geo informatics approach. *Remote Sensing Applications: Society and Environment, 20*, 100375.

Mondal, A., Khare, D., & Kundu, S., (2018). A comparative study of soil erosion modeling by MMF, USLE and RUSLE. *Geocarto International, 33*(1), 89–103.

Morgan, R. P. C., (2009). *Soil Erosion and Conservation*. John Wiley & Sons.

Morgan, R., Quinton, J., Smith, R., Govers, G., Poesen, J., Auerswald, K., et al., (1998). The European soil erosion model (EUROSEM): A dynamic approach for predicting sediment transport from fields and small catchments. *Earth Surface Processes and Landforms: The Journal of the British Geomorphological Group, 23*(6), 527–544.

Nieder, R., Benbi, D. K., & Reichl, F. X., (2018). *Soil Components and Human Health*. Springer.

Norouzi, B. Y., Bathurst, J., & Walling, D., (2004). Use of caesium-137 data to evaluate SHETRAN simulated long-term erosion patterns in arable lands. *Hydrological Processes, 18*(10), 1795–1809.

Nwokocha, O. U., (2021). *Assessment of the Impacts of Gully Erosion on Livelihood of Residents of Ohafia*. Abia State.

Pandey, A., Himanshu, S. K., Mishra, S. K., & Singh, V. P., (2016). Physically based soil erosion and sediment yield models revisited. *Catena, 147*, 595–620.

Pierson, F. B., Bates, J. D., Svejcar, T. J., & Hardegree, S. P., (2007). Runoff and erosion after cutting western juniper. *Rangeland Ecology & Management, 60*(3), 285–292.

Raj, A. D., Kumar, S., & Sooryamol, K., (2022). Modeling climate change impact on soil loss and erosion vulnerability in a watershed of Shiwalik Himalayas. *CATENA, 214*, 106279.

Ribolzi, O., Patin, J., Bresson, L. M., Latsachack, K., Mouche, E., Sengtaheuanghoung, O., et al., (2011). Impact of slope gradient on soil surface features and infiltration on steep slopes in northern Laos. *Geomorphology, 127*(1, 2), 53–63.

Rudra, R., Dickinson, W., Clark, D., & Wall, G., (1986). GAMES—A screening model of soil erosion and fluvial sedimentation on agricultural watershed. *Canadian Water Resources Journal, 11*(4), 58–71.

Shi, H., Li, T., Wang, K., Zhang, A., Wang, G., & Fu, X., (2016). Physically based simulation of the streamflow decrease caused by sediment-trapping dams in the middle Yellow River. *Hydrological Processes, 30*(5), 783–794.

Silburn, D., & Loch, R., (1989). Evaluation of the CREAMS model. I. Sensitivity analysis of the soil erosion sedimentation component for aggregated clay soils. *Soil Research, 27*(3), 545–561.

Singh, R. L., & Singh, P. K., (2017). Global environmental problems. *Principles and Applications of Environmental Biotechnology for a Sustainable Future* (pp. 13–41). Springer.

Smith, H., (1999). Application of empirical soil loss models in southern Africa: A review. *South African Journal of Plant and Soil, 16*(3), 158–163.

Sun, L., Guo, H., Liu, B., Wu, S., Weckler, P. R., & Yang, J., (2021). Characterizing erosion processes on a convex slope based on 3D reconstruction method. *Geoderma, 402*, 115364.

Supratak, A., Dong, H., Wu, C., & Guo, Y., (2017). DeepSleepNet: A model for automatic sleep stage scoring based on raw single-channel EEG. *IEEE Transactions on Neural Systems and Rehabilitation Engineering, 25*(11), 1998–2008.

Tesfahunegn, G. B., Tamene, L., & Vlek, P. L., (2014). Soil erosion prediction using Morgan-Morgan-Finney model in a GIS environment in Northern Ethiopia catchment. *Applied and Environmental Soil Science, 2014.*

Tully, K., Sullivan, C., Weil, R., & Sanchez, P., (2015). The state of soil degradation in Sub-Saharan Africa: Baselines, trajectories, and solutions. *Sustainability, 7*(6), 6523–6552.

Walling, D. E., (2017). Measuring sediment yield from river basins. *Soil Erosion Research Methods* (pp. 39–82). Routledge.

Walling, D., He, Q., & Whelan, P., (2003). Using 137Cs measurements to validate the application of the AGNPS and ANSWERS erosion and sediment yield models in two small Devon catchments. *Soil and Tillage Research, 69*(1, 2), 27–43.

Wang, G., Fu, X., Shi, H., & Li, T., (2015). Watershed sediment dynamics and modeling: A watershed modeling system for yellow river. *Advances in Water Resources Engineering* (pp. 1–40). Springer.

Wang, Y., & Dai, E., (2020). Spatial-temporal changes in ecosystem services and the trade-off relationship in mountain regions: A case study of Hengduan Mountain region in Southwest China. *Journal of Cleaner Production, 264*, 121573.

Williams, J., Renard, K., & Dyke, P., (1983). EPIC: A new method for assessing erosion's effect on soil productivity. *Journal of Soil and Water Conservation, 38*(5), 381–383.

Yuan, L., Sinshaw, T., & Forshay, K. J., (2020). Review of watershed-scale water quality and nonpoint source pollution models. *Geosciences, 10*(1), 25.

Zema, D. A., Nunes, J. P., & Lucas-Borja, M. E., (2020). Improvement of seasonal runoff and soil loss predictions by the MMF (Morgan-Morgan-Finney) model after wildfire and soil treatment in Mediterranean forest ecosystems. *Catena, 188*, 104415.

Zhang, S., Hou, D., Wang, C., Pan, F., & Yan, L., (2020). Integrating and managing BIM in 3D web-based GIS for hydraulic and hydropower engineering projects. *Automation in Construction, 112*, 103114.

CHAPTER 5

Deep Learning-Based Approach for Disease Detection in the Changing Global Ecosystem

OWAIS BASHIR,[1] SHABIR AHMAD BANGROO,[1] NASIR BASHIR NAIKOO,[1] NAZIR AHMAD GANIE,[2] ZOHRA SHABIR,[3] SHEIKH ASIF,[4] RAZEEF MOHD.,[5] SHAFEEQ UR REHMAN,[6] FAZIL FAYAZ,[3] and KHALID REHMAN HAKEEM[7]

[1]*Division of Soil Science, Sher-e-Kashmir University of Agricultural Sciences and Technology of Kashmir, Kashmir, India*

[2]*Division of Animal Genetics and Breeding, Sher-e-Kashmir University of Agricultural Sciences and Technology of Kashmir, Kashmir, India*

[3]*Division of Plant Pathology, Sher-e-Kashmir University of Agricultural Sciences and Technology of Kashmir, Kashmir, India*

[4]*Division of Entomology, Sher-e-Kashmir University of Agricultural Sciences and Technology of Kashmir, Kashmir, India*

[5]*NAHEP, Sher-e-Kashmir University of Agricultural Sciences and Technology of Kashmir, Kashmir, India*

[6]*School of Environment and Civil Engineering, Dongguan University of Technology, Dongguan, Guangdong, China*

[7]*Department of Biological Sciences, Faculty of Science, King Abdulaziz University, Jeddah, Saudi Arabial:*

Soil, Agriculture, and Ecosystem Modelling: Smart Technologies for Sustainable Solutions.
Owais Bashir, Khalid Rehman Hakeem, Surinder S. Kukal, and Razeef Mohd (Eds.)

ABSTRACT

Plant disease threatens food security as it deteriorates the quality and quantity of harvest. Therefore, diagnosing plant illnesses as early as possible is essential to prevent significant crop damage. However, an incorrect diagnosis of the condition leads to the wrong application of pesticides. To improve the quality and quantity of production, a deep learning (DL)-based method is developed for detecting and classifying plant diseases using various datasets. Plant diseases are significant determinants of plant output and quality. By utilizing digital image processing, plant diseases and pests may be identified. In recent years, DL has achieved significant advances in digital image processing, far outperforming conventional approaches. Image acquisition is the first step in computer vision-based systems. Next, different image-processing tasks, such as scaling, filtering, segmentation, selection, and feature extraction, are performed. Finally, machine learning (ML)- or DL-based algorithms are used for recognition and classification. How to employ DL technology to investigate the identification of plant diseases and pests has become a major worry for academics. This chapter defines the plant disease and pest detection challenge and compares it to conventional plant disease and pest detection techniques. This chapter examines the research on plant disease and pest detection based on DL in recent years from the three perspectives of classification network, detection network, and segmentation network and summarizes the merits and disadvantages of each technique. This chapter explores potential obstacles in practical implementations of DL-based plant disease and pest detection. In addition, potential remedies and research ideas for the issues are suggested, and numerous proposals are made. This chapter concludes with an analysis and forecast of the future trend of DL-based plant disease and pest detection.

5.1 INTRODUCTION

Modern technology has enabled humanity to generate enough food to feed over 7 billion people. According to the United Nations Food and Agriculture Organization (FAO), the global population is estimated to reach 9 billion by 2050 (Dawson et al., 2016; Kim & Kwak, 2020; Kwak, 2019). To meet the nutritional demands of such a large population, the food growth rate should be boosted to 70% by 2050 (Searchinger et al., 2019). However, growing food production is hampered by several problems, such as the lack of sizeable cultivable land, plant diseases and other resources. In addition, plant diseases significantly reduce the amount and quality of harvests. These damages significantly impact the economy, resulting in a decline in farmer income and food costs for consumers (Mesterházy et al., 2020). Moreover, it may result in severe food shortages following extreme hunger and famine, particularly in underdeveloped nations with scarce preventative measures (Manirambona et al., 2021). Human resources typically identify agricultural diseases by physically inspecting crop fields. To examine every plant, though, would take much time and be nearly impossible for humans (Rehman et al., 2022; Sinha & Dhanalakshmi, 2022). As a result of the limited labor available, crop disease analysis procedures frequently experience delays. Therefore, early detection of agricultural diseases is essential since it increases food output and shields farmers from expensive spraying procedures (Bieganowski et al., 2021). The scientific community has switched its attention to introducing and analyzing computer-aided technologies to simplify this complicated work and to build efficient automated crop analysis techniques to cope with the issues of physical plant disease identification (Albattah et al., 2022). Genomic biology and immunology methods were initially used to identify crop diseases. However, these strategies needed to be formed; they needed much money and human expertise (Bevan et al., 2017; Sundin et al., 2016). New machine learning (ML) based hand-coded algorithms are currently used in agriculture to improve decision-making capability (Thakur & Rane, 2021). A significant quantity of data is now being captured in real-time thanks to the development of digital techniques; this data is then used to create ML-based approaches that help people make better decisions (Ayvaz & Alpay, 2021). These techniques, including decision trees (DT), support vector machines (SVM), deep learning (DL), convolution neural networks, K-nearest neighbors (KNN), and Gaussian frameworks, have also undergone extensive testing for the diagnosis of agricultural diseases (Iqbal et al., 2021; Pattnaik & Parvathy, 2022; Swain et al., 2020).

A subtype of ML called "deep learning" was created in 1943 with the inclusion of threshold logic to create a computer model that closely resembles the biological pathways of humans. This subject of study is continually growing; it may be separated into two periods: 1943–2012 and 2012–Present (Arabi, 2021; Xu et al., 2021). During the initial phase, several innovations were made, including back propagation, chain rules, Neocognitron, handwritten text identification (LeNET architecture), and solving the training problem. However, in the second phase, through picture recognition, cutting-edge algorithms/architectures were created for various applications, including self-driving cars, earthquake forecasts, marketing, finance, rainfall prediction, and disease prediction (Arabi, 2021; Darwish et al., 2020; Pandey &

Janghel, 2019). The effectiveness of DL-dependent approaches such as CNN, deep belief networks and recurrent neural networks (RNNs) in various application areas such as image segmentation, classification, and change detection has revolutionized the agriculture sector. DL-dependent techniques, for example, CNN, can instantaneously recognize the optimum solution main facts from the input data without human experts (Abdel-Basset et al., 2020; Hakim et al., 2022). The DL-based framework's operation mimics how the human brain processes information. By looking at several object samples, people can visually recognize and locate things. These methods use the same process to recognize objects and patterns. DL architectures produce more accurate findings than customized ML-based techniques (Albattah et al., 2022; Deng et al., 2021; Garcia-Molsosa et al., 2021), enabling better prediction in agriculture and allied sectors. Due to the rapid advancement of hardware, DL frameworks are being extensively researched to find solutions to difficult issues in a respectably short amount of time. In agriculture, DL-based approaches demonstrate cutting-edge precision and are adapted well to various jobs. Deep neural networks (DNNs) in various forms have demonstrated noteworthy performance compared to hyperspectral analysis. GoogLeNet, AlexNet, VGG, and ResNet are a few of the well-known CNN models that did well in crop-related classification tasks such as yield amount prediction, crop tops identification, fruit enumeration, detection of plant diseases, classification, and many more (Li & Du, 2022; Mahmud et al., 2021; Shaikh et al., 2022). These frameworks can produce reliable results with the least amount of preparation and computing labor.

Additionally, DL-based algorithms are resistant to post-processing assaults such as rotation, translation, etc., and provide higher performance for object localization by utilizing the topological data from the input pictures (Sarkar, 2021). With pre-trained frameworks, these approaches also provide an output of higher quality. There is always potential for improvement, even though much work has been done on agricultural disease detection and categorization. In comparison to human brain intelligence, ML techniques are remarkably better at solving complex real-world situations. Since these frameworks generate lengthy codes that increase computational complexity, the main problems with ML-based approaches to crop disease detections are their low efficacy and lengthy processing times. However, at the expense of increased code complexity, DL approaches have been developed to address the problem of lengthy codes (Gupta et al., 2021; Jiang & Li, 2020; Vamathevan et al., 2019). Additionally, current approaches cannot be used to discover several cases of the same disease in a picture or to identify multiple diseases across one sample due to poor generalization to real-world circumstances.

As DL architectures began to grow, researchers used this technology to categorize and recognize photos. In DL, the three main types of networks used nowadays are RNNs, multilayered perceptron, and CNN (RNN). CNN is the most used technique for categorizing plant diseases. Since they are often used for image analysis or medical problems, DL models like deconvolutional networks and fully convolution networks (FCNs) are not used to classify plant leaf diseases (Lu et al., 2021; Singh et al., 2018). CNN typically consists of convolution layers, pooling, and fully linked networks. The convolution layer extracts features based on the spatial correlations of the image's data (Shi et al., 2022; Xiong & Zuo, 2022). The procedure is continued until all potential sites in the picture are filtered, after which the kernel is shifted forward by one pixel. In addition to making the model robust to translations, rotation, and scaling, the pooling layer samples features from the higher layer feature map (Chaman & Dokmanic, 2021; Li et al., 2021b). The highest or average pooling is the most popular (Figure 5.1). The result of average pooling is the average across all regions by dividing the input picture into several rectangular areas dependent on the filter's size, maximum pooling outputs, and the highest value for each zone. The multimodal features are merged and reduced to one-dimensional characteristics in the classifier to classify or detect tasks since every neuron in the fully connected layers is linked to the top neuron (Chen et al., 2022; Sánchez-Reolid et al., 2022; Sharif et al., 2022).

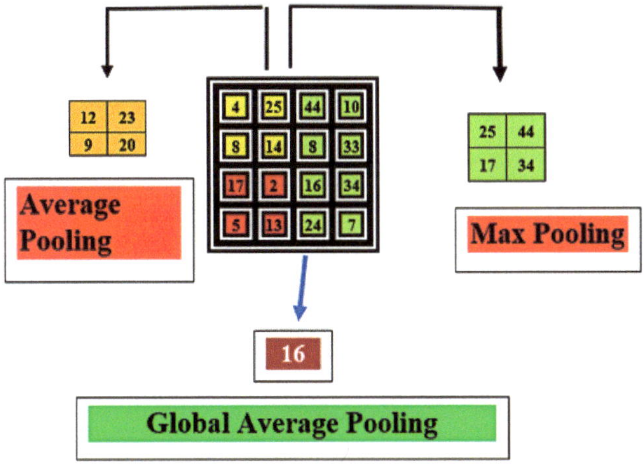

FIGURE 5.1 Three types of pooling operations.

As a result of this approach, we advise using a more holistic approach to provide a more suitable foundation to build a comprehensive understanding of DL. This is because a thorough strategy will likely account for all pertinent elements. This research aims to offer a more thorough analysis of the most crucial components of DL in disease categorization, including the current advancements made in this area (Alzubaidi et al., 2021; Ren et al., 2021). The importance of DL, various DL techniques, implementation of DL in various crops, and the networks that enable them are all thoroughly covered in this study. The most prevalent kind of DL networks, convolutional neural networks (CNNs), are next covered, along with their main properties and the development of CNN designs (Alzubaidi et al., 2021; Li et al., 2021b).

5.2 PLANT DISEASES DETECTION TECHNIQUES USING DEEP LEARNING (DL)

This section summarizes approaches based on DL to combat plant diseases. The use of conventional networks in agriculture is demonstrated by implementing DL-based plant disease methods. In accordance, the network can be segmented into detection, classification, and segmentation networks.

5.2.1 CLASSIFICATION NETWORK

Plant diseases are difficult to identify naturally because of their profound changes in size, shape, texture, color, background, arrangement, and imaging lighting. CNN's robust feature extraction capacity has led to the widespread deployment of CNN-based classification networks for the categorization of crop diseases (DAGIM, 2021; Valicharla, 2021). A complete connection layer (or average pooling layer) and a SoftMax structure follow the feature extraction stage of a CNN classification network, typically consisting of a cascaded convolution layer and a pooling layer. Existing network architectures for crop diseases categorization include AlexNet, Inception V4, DenseNets, MobileNet, GoogleLeNet, VGGNet, SqueezeNet, ResNet, and SqueezeNet (Baht & Domínguez, 2022; Falaschetti et al., 2021; Liu & Wang, 2021). Several researchers have also built network architectures based on real situations. The network evaluates the image and produces a categorization label by providing a test picture to the classification model. Based on the various tasks that the classification network technique completes, it may be divided into three subcategories: using the networks as feature representation, using the networks for classification effectively, and using the networks to locate lesions (Lai et al., 2015; Yang et al., 2016).

5.2.1.1 USING THE NETWORK AS FEATURE REPRESENTATION

In the earliest study on DL-based classification systems for crop diseases, numerous researchers used CNN's full feature extraction capacity and merged it with conventional classifiers. To characterize the images, the photographs were first put into a pertained CNN network. Then, for classification, these traits are loaded into a typical ML classifier (such as an SVM) (Pandey et al., 2022; Wang et al., 2022b). Yalcin & Razavi (2016) developed a DNN architecture using polynomial kernels with SVM classifiers and used feature modifiers like GIST and LBP to retrieve the picture features. The results of the experiments proved that the strategy was effective. After traversing through the meta-architecture, the input photos of healthy and diseased plants were classified into the appropriate groups. Fuentes et al. (2018) suggested a CNN-based meta-architecture comprising various feature extractors. Hassan et al. (2021) categorized eight different kinds of rice diseases with a 97.5% accuracy rate utilizing attributes extracted from the DCNN technique and submitted them into the SVM.

5.2.1.2 USING THE NETWORK FOR CLASSIFICATION ACCURACY

The oldest widespread application of CNN in crop disease detection is categorizing lesions directly using classification networks. On the basis of the features of existing research, it is further subdivided into original picture classification; classification after identifying the region of interest (ROI), and multi-category classification (Zheng et al., 2022).

5.2.1.3 USING THE NETWORK FOR LESIONS LOCATION

Typically, a classification system can only classify images at the tag level. It can detect the site of lesions and classify images pixel-by-pixel by integrating several techniques and procedures. Based on the features of current studies, it can be further separated into original picture categorization, multi-category classification, and classification after locating the Regions of interest (ROI) (Huang et al., 2019; Jia et al., 2022). It can be further divided into three categories based on the applied methods: multi-task learning networks, sliding windows, and heatmap.

5.2.2 DETECTION NETWORK

Object placement is one most the major computer vision challenges. Its role is most similar to traditional plant disease and pest detection. Its goal is to determine the object's precise position and categorization. Currently, DL-based object recognition systems are constantly evolving (Fadlullah et al., 2017; Fridman et al., 2017). A network for identifying crop disease depending on DL may generally be categorized into two groups: a two-stage network characterized by Faster R-CNN and a one-stage model described by SSD and YOLO (Nasirahmadi et al., 2021; Roy et al., 2022). Before running the object detection algorithm, the two-stage network must develop a candidate box (proposal) containing lesions. In comparison, the one-stage network uses the network's retrieved properties to forecast the position and kind of lesions (Nazir et al., 2021).

5.2.2.1 PLANT DISEASE DETECTION IS BASED ON TWO STAGES OF NETWORK

Two-stage detection network (Faster R-CNN) collects the feature map of the input image using the backbone network, evaluates the anchor box confidence using RPN, and then provides a suggestion (Vecvanags et al., 2022). After ROI pooling, enter the suggested area's feature map into the network, refine the initial detection results, and obtain lesion localization and classification (Yu et al., 2022). When detecting plant diseases, common ways to improve the layered structure anchor ratio, feature map, ROI pooling, and loss function. Fuentes et al. (2018) worked on the tomato crop using Faster R-CNN to identify crop diseases directly. Using DL approaches such as ResNet and VGG-Net, the map value of a dataset comprising 5,000 tomato diseases improved by 85.98%. Ozguven & Adem (2019) designed Faster R-CNN architecture for unsupervised leaf spot identification in beets. This method's total classification accuracy is 95.4%. Zhao et al. (2021) employed Faster R-CNN and FCM-KM to rapidly diagnose rice diseases. Bacterial leaf blight, rice blast, and sheath blight accuracy rates were found to be 96.71%, respectively, based on 3,010 images (Zhao et al., 2021). Xie et al. (2020) suggested a Faster DR-IACNN model, based on the Faster R-CNN detection technique and self-built grape leaf disease dataset (GLDD), using Inception-ResNet-v2, Inception-v1, and SE. The suggested model has better feature extraction, 81.1% mAP accuracy, and 15.01 FPS detection speed.

The two-stage detection system has increased the detection performance to improve real-time and usability, but it still isn't concise, and the inference speed is too sluggish.

5.2.2.2 PLANT DISEASES AND PESTS DETECTION BASED ON ONE-STAGE NETWORK

The one-stage object detection method omits the region proposal stage. It puts the detection head straight into the regression and classification backbone networks, resulting in a significant increase in inference speed. Single-stage detection networks, SSD and YOLO, take the complete picture as input and return the bounding box location and category at the output layer. SSD uses VGG16 as the network's trunk and a feature pyramid network to forecast from multiple levels (Pacal et al., 2020). Singh et al. (2020) designed PlantDoc to identify diseases. The combination of MobileNets with SSD is often used in computer vision applications, especially for on-device or edge-based object detection due to the efficiency of MobileNets and the accuracy of SSD in detecting objects within images or video frames. Wu et al. (2021) introduced a CNNs-based multi-scale feature fusion method to identify maize leaf blight under a complicated backdrop. Data preparation, feature fusion, feature sharing, and illness diagnosis were included. The new model's mAP is greater than the old SSD model's (71.80% to 91.83%). The updated model's FPS (from 24 to 28.4) meets true detection standards. For CNN network detection from beginning to end, YOLO sees detection as a regression problem and leverages metadata to explicitly forecast the object's boundaries and categories. YOLO improves detection speed and accuracy globally. Bhatt et al. (2019) developed a technique for diseases in tea gardens. YOLOv3 detected diseases and pests. With 50% IOU, 86% mAP was obtained while assuring real-time system availability. Zhang et al. (2021) coupled the pooling of improved YOLOv3 with the spatial pyramids by juxtaposing up-sampling and convolution procedures, allowing the algorithm to efficiently identify tiny crop pest specimens in the picture and reduce the issue of low recognition rate due to plant pest attitudes and scales. Around 20 classes of real-world pests have a recognition accuracy of 88.07%. Plant disease detection models will become more inventive with the progression of object identification systems in computer vision. In the sector of detection of plant diseases that emphasizes precision, more two-stage models were applied, while in the sector that prioritizes time, more one-stage models are employed.

Can the classification network replace the detection network? Detection networks find plant diseases. The classification network classifies crop diseases. The physically disguised data of a detection system is composed of category data, i.e., the categories of crop diseases that need to be found and the related annotation information. The detection network does seem to contain the classification network's processes; it can respond to "what crop diseases are where." Misconception: "what sort of plant illnesses and pests" is given a priori; what is identified throughout training is not always the actual outcome. In cases of significant model differentiation, where the detection network gives correct findings, it may answer "what crop diseases and pests exist where." In the actual world, it cannot uniquely express the distinctiveness of crop disease categories, merely answering "what sort of crop diseases may be in what area," thus a classification network is needed. Detection cannot substitute categorization.

5.2.3 SEGMENTATION NETWORK

Using a segmentation network, the morphological and even occurrence classification of lesions and healthy zones represent plant disease (Dhingra et al., 2018). It divides the lesion zone precisely and establishes its position, group, and spatial relationships (dimensions, region, form, axis, etc.). Fully convolutional networks (FCNs) and Mask R-CNN are promising segmentation networks (Lin et al., 2020; Qadir et al., 2019).

5.2.3.1 FCN

A full convolution neural network (FCN) is the foundation for segmenting picture semantics. Currently, FCN is the foundation of almost all semantic segmentation systems. FCN uses convolution to extract and decode the information from the image initially, and deconvolution or upsampling is used to gradually recover the featured image's size with that of the input image (Zhao, 2022; Zhou et al., 2022). Based on the variations in FCN network topology, the segmentation techniques for plant diseases may be separated into traditional FCN, U-net, and SegNet (Sodjinou et al., 2022).

5.2.3.2 CONVENTIONAL FCN

Wang et al. developed a new maize leaf disease segmentation technique based on a complete convolution neural network with a recognition accuracy of 96.26%. Wang et al. (2019a) developed a segmentation approach for crop diseases based on enhanced FCN. In this technique, multi-layer feature information was extracted from the input maize leaf lesion picture using a convolution layer, and the width and resolution of the original image were recovered using a deconvolution operation. In contrast to the initial FCN approach, the segmentation of smaller lesion areas was highlighted, having a prediction accuracy of 95.87%.

5.2.3.3 U-NET

U-net is an integration of the encoder-decoder structure with the conventional FCN structure. It stands out because it incorporates a layer-hopping connection that combines the attributes that were extracted during the coding and decoding processes to make it easier to retrieve segmentation data. To separate 50 naturally acquired cucumber powdery mildew leaves (Lin et al., 2009), used a U network-shaped CNN. A batch data normalization phase was inserted every after convolution layer, compared to the earlier U-net, making the neural network resilient to weight initialization. The research compares favorably to the Random-forest, K-means, and GBDT methods presently in use; the U-net CNN can discriminate powdery mildew on cucumber leaves with an image pixel correctness of 96.08%. Whilst retaining high exactness with smaller samples amid a complex background, the U-net technique can precisely partition the lesion area (Haq, 2022).

5.2.3.4 SEGNET

It also possesses a conventional encoder-decoder structure. Its distinctive feature would be that the up-sampling operation in the decoder uses the encoder's largest pooling index (Gao & Ji, 2019). Kerkech et al. (2020) introduced an unmanned aerial vehicle (UAV) picture segmentation algorithm. SegNet distinguished four categories from visible and infrared photos: shadows, symptomatic ground, and healthy grape plants. On grapevines and leaves, the detection rates of the suggested approach were 92% and 87%, respectively.

5.2.3.5 MASK R-CNN

Mask R-CNN has become one of the most used visual scenario segmentation methods using a network-based multifunctional learning strategy for classification and extraction (Kadam et al., 2022). While several lesions of a similar kind display adhesion or overlap, the segmentation of incidences get separated and may be identified as individual lesions. Nonetheless, semantic segmentation frequently interprets several lesions of a similar kind as a single entity (Liu et al., 2022). In contrast, some study integrates the Mask R-CNN structure alongside object recognition systems for diagnosing plant diseases. Wang et al. (2022c) used two different algorithms, Faster R-CNN and Masked R-CNN, in which Faster R-CNN was used to classify the category of tomatoes diseases, and Mask R-CNN was used to recognize and identify the location and shape of the affected region. The results demonstrated proposed model could promptly and accurately identify 12 kinds of tomato diseases and categorize the positional accuracy of afflicted areas. Mask R-CNN achieved a 99.64% detection for all classes of tomato diseases. Segmentation has advantages over classifying and network detection techniques for gathering lesion data. However, comparable to the detection network, it needs a large amount of annotating input. Its annotated input is pixel-by-pixel, typically demanding a significant amount of money and time.

5.3 PLANT DISEASE DETECTION BY WELL-KNOWN DL ARCHITECTURES

Multiple cutting-edge DL algorithms and frameworks for visual recognition, segmentation, and classification have evolved since the introduction of AlexNet (Sharma & Guleria, 2022). These investigations have been presented using well-known DL structures to classify and diagnose plant diseases. To get better results, numerous related publications also included unique visualization techniques and improved/modified DL architectures. The Plant Village collection has been extensively used, and the chosen DL models were assessed using various performance factors (Albattah et al., 2022; Hanh et al., 2022).

5.3.1 WITHOUT VISUALIZATION TECHNIQUE

To categorize diseases in plants, CNN was used in the study, and histogram techniques were used to highlight the model's significance. Basic CNN architectures such as GoogLeNet, AlexNet, and ResNet were applied to diagnose tomato leaf diseases (Aquil & Ishak, 2021). Training and validation reliability have been used to demonstrate the model's effectiveness; ResNet was judged to be the best CNN architecture. Although CA and F1-score were used to assess the model in color and grayscale modes (Saleem et al., 2019), LeNet architecture was used to detect diseases in banana leaves (Dhaka et al., 2021). Five DL architectures were utilized, with VGG surpassing all other models (AlexNet, GoogLeNet, AlexNetOWTbn, Overfeat, and VGG) (Wang et al., 2022a). In another instance, to recognize eight different plant diseases, cutting-edge DL models including ResNet-50, ResNet-101, InceptionResNetv2, GoogLeNet, Inception-v3, and SqueezeNet were integrated with SVM, extreme learning machines (ELM), and KNN classifiers (Indrakumari et al., 2021).

Regarding reliability criteria such as sensitivities, specificity, and F1-score, ResNet-50 with an SVM Classification model beat the competing models. A new DL model named Inception-v3, the model was used to determine cassava disease (Singh et al., 2018). The two basic CNN versions were used to classify plant diseases in cucumbers, and the accuracy score was 0.823. The super-resolution convolutional neural network (SRCNN) has replaced the traditional method of identifying and categorizing plant diseases (SRCNN). The classification of tomato plant disease using the AlexNet and SqueezeNet v1.1 models was performed, with AlexNet proving to be the more accurate DL model overall. A comparative study was conducted to determine the ideal DL design for detecting plant diseases. The study detected six tomato plant diseases using the VGG-16 DL and AlexNet, and classification accuracy was compared in depth (Hong et al., 2020).

5.3.2 WITH VISUALIZATION TECHNIQUES

The techniques mentioned above used DL models and visualization tools to get a better knowledge of plant diseases. For instance, the saliency map for visualizing plant disease symptoms is superior to the prior technique, such as SVM (Kumar & Kukreja, 2022). Furthermore, different filters are utilized to identify the disease patches. To identify olive plant disease, a modified LeNet model was utilized (Saleem et al., 2019). To diagnose plant diseases, edge and segmentation maps are used. Four cucumber diseases were detected, and their accuracy was compared to SVMs, random forest, and AlexNet models (Ma et al., 2018).

Furthermore, the picture segmentation approach is applied to view plant disease signs. A new DL model termed teacher/student network was developed, along with a unique visualization approach for identifying plant disease areas. DL models with specific detection sensors are implemented, and plant diseases are tagged using their respective forecast percentages. SSD, RFCN, and Faster-RCNN were the three detectors utilized with a very sound architecture such as GG, ZFNet, ResNet-50, AlexNet, V ResNet-101, GoogLeNet, and ResNetXt-101 for the comparative study to determine the best architecture between all those studied. The best results were obtained using ResNet-50 and the detector R-FCN (Saleem et al., 2019).

Additionally, a bounding box was constructed to locate the type of plant disease. Banana leaves were analyzed using several CNN models (Inception-V2, ResNet-50 and MobileNet-V2) with Faster-RCNN and SSD detectors to detect disease and pests (Ukwuoma et al., 2022). According, various CNN combinations were utilized, and heat maps were supplied as primary data to the visuals of diseased plants, providing the chance of a certain kind of disease emerging. Furthermore, the ROC curve assesses the model's performance. In a comparative study of GoogLeNet and AlexNet architectures for tomato diseases, the GoogLeNet surpasses AlexNet.

Furthermore, occlusion methods were presented to identify disease zones. VGG-CNN and VGG-FCN models have been utilized to identify wheat diseases and visualize characteristics in each block (Lu et al., 2017). In a study, the VGG-CNN model was employed to determine Fungal wilt in turnip (Arinichev et al., 2021), while the K-means clustering approach was employed to identify disease symptoms. A CNN-based semantic segmentation method was developed to identify the disease in cucumbers (Mia et al., 2021). Another technique depending on specific disease symptoms in crops was developed that used a DL model for identifying plant diseases. A Deep CNN framework was designed to detect, classify, and assess nine soybean stresses (Singh et al., 2018). CNN recognized paddy diseases, and its disease spots were identified using feature maps. To detect cucumber plant diseases, a dilation CNN was utilized. Correlation coefficient and DL models such as AlexNet and VGG-16 architectures were offered as cutting-edge visualization tools for detecting diseases in grapes (LeNet).

After applying simple grouping to the renewed image's channels, a narrow band heat map was produced. A simple binary cutoff approach was employed to detect plant disease. Comparing the new method to others showed its usefulness. LRP-Epsilon, LRP-Z, and gradient missed left-side plant diseases. DL produced better findings while suggesting leaf disease. On the right, a decoder in the recommended strategy corrected grad-inaccurate cam's disease localization. Feature maps were used to evaluate CNN architectures' ability to differentiate plant diseases. The results show that the CNN model identifies plant disease well. Segmentation and bordered maps identified plant diseases. The segmentation map shows the affected leaf as a yellow region with a white surface (Saleem et al., 2019).

5.4 HYPER-SPECTRAL IMAGING (HSI) WITH DL MODELS

During the growing process, crops may be impacted by many diseases at once, and certain microorganisms may create symptoms and indications that are identical. Additionally, early symptoms of plant diseases are not obvious, making it challenging to identify plant diseases with the naked eye or simple computer vision (Brahimi et al., 2018). The electromagnetic spectrum bandwidth of multispectral images is mostly centered on the visible and near-infrared (400–1,100 nm), with the occasional inclusion of shortwave infrared. This sensor is capable of collecting spectral data from a large number of small wavelength bands. These small bands are sensitive to minor plant leaf alterations generated by disorders, allowing for quick diagnosis of asymptomatic disease (Cordon et al., 2022). Qin & Zhang (2005) proposed a novel strategy for generating feature bands from data on illness differences. Using an improved combination of adaptive weighting method (CARS) and consecutive projection technique, a model for early detection of cucumber downy mildew was built (SPA). Abdulridha et al. (2020) employed hyperspectral scanning techniques and the MLP classification approach to determine the four phases of tomato bacterial spot disease. The classification performance of hyper-spectral frequencies was assessed using a saliency map-based visualization technique. The classification model performance 3D-CNN was 95.73% based on a hyper-spectral scan of the inoculated and simulated infected stem pictures, and the accuracy of the infected group was 87%. The use of DL to characterize spectral data for the detection of potato viruses gave satisfactory precision (78%) and recall levels (88%) (Polder et al., 2019). A new hyper-spectral analysis proximal sensing technique based on generative adversarial networks (GAN), dubbed outliers removal auxiliary classifier generative adversarial nets (OR-AC-GAN) (Wang et al., 2019b), was developed to identify tomato plant disease before the emergence of obvious symptoms. Before leaf-level symptoms appeared, the categorization accuracy was 96.25%.

5.5 DEEP LEARNING (DL) APPROACHES FOR DISEASE IDENTIFICATION IN DIFFERENT CROPS

Early detection of plant diseases can decrease the consumption of potentially dangerous herbicides and labor expenses. Due to the large scale of several fields, even the most skilled farmers have difficulties recognizing plant diseases. Consequently, an automatic disease detection method will majorly contribute to farmers' work and talents. Researchers have already built feature extraction and **ML**-based automated recognition and diagnostic systems (Asiri et al., 2019; Mansour, 2018).

5.5.1 DEEP LEARNING (DL) FOR WHEAT DISEASES

Among the world's most productive and significant crops, wheat (*Triticum aestivum* L.) is essential for food security. Currently, there are many obstacles to wheat production, with illnesses among the most significant (Bolton et al., 2008). Wheat illnesses might decrease grain quality or cause grain contamination due to toxins that are released by infections, which would also reduce production. Rusty leaves Common diseases that influence wheat production include powdery mildew, loose smut, brown rust, stripe rust/yellow rust, black rust, flag smut, hill bunt or stinking smut, Karnal bunt, and tan spot. These diseases can reduce wheat output by 10–40%. The application of DL methods based on neural network algorithms has improved wheat disease monitoring using digital RGB images. Multilayered neurons separate DL neural networks from other methodologies. CNNs are successful photo labeling systems (Mane & Kulkarni, 2020). AlexNet and VGG were early CNN concepts. These solutions improved algorithm convergence (ResNet network), parameter count reduction (MobileNet network), and model training (SENet network). These enhancements have boosted DL's utilization. They have also mastered plant phenotyping. DL neural networks identify wheat diseases successfully. DL outperforms SVM, random forest, and stochastic gradient descent. Constricted data and Bayesian DL improved plant disease diagnosis (Ho et al., 2021). Using digital RGB photos, DL algorithms have successfully diagnosed wheat illnesses. Barbedo et al. (2016) utilized CNN to identify wheat illnesses in leaf photographs. Picon et al. (2019) used deep CNNs to identify wheat illnesses (Septoria, tan spot, and two forms of rust). Lu et al. (2017) developed a technique for identifying six wheat illnesses. Using DL to identify wheat leaf fungal disease is still under development.

Over the past few years, DL has encountered substantial progress in classification tasks, with the CNNs possessing the center idea. The CNNs made it possible to use algorithms for automated feature extraction that do not need domain expertise. In-depth features (DFs; features automatically retrieved by CNNs) have recently been applied to agricultural disease diagnosis, and there is little research in this area. Lu et al. (2017) extracted DFs to distinguish between wheat illnesses such as leaf blotch, smut, stripe rust, and black chaff, but they did not describe how well the algorithms performed at determining the severity of the diseases. Additionally, DFs have been utilized to identify rice leaf disease and apple scab disease (Khan et al., 2018). However, only a few studies have shown the use of DFs to distinguish between and gauge the severity of wheat leaf rust and tan spot illnesses. It was also impossible to get the diagnostic accuracy estimates based on certain characteristics (HFs and DFs), which affect classification performance. A decent sickness detection algorithm requires many annotated photos. Publicly available data supports modern picture recognition systems for plant characterization and disease identification. Second, network design affects sickness recognition. Improved DL architectures identify plant diseases. Modern illness detection is accurate. EfficientNet is a labeling network architecture. Swish replaces ReLU in CNN models. EfficientNet (B0 through B7) is a series of CNNs with similar architectures that differ in layer depth, layer width, and input picture size. With more models, more parameters are estimated. On ImageNet, the 66M-parameter EfficientNet-B7 model was 84.3% accurate. This architecture beat ConvNet for fruit identification. On PlantVillage, EfficientNet beat VGG16, Inception V3, AlexNet, and ResNet (Bande and Dewangan).

5.5.2 DEEP LEARNING (DL) IN RICE DISEASES

For decades, agronomists have utilized computer vision to predict crop yields, diagnose nutrient insufficiency, estimate geometric crop sizes, and identify weeds (Jiang & Li, 2020). AI is frequently utilized in agricultural research for plant recognition and pathology, emphasizing creating of DL/ML architectures to address varied difficulties. Supervised ML identifies crops, forecasts production, detects diseases and identifies deficiencies. Abbaspour-Gilandeh et al. studied 13 southeast rice types using DA and ANN. ELM trumps support vector regression (SVR) and various techniques for wheat yield predictions. K-means and SVM obtained 85.06 to 93% accuracy for classifying five rice deficits. DL outperforms ANN, SVM, and KNN in diagnosing nutritional deficits in maize (Leena & Saju, 2018).

CNN-based DL models are used for identifying deficiencies from visual data. Convolutional encoders outperformed RAN-CNN in detecting soybean iron deficiency (Sharma et al., 2022). ResNet-50 detected seven nutritional deficits in 4,088 black gram pictures with 65.4% accuracy. In another experiment, Inception-ResNet and auto-encoder accurately classified three nutritional deficits in 571 tomato photos. Fine-tuning CNN models improve plant nutrient deficit identification. TL outperforms CNN. CNN models ResNet50, InceptionV3, and DenseNet121 were used to identify 11 kinds of nutritional deficits in rice plant leaves, with DenseNet121 having the greatest accuracy (97.44%). CNNs have demonstrated good efficacy in identifying nutrient insufficiency in rice plants, although region-based CNN lacked detail on rice crops. The pre-trained CNN model (MobileNetV2) improves the accuracy of rice leaf disease diagnosis in low-quality test images. Deep residual CNNs can be utilized with distinctive block pre-processing to identify and classify the rice yellow vein disease (Chen et al., 2021). The bagged tree classification technique and Complex tree classifier outperformed other models in a classification system for rice plant diseases. In forecasting cereal disease, the pre-trained and fine-tuned custom-developed convolution classifier which was very inexpensive. Numerous research on crop management revealed that DL models (InceptionResNetV2, EfficientNet, InceptionV3, MobileNet, DenseNet, and others) performed better in terms of precision and training time.

5.5.3 DEEP LEARNING (DL) IN MAISE DISEASE

The signs and symptoms of corn leaf disease vary. Less-experienced farmers may have more difficulty diagnosing diseases than experts. We can identify and classify leaf diseases in the maze using SVMs, digital image processing, and neural networks, Wong et al. suggested an SVM-based multi-classifier for identifying maize leaf diseases having 89.6% of recognition (Wong et al., 2004). The SVM classification approach is suitable for only a few samples and cannot achieve high recognition accuracy with large samples. Chen et al. (2014) suggested using a probabilistic neural network and image processing to identify maize leaf diseases (PNN). Around 90.4% was the greatest detection frequency attained with this strategy. The PNN classifier loses precision and speed as training data rises. Xu et al. (2011) suggested a weighted adaptive multi-classifier fusion approach for identifying maize leaf diseases. This technique investigated seven common maize leaf diseases having average accuracy of 94.71%. To diagnose maize leaf disease, they suggested applying digital image processing techniques relying on the Retinex algorithm, Fisher discriminant and coupled with quantum neural network (QNN), PCA, and SVM and other combining features. These investigations had the best detection accuracy (95.3%) while using fewer maize illnesses. Maximum identification accuracy was 95.3%, which does not meet modern criteria for detection performance.

Consequently, future investigations should concentrate on improving identification precision. Significant progress has been made in DL in recent years. It can currently produce meaningful representations of image features from a huge number of images. DL enables detectors to rapidly and accurately identify maize diseases, hosting plant protection and expanding the computer vision application for precision agriculture. Researchers employed various pooling techniques, filter sizes, and algorithms to detect 11 maize diseases (Nagaraju & Chawla, 2020). The accuracy of the suggested CNN model was 95.48%. DeChant et al. (2017) trained CNNs to detect maize leaf blight. Using several convolution neural network models and adjusting the training-to-testing set size ratio improved the diagnostic precision of illness in some experiments. More components and longer training convergence periods have a detrimental effect on the recognition rate. To accurately detect corn leaf diseases, it is necessary to design a recognition system with fewer features and improved identification precision.

5.5.4 DEEP LEARNING (DL) IN POTATO DISEASES

Deep learning (DL) models have recently enhanced picture identification precision, and these approaches have also been implemented in agricultural settings. Islam et al. (2017) suggested combining image analysis and ML to identify leaf image diseases. Plant Village is a computerized classification method for potato plant illnesses and healthy leaves. The accuracy of the suggested model for picture classification and segmentation using a SVM (95%). Consequently, the suggested method enables automated large-scale plant disease detection. Multiclass SVM image segmentation is utilized to develop an automated and user-friendly method (Varshney & Arora, 2004). Primitive potato diseases, like Late Blight and Early Blight, require minimal computing effort. The objective is to give farmers a practical, trustworthy, and time-saving disease detection tool. They exhibited techniques for recognizing plant diseases and leaf texture patterns. The segmentation of leaves is then studied using an image segmentation high-pass filter. The end product is a unique leaf texture. Local or regional fractal features supply an adequate texture model. Each illness would have a different texture. Multiclass SVM is used to classify the extracted texture pattern. Multiclass SVM implementations are developed for automatic DSS (Decision Support Systems) and growers to diagnose diseases in grape plants (Chougule et al., 2019). The Scheme segmented and analyzed a single leaf, and the pass band found a diseased area on the leaf. Using fractals, the robust texture feature allows the recovery of segmentation leaf surface function invariants in nature. The texture reduction pattern is then categorized as an SVM multiclass into the healthy or unhealthy classes. Downy mildew and black red, two prominent grass diseases, are the focus of the inquiry. The suggestion method gives producers 96.6% accurate professional agricultural advice immediately. The breakthroughs in leaf-based image identification algorithms utilized in 2018 yield remarkable results. This chapter uses random forests for data collection to identify healthy and diseased leaves.

5.5.5 DEEP LEARNING (DL) IN APPLE DISEASES

Apple is an extensively grown, plentiful, and popular fruit. Apple plantations' economic success is impacted by fruit quality. Diseases have remained an important cause of declining apple quality and output, which affects agriculture (Ishii, 2006). Identifying and treating apple diseases precisely can reduce agricultural losses and enhance economic growth. Unlike ML methods, CNNs can effectively learn high-level disease feature representations from photos (Allugunti, 2022; Yin et al., 2022). CNNs have succeeded in numerous visual tasks because of their richer and better high-level feature representations. In recent years, powerful DL architectures like ResNet, ResNet, and DenseNet have improved ImageNet's recognition accuracy and speed. Scholars have developed lightweight designs such as Xception, MobileNet, ShuffleNet, and so on to overcome the mobile deployment challenge. For stable, efficient, low-cost, and intelligent disease identification, Chao et al. (2020) suggested combining ZDNet, DenseNet, and Xception to improve model feature extraction. Five apple leaf diseases were identified with 98.82% accuracy using fewer criteria. Liu et al. (2020) used inception and a dense connection technique to develop a real-time neural network model that accurately identified six grape leaf diseases. Ramcharan et al. (2019) also trained a mobile cassava disease identification algorithm. Complicated variables, such as angles, brightness, and occlusion, might alter the model's performance in the field, proving that picture classification underneath the complex background is difficult.

Attention can improve feature extraction. The attention mechanism assigns bigger weights to regions of interest and lesser weights to backgrounds to improve the model and generate more accurate judgments. In other investigations, attention mechanisms excelled at categorization, recognition, and segmentation. This work offers a CNN for apple disease identification. These are the study's main contributions and innovations:

1. Using image generation, a new ALDID is created. Image enrichment techniques are utilized to enlarge the data set and imitate apple leaf disease photos acquired under diverse situations to train the algorithm.
2. An attention-based approach for apple leaf disease identification is suggested. The optimal network depth, breadth, and input picture resolution are determined using a network search approach. Deep segmented convolution is performed to the CA-Conv infrastructure to decrease parameters and avoid overfitting. Finally, a coordination attention block is incorporated into the infrastructure to integrate distinctive channel and geospatial attention and increase the model's lesion-area learning capacity.

5.5.6 DEEP LEARNING (DL) FOR BANANA DISEASE

Banana leaf disease identification and classification have been a concern for a long time (Singh & Misra, 2017). Many studies have been done on this topic, with truthful outcomes. Various researchers published the ML-based banana disease and pest detection system. The technique utilized is a deep CNN for disease identification (Jiang et al., 2019). The author collected data sets for eight distinct banana diseases using 30,000 photos, and the proposed system is 90% accurate. **ML** algorithms were created for identifying and classifying plant diseases, indicating research needs for recognizing the disease in plants before the symptoms are visible. DL-based banana leaf disease classification using LeNet architecture as a CNN. This research shows its usefulness in diverse backgrounds, sizes, and orientations. In 25 cycles, this approach stabilizes and reaches high accuracy. Using the SVM classifier and **ML**, banana diseases may be detected early. Close-range hyperspectral remote sensing photos are employed. The classifiers' outcomes are assessed by overall accuracy, which is 96% in early diagnosis, 90% in mid-identification, and 92% in late identification (Hassanzadeh, 2022).

Various agriculture scientists and plant pathologists propose picture manufacturing banana leaf disease diagnosis. Images are first obtained, the RGB model is turned into an HSI color model, then segmented using thresholding, and histogram equalization is determined for the HSI image (Kumar & Raghavendra, 2019). The classification is evaluated using backpropagation neural networks, SVM, and PCA (PCA). An artificial neural network (ANN)-based banana leaf disease identification and classification method are proposed by many pathologists (Narayanan et al., 2022). The image is acquired and pre-processed, the color and HOT (Histogram of Template) features are extracted, and the data set is trained using the ANN. The query images are graded based on the total percentage of the affected area. The picture is then categorized by ailment. Not only for banana leaves but also for rice, maize, apple, cheerful, and other common plants. Below are some findings. Using upgraded CNNs classifies apple and cherry plant diseases. The CNN is improved by integrating inception, excitation, and a global pooling layer. This method's test data accuracy is 91.7%. Deep CNNs and object detection techniques based on tomato disease diagnosis have been implemented with two different techniques: Faster R–CNN, which is used for identifying the type of tomato disease, and Mask R–CNN, which is used to find and segment the location and shape of infected areas (Wang et al., 2019c). The experiments demonstrated that the proposed technique provided an accuracy of 90% in detecting the disease and 99.9% in identifying the shape of infected areas.

5.6 DEEP LEARNING (DL) CHALLENGES

Deep learning (DL) has attained excellent performance in various tasks, such as visual identification, language, and voice detection systems, in addition to attracting focus to its research locations and significant advancements (Gu et al., 2018; Hassanzadeh, 2022; Nagaraju & Chawla, 2020; Singh & Misra, 2017; Xu et al., 2022; Yin et al., 2022). In contrast, because of the dearth of publicly accessible data and the complexity of several topics, DNNs have not yet addressed them. Therefore, a rich environment creates significant prospects for future study opportunities. This section will examine the significant DL difficulties and potential alternatives, including hardware configuration options. Using DL, identifying and categorizing plant diseases pose similar difficulties.

The absence of sufficient labeled training examples is still the most challenging part of DL strategies. Regular contributions of petabytes are made, in addition to the zettabytes that already exist. This enormous progress is accumulating data that cannot be labeled without human assistance (Kashyap et al., 2015; Razzak et al., 2020). The recent performance of supervised learning approaches is often attributable to the current availability of massive datasets and labels. In contrast, unsupervised learning approaches will become the primary solution explored when data intricacy and size increase rapidly (Islam et al., 2017; Kumar & Raghavendra, 2019; Liu et al., 2020).

Moreover, growing difficulties like complex datasets, missing data, and data sparsity necessitate a revision of the present DL models by requiring the capture of approximation information throughout observations rather than training. In addition, reduced, unstructured, elevated, diverse, and unfinished datasets are now accessible to DL methods (Singh et al., 2020; Xie et al., 2020; Zhao, 2022). Due to the DNN's inherent dubious black-box character, the sole potential to deal with unsupervised data is supplied, which is very motivating. Numerous upgraded DL models are being developed to manage dirty and noisy data (Ozguven & Adem, 2019). As in Torralba et al. the complicated database containing low-resolution RGB photographs was populated with 80 million tiny pictures. A novel adaptive function was created to minimize the number of chaotic tags within the dataset. In addition, other applications, such as social networking sites, XML files, Genetic code sequencing, and time series, utilize vast amounts of data in a streaming-live manner. These volumes of data are plagued by imbalance, heterogeneity, and incompleteness. How DL techniques learn in such settings is a topic of interest and is currently the subject of debate (Bhatt et al., 2019; Qadir et al., 2019; Zhao et al., 2021).

Disappearing and bursting Other key obstacles for DL approaches include gradients. Each layer of a DL network computes derivatives (or gradients) in a cascading fashion due to the network's fundamental architecture (layer by layer). In contrast, variables are progressively increased or decreased at each level. This method is susceptible to bursting (or disappearing) gradients. However, weights are decreased/increased depending on gradients to reduce the error or cost function (Gao & Ji, 2019; Haq, 2022). Extremely slight gradients may cause the network's training to take longer, whereas huge gradients may cause the training to diverge or overshoot. In addition, weight change has a negligible effect on output training and may need additional time. Linear activation functions, such as correct weight normalization and ReLU, can mitigate this issue's impact.

One of the most prevalent concerns with DL models is overfitting. Adding additional neurons to a DL network will undoubtedly result in the formation of a network capable of tackling more complex problems. The DL network is capable of providing itself with high flexibility in the data training process. During the dataset learning phase, there is a greater risk of generalization owing to disturbance and aberrations. This issue leads to a delay in testing and training and a decline in the accuracy of predictions on actual test data. For instance, overfitting provides a high-order polynomial output in clustering and classification issues. This output identifies the assessment limit for the training dataset, which requires additional time and reduces the accuracy of the results for most of the test datasets. One approach to the overfitting issue is the intelligent selection of the multiple numbers of neurons that match the situation's type and scale. Several strategies are available for estimating the appropriate number of neurons (Madhiarasan & Deepa, 2017). Unfortunately, there is no silver bullet; the best solution is to experiment with each use case to determine the optimal value.

Using the smallest amount of input dataset while achieving the highest productivity is a further challenge of DL that interacts with the issue of computing efficiency (Goh et al., 2021; Madhiarasan & Deepa, 2017). Current DL frameworks require considerable computer resources to approach current performance levels. Using reservoir computation is one method for overcoming this difficulty. Another solution is incremental techniques, which employ big and medium databases with offline training. Several academics have recently changed their approaches to developing scalable and parallel deep-learning frameworks. Recent modifications to their concept involve transferring the learning process to GPUs. GPUs are notorious for their leakage currents, precluding any practical implementation of deep-learning models on mobile devices. FPGAs (Field-Programmable-Gate-Arrays) provide an alternative approach. FPGAs have been employed as deep-learning accelerators to optimize data-access pipelines to get significantly improved results. Wang et al. (2016) utilized a scalable design called DLAU (Deep Learning Accelerator Unit). They were utilizing three pipelined processing units, the DLAU. Using locality and tile approaches, they accomplished results 361 times quicker than CPUs while spending 234 mW. Another method obtained a detection rate of 97% by employing an architecture based on low-end FPGAs with leakages, arc losses, and "and so on and so forth."

In comparison to software implementation, they made 7.5 times faster progress. FPGAs may be combined on a motherboard and spend less energy than GPUs for the same level of performance. In contrast, GPUs give the best performance in floating-point operations. Zhang et al. (2018) developed a novel implementation strategy for CNNs based on a roofline model. They utilized loop tiling to estimate the necessary memory bandwidth, a crucial aspect of FPGAs. Their design significantly reduces power consumption and achieved 61.62 gigaflops under 100 MHz. A gigaflop is the unit of measurement for the performance of a floating-point processing unit. Currently, there are no FPGA-based DL testbeds. Consequently, only FPGA design professionals can investigate this topic.

5.7 RESEARCH DIRECTIONS AND FUTURE ASPECTS OF DEEP LEARNING (DL)

Current methods have established a solid foundation for DL, but many areas need improvement in the DL of plant diseases:

1. **Data Annotation Automation:** Most DL methods for plant diseases are developed using publicly available, significantly explained data sets. Collecting original information from suitable sources to build a new design problem for plant diseases is of primary importance. Annotations of original information, such as categorization, marking, or tagging, are essential for developing prejudiced DL models for plant diseases. Especially for large datasets, automatic and dynamic data annotation may be higher productive for supervised learning and reducing manual input than manual annotations or hiring annotators. In-depth examination of data collection and annotation approaches or developing an unsupervised learning-dependent solution might become one of the most important study topics in plant disease using a DL approach.
2. **Data Quality Preparation:** As previously explained in the chapter, DL approaches influence the data quality, training accessibility, and the final model for a particular plant disease. DL methods could become ineffective or even less reliable if the training dataset is low quality, confused, loud, unbalanced, inappropriate, irregular, insufficient, etc. These data problems in the plant diseases may result in poor handling and inaccurate conclusions, a significant worry when attempting to get insight into the data. DL systems for plant disease must adjust to growing data difficulties to calculate knowledge from

various observations. Efficient information pretreatment methods are designed to establish the data issues and characteristics, which may be an additional area of research in plant disease (Li et al., 2021a; Shi et al., 2019).

3. **Black-Box Perception, DL/ML Algorithm Selection:** It is difficult to explain how DL outcomes and model choices are generated. DL models achieve great results when studying large datasets. However, inherent "black-box" perspective indicates poor statistical comprehensibility, which could be a significant negative (Mathews, 2019; Zihni et al., 2020). ML technologies, particularly rule-based DL systems, provide unambiguous logical principles for producing simpler judgments to comprehend, amend, or remove. If the wrong learning approach is used in plant disease, unanticipated repercussions may develop, lowering the model's effectiveness and precision. Assessing efficiency, complexity, model accuracy, and adaptability, selecting a model for the intended application is challenging and necessitates in-depth research for improved comprehension and decision-making (Ishii, 2006; Ramcharan et al., 2019; Wang et al., 2019c).
4. **Deep Networks for Supervised or Discriminative Learning:** CNN, MLP, and RNN, as well as their variants, constitute prejudiced architectures. Developing new methods or variants of these discriminative techniques for plant disease takes into account parameterization, efficiency, and adaptability depending on the intended practical applications and data quality. This would be a novel innovation and a key future aspect in automated or exclusionary training of the plant diseases in the DL.
5. **Deep Learning (DL) Networks:** In the case of plant diseases, supervised algorithms enable us to detect higher predictive traits or characteristics in data or develop a novel set of information via exploratory investigation. Pre-processing guarantees training correctness and modeling effectiveness for supervised, discriminatory, and quasi-training sets in plant diseases. SOM, RBM, GAN, AE, DBN, and their variations are generative techniques used in plant disease detection. Developing unique data modeling or visualization solutions for real-world applications might be a novel addition to supervised or spontaneous learning.

6. **Hybrid/Ensemble Modeling and Uncertainty:** As per DL architecture, hybridization is yet another essential aspect of the DL process. Competent hybridization may exceed everyone else in terms of reliability and ambiguity handled in high-risk circumstances. DL with DL/ML may play a major role in the hybridization area. Designing effective hybridization discriminative and generative models in plant disease instead of native techniques can be a research possibility to overcome semi-supervised learning issues, which reflects more uncertainty.
7. **Selecting Threshold/Hyper-Parameter Values and Network Structures:** DL models and applications focus on performance, model complexity, and computing needs. Combining algorithmic breakthroughs with greater accuracy and retaining the computing economy might boost the usefulness of DL models in future real-world applications. Due to the nature of the intended applications, progressive or recency-based learning may be advantageous for plant diseases. Considering network architectures with a fixed number of nodes and layers, hyper-parameter values and threshold settings may not be applicable in many situations because data is constantly changing. Regarding reliability and actual practicality, a data-driven method that dynamically selects them may be more advantageous when creating a DL model in plant disease. This data-driven mechanization may lead to much more intelligent DL models in the future, which might be an essential future aspect and area of plant disease research.
8. **Lightweight Deep Learning (DL) Modeling for Smart Devices and Apps:** In current years, the internet of things (IoT) has gained popularity for monitoring and collecting human and ecological data (such as bio-data, human behaviors, geo-information, plant diseases and weather parameters, etc.) for smart technologies and services. Smartphones generate large amounts of data daily, necessitating rapid processing on smartphones. DL could be utilized to explore fundamental characteristics and manage huge quantities of sensor information for IoT applications like health monitoring and disease analysis, intelligent transportation, road traffic forecasting and surveillance, transportation, fabricating inspection, fault evaluation, smart industry or Industry 4.0, and more. Due to serious learning's computational cost and storage overhead, lightweight modeling is

essential for resource-constrained devices. In some situations, optimization, simplifying, compressing, thinning, generalizations, and extraction of features may be beneficial. Future emphasis may be placed on developing lightweight DL algorithms based on a basic network architecture in order to adapt the DL model for next-generation mobile, IoT, or resource-constrained devices and applications.

9. **Designing a Deep Learning (DL) Framework for Applications:** Developing a universal framework for DL-dependent solutions is a comprehensive research path. The whole model facilitates a learning algorithm to concentrate on the most critical pieces of incoming input and the ability to gather hidden features to retain unique and relevant elements. The concentration model can be utilized in computer vision, speech recognition, text or image categorization, computational linguistics, recommendation systems, behavior analysis, etc., because of its intuitiveness, flexibility, and comprehensibility. Supervised learning may be utilized to find an attention mechanism's first and most helpful feature (Conneau et al., 2017; Gilpin et al., 2018; Mathews, 2019; Viola & Jones, 2001; Zhang et al., 2019). CNN could be combined with proper concentration processes to create a comprehensive categorization foundation for gathering attributes at numerous levels and regions. Another addition would be designing a ubiquitous DL model that considers attention and underlying characteristics for different applications would be another addition.

DL is a broad open subject towards which researchers may contribute by developing new techniques or upgrading existing ones to handle real-world difficulties in various application areas. This lets researchers assess the application's underlying and unanticipated hurdles for more reliable and authentic outcomes. Addressing the above-mentioned issues and finding efficient and inexpensive alternatives could lead to "Future Generation DL" modeling and more automated applications.

5.8 CONCLUSION

The initial step in computer vision involves acquiring images, followed by diverse image-processing tasks like resizing, filtering, segmentation, and feature extraction. ML or DL algorithms are subsequently employed for recognition and classification. Further advancements include sophisticated image enhancement, intricate segmentation, and innovative feature engineering. Integrating these advancements with ML/DL enhances accuracy and efficiency in computer vision tasks. We also reviewed recent papers that identified and analyzed plant diseases using **ML** or DL-based techniques. The maximum accuracy attained by several classifiers, the type of crop used, the classifiers employed, and the datasets' nature were further reported parameter-wise. We provided a thorough review of recent **ML** and DL studies to identify and categorize plant diseases. According to our research, 53% of DL-based studies used pretrained models for classifying images of plant leaves.

In contrast, about 33% of studies used DL models for segmentation or optimization. DL models are a better option for image categorization problems than **ML**-based methods. However, some problems still exist because the training of the underlying model used for classification purposes significantly impacts how well **ML** and DL models perform. We also highlighted some critical problems in identifying and classifying plant diseases that may greatly impact the model's performance. Through a real-time-based system, this effort will direct researchers to investigate new features in the agricultural disease recognition and classification sector. Researchers will be guided to identify the factors which may significantly affect the performance. The major objective of the upcoming work will be to create a comprehensive system made up of server-side components with a trained model and an application for smart mobile devices with capabilities like displaying detected plant diseases based on leaf photos taken with the camera of the mobile phone. Regardless of their experience level, farmers will benefit from this program since it will help them identify plant diseases quickly and effectively and decide whether to use chemical pesticides. Future studies will also expand the model's application by training it to recognize plant diseases over a larger range of geographical areas, merging drone-taken aerial images of orchards and vineyards with convolution neural networks for object detection, and more. The authors want to improve crop quality for future generations by furthering their research to contribute significantly to sustainable development.

ACKNOWLEDGMENTS

The author expresses gratitude to S. S. Kukal for his significant contribution in preparing this chapter.

KEYWORDS

- **classification network**
- **deep learning**
- **detection network**
- **hyper-spectral imaging**
- **maize disease**
- **one stage network**
- **plant disease detection**
- **visualization technique**

REFERENCES

Abdel-Basset, M., Chang, V., Hawash, H., Chakrabortty, R. K., & Ryan, M., (2020). Deep-IFS: Intrusion detection approach for industrial internet of things traffic in fog environment. *IEEE Transactions on Industrial Informatics, 17*, 7704–7715.

Abdulridha, J., Ampatzidis, Y., Kakarla, S. C., & Roberts, P., (2020). Detection of target spot and bacterial spot diseases in tomato using UAV-based and benchtop-based hyperspectral imaging techniques. *Precision Agriculture, 21*, 955–978.

Albattah, W., Nawaz, M., Javed, A., Masood, M., & Albahli, S., (2022). A novel deep learning method for detection and classification of plant diseases. *Complex & Intelligent Systems, 8*, 507–524.

Allugunti, V. R., (2022). A machine learning model for skin disease classification using convolution neural network. *International Journal of Computing, Programming and Database Management, 3*, 141–147.

Alzubaidi, L., Zhang, J., Humaidi, A. J., Al-Dujaili, A., Duan, Y., Al-Shamma, O., Santamaría, J., et al., (2021). Review of deep learning: Concepts, CNN architectures, challenges, applications, future directions. *Journal of Big Data, 8*, 1–74.

Aquil, M. A. I., & Ishak, W. H. W., (2021). Evaluation of scratch and pre-trained convolutional neural networks for the classification of tomato plant diseases. *IAES International Journal of Artificial Intelligence, 10*, 467.

Arabi, A. A., (2021). Artificial intelligence in drug design: Algorithms, applications, challenges, and ethics. *Future Drug Discovery, 3*, FDD59.

Arinichev, I. V., Polyanskikh, S. V., Volkova, G. V., & Arinicheva, I. V., (2021). Rice fungal diseases recognition using modern computer vision techniques. *International Journal of Fuzzy Logic and Intelligent Systems, 21*, 1–11.

Asiri, N., Hussain, M., Al Adel, F., & Alzaidi, N., (2019). Deep learning-based computer-aided diagnosis systems for diabetic retinopathy: A survey. *Artificial Intelligence in Medicine, 99*, 101701.

Ayvaz, S., & Alpay, K., (2021). Predictive maintenance system for production lines in manufacturing: A machine learning approach using IoT data in real-time. *Expert Systems with Applications, 173*, 114598.

Baht, N. M., & Domínguez, E., (2022). Detection of plant diseases based on convolutional neural network approach: A review. In: *2022 14th International Conference on Electronics, Computers and Artificial Intelligence (ECAI)* (pp. 1–4). IEEE.

Bande, P., & Dewangan, M. K., (2022). Plant disease detection using deep learning. *International Journal for Research in Applied Science and Engineering Technology (IJRASET).* ISSN: 2321-9653; IC Value: 45.98 Volume 10 Issue VI. https://www.academia.edu/download/88151689/Plant_Disease_Detection_using_Deep_Learning.pdf (accessed on 30 November 2023).

Barbedo, J. G. A., Koenigkan, L. V., & Santos, T. T., (2016). Identifying multiple plant diseases using digital image processing. *Biosystems Engineering, 147*, 104–116.

Bevan, M. W., Uauy, C., Wulff, B. B., Zhou, J., Krasileva, K., & Clark, M. D., (2017). Genomic innovation for crop improvement. *Nature, 543*, 346–354.

Bhatt, P. V., Sarangi, S., & Pappula, S., (2019). Detection of diseases and pests on images captured in uncontrolled conditions from tea plantations. In: *Autonomous Air and Ground Sensing Systems for Agricultural Optimization and Phenotyping IV* (pp. 73–82). SPIE.

Bieganowski, A., Dammer, K. H., Siedliska, A., Bzowska-Bakalarz, M., Bereś, P. K., Dąbrowska-Zielińska, K., Pflanz, M., et al., (2021). Sensor-based outdoor monitoring of insects in arable crops for their precise control. *Pest Management Science, 77*, 1109–1114.

Bolton, M. D., Kolmer, J. A., & Garvin, D. F., (2008). Wheat leaf rust caused by *Puccinia triticina. Molecular Plant Pathology, 9*, 563–575.

Brahimi, M., Arsenovic, M., Laraba, S., Sladojevic, S., Boukhalfa, K., & Moussaoui, A., (2018). Deep learning for plant diseases: Detection and saliency map visualization. *Human and Machine Learning* (pp. 93–117). Springer.

Chaman, A., & Dokmanic, I., (2021). Truly shift-invariant convolutional neural networks. In: *Proceedings of the IEEE/CVF Conference on Computer Vision and Pattern Recognition* (pp. 3773–3783).

Chao, D., Zhou, W., Xie, F., Ye, C., Li, H., Jaroniec, M., & Qiao, S. Z., (2020). Roadmap for advanced aqueous batteries: From design of materials to applications. *Science Advances, 6*, eaba4098.

Chen, J., Zhang, D., Zeb, A., & Nanehkaran, Y. A., (2021). Identification of rice plant diseases using lightweight attention networks. *Expert Systems with Applications, 169*, 114514.

Chen, S., Yu, J., & Wang, S., (2022). One-dimensional convolutional neural network-based active feature extraction for fault detection and diagnosis of industrial processes and its understanding via visualization. *ISA Transactions, 122*, 424–443.

Chen, X., Zhou, J., Xiao, H., Wang, E., Xiao, J., & Zhang, H., (2014). Fault diagnosis based on comprehensive geometric characteristic and probability neural network. *Applied Mathematics and Computation, 230*, 542–554.

Chougule, A., Jha, V. K., & Mukhopadhyay, D., (2019). Decision support for grape crop protection using ontology. *International Journal of Reasoning-based Intelligent Systems, 11*, 24–37.

Conneau, A., Kiela, D., Schwenk, H., Barrault, L., & Bordes, A., (2017). *Supervised Learning of Universal Sentence Representations from Natural Language Inference Data.* arXiv preprint arXiv:1705.02364.

Cordon, G., Andrade, C., Barbara, L., & Romero, A. M., (2022). Early detection of tomato bacterial canker by reflectance indices. *Information Processing in Agriculture, 9*, 184–194.

Dagim, F. Y., (2021). *Image Based Sorghum Leaf Disease Classification Using Deep Learning Approach.* In.

Darwish, A., Hassanien, A. E., & Das, S., (2020). A survey of swarm and evolutionary computing approaches for deep learning. *Artificial Intelligence Review, 53*, 1767–1812.

Dawson, T. P., Perryman, A. H., & Osborne, T. M., (2016). Modeling impacts of climate change on global food security. *Climatic Change, 134*, 429–440.

DeChant, C., Wiesner-Hanks, T., Chen, S., Stewart, E. L., Yosinski, J., Gore, M. A., Nelson, R. J., & Lipson, H., (2017). Automated identification of northern leaf blight-infected maize plants from field imagery using deep learning. *Phytopathology, 107*, 1426–1432.

Deng, J., Lu, Y., & Lee, V. C. S., (2021). Imaging-based crack detection on concrete surfaces using you only look once network. *Structural Health Monitoring, 20*, 484–499.

Dhaka, V. S., Meena, S. V., Rani, G., Sinwar, D., Ijaz, M. F., & Woźniak, M., (2021). A survey of deep convolutional neural networks applied for prediction of plant leaf diseases. *Sensors, 21*, 4749.

Dhingra, G., Kumar, V., & Joshi, H. D., (2018). Study of digital image processing techniques for leaf disease detection and classification. *Multimedia Tools and Applications, 77*, 19951–20000.

Fadlullah, Z. M., Tang, F., Mao, B., Kato, N., Akashi, O., Inoue, T., & Mizutani, K., (2017). State-of-the-art deep learning: Evolving machine intelligence toward tomorrow's intelligent network traffic control systems. *IEEE Communications Surveys & Tutorials, 19*, 2432–2455.

Falaschetti, L., Manoni, L., Rivera, R. C. F., Pau, D., Romanazzi, G., Silvestroni, O., Tomaselli, V., & Turchetti, C., (2021). A low-cost, low-power and real-time image detector for grape leaf esca disease based on a compressed CNN. *IEEE Journal on Emerging and Selected Topics in Circuits and Systems, 11*, 468–481.

Fridman, L., Brown, D. E., Glazer, M., Angell, W., Dodd, S., Jenik, B., Terwilliger, J., et al., (2017). *MIT Autonomous Vehicle Technology Study: Large-Scale Deep Learning-Based Analysis of Driver Behavior and Interaction with Automation.* arXiv preprint arXiv:1711.06976, 1.

Fuentes, A. F., Yoon, S., Lee, J., & Park, D. S., (2018). High-performance deep neural network-based tomato plant diseases and pests diagnosis system with refinement filter bank. *Frontiers in Plant Science, 9*, 1162.

Gao, H., & Ji, S., (2019). Graph u-nets. In: *International Conference on Machine Learning* (pp. 2083–2092). PMLR.

Garcia-Molsosa, A., Orengo, H. A., Lawrence, D., Philip, G., Hopper, K., & Petrie, C. A., (2021). Potential of deep learning segmentation for the extraction of archaeological features from historical map series. *Archaeological Prospection, 28*, 187–199.

Gilpin, L. H., Bau, D., Yuan, B. Z., Bajwa, A., Specter, M., & Kagal, L., (2018). Explaining explanations: An overview of interpretability of machine learning. In: *2018 IEEE 5th International Conference on Data Science and Advanced Analytics (DSAA)* (pp. 80–89). IEEE.

Goh, G. D., Sing, S. L., & Yeong, W. Y., (2021). A review on machine learning in 3D printing: Applications, potential, and challenges. *Artificial Intelligence Review, 54*, 63–94.

Gu, J., Wang, Z., Kuen, J., Ma, L., Shahroudy, A., Shuai, B., Liu, T., et al., (2018). Recent advances in convolutional neural networks. *Pattern Recognition, 77*, 354–377.

Gupta, R., Srivastava, D., Sahu, M., Tiwari, S., Ambasta, R. K., & Kumar, P., (2021). Artificial intelligence to deep learning: Machine intelligence approach for drug discovery. *Molecular Diversity, 25*, 1315–1360.

Hakim, M., Omran, A. A. B., Ahmed, A. N., Al-Waily, M., & Abdellatif, A., (2022). A systematic review of rolling bearing fault diagnoses based on deep learning and transfer learning: Taxonomy, overview, application, open challenges, weaknesses, and recommendations. *Ain Shams Engineering Journal*, 101945.

Hanh, B. T., Van, M. H., & Nguyen, N. V., (2022). Enhancing the performance of transferred efficient net models in leaf image-based plant disease classification. *Journal of Plant Diseases and Protection, 129*, 623–634.

Haq, I. U., (2022). *An Overview of Deep Learning in Medical Imaging*. arXiv preprint arXiv:2202.08546.

Hassan, S. M., Maji, A. K., Jasiński, M., Leonowicz, Z., & Jasińska, E., (2021). Identification of plant-leaf diseases using CNN and transfer-learning approach. *Electronics, 10*, 1388.

Hassanzadeh, A., (2022). *On the Use of Imaging Spectroscopy from Unmanned Aerial Systems (UAS) to Model Yield and Assess Growth Stages of a Broadacre Crop*. In: Rochester Institute of Technology.

Ho, W. K., Tang, B. S., & Wong, S. W., (2021). Predicting property prices with machine learning algorithms. *Journal of Property Research, 38*, 48–70.

Hong, H., Lin, J., & Huang, F., (2020). Tomato disease detection and classification by deep learning. In: *2020 International Conference on Big Data, Artificial Intelligence, and Internet of Things Engineering (ICBAIE)* (pp. 25–29). IEEE.

Huang, Y., Han, L., Dou, H., Luo, H., Yuan, Z., Liu, Q., Zhang, J., & Yin, G., (2019). Two-stage CNNs for computerized BI-RADS categorization in breast ultrasound images. *Biomedical Engineering Online, 18*, 1–18.

Indrakumari, R., Poongodi, T., Khaitan, S., Sagar, S., & Balamurugan, B., (2021). A review on plant diseases recognition through deep learning. *Handbook of Deep Learning in Biomedical Engineering*, 219–244.

Iqbal, S., Siddiqui, G. F., Rehman, A., Hussain, L., Saba, T., Tariq, U., & Abbasi, A. A., (2021). Prostate cancer detection using deep learning and traditional techniques. *IEEE Access, 9*, 27085–27100.

Ishii, H., (2006). Impact of fungicide resistance in plant pathogens on crop disease control and agricultural environment. *Japan Agricultural Research Quarterly: JARQ, 40*, 205–211.

Islam, M., Dinh, A., Wahid, K., & Bhowmik, P., (2017). Detection of potato diseases using image segmentation and multiclass support vector machine. In: *2017 IEEE 30th Canadian Conference on Electrical and Computer Engineering (CCECE)* (pp. 1–4). IEEE.

Jia, R., Nahli, A., Li, D., & Zhang, J., (2022). A multi-feature extraction-based algorithm for stitching tampered/untampered image classification. *Applied Sciences, 12*, 2337.

Jiang, P., Chen, Y., Liu, B., He, D., & Liang, C., (2019). Real-time detection of apple leaf diseases using deep learning approach based on improved convolutional neural networks. *IEEE Access, 7*, 59069–59080.

Jiang, Y., & Li, C., (2020). Convolutional neural networks for image-based high-throughput plant phenotyping: A review. *Plant Phenomics, 2020.*

Kadam, K. D., Ahirrao, S., & Kotecha, K., (2022). Efficient approach towards detection and identification of copy move and image splicing forgeries using mask R-CNN with MobileNet V1. *Computational Intelligence and Neuroscience, 2022.*

Kashyap, H., Ahmed, H. A., Hoque, N., Roy, S., & Bhattacharyya, D. K., (2015). *Big Data Analytics in Bioinformatics: A Machine Learning Perspective*. arXiv preprint arXiv:1506.05101.

Kerkech, M., Hafiane, A., & Canals, R., (2020). Vine disease detection in UAV multispectral images using optimized image registration and deep learning segmentation approach. *Computers and Electronics in Agriculture, 174*, 105446.

Khan, M. A., Akram, T., Sharif, M., Awais, M., Javed, K., Ali, H., & Saba, T., (2018). CCDF: Automatic system for segmentation and recognition of fruit crops diseases based on correlation coefficient and deep CNN features. *Computers and Electronics in Agriculture, 155*, 220–236.

Kim, H. S., & Kwak, S. S., (2020). *Crop Biotechnology for Sustainable Agriculture in the Face of Climate Crisis* (pp. 139–141). In: Springer.

Kumar, D., & Kukreja, V., (2022). Deep learning in wheat diseases classification: A systematic review. *Multimedia Tools and Applications*, 1–45.

Kumar, S. S., & Raghavendra, B., (2019). Diseases detection of various plant leaf using image processing techniques: A review. In: *2019 5th International Conference on Advanced Computing & Communication Systems (ICACCS)* (pp. 313–316). IEEE.

Kwak, S. S., (2019). *Biotechnology of the Sweetpotato: Ensuring Global Food and Nutrition Security in the Face of Climate Change* (pp. 1361–1363). In: Springer.

Lai, S., Xu, L., Liu, K., & Zhao, J., (2015). Recurrent convolutional neural networks for text classification. In: *Twenty-ninth AAAI Conference on Artificial Intelligence.*

Leena, N., & Saju, K., (2018). Classification of macronutrient deficiencies in maize plant using machine learning. *International Journal of Electrical and Computer Engineering (IJECE), 8*, 4197–4203.

Li, D., & Du, L., (2022). Recent advances of deep learning algorithms for aquacultural machine vision systems with emphasis on fish. *Artificial Intelligence Review, 55*, 4077–4116.

Li, L., Rong, S., Wang, R., & Yu, S., (2021a). Recent advances in artificial intelligence and machine learning for nonlinear relationship analysis and process control in drinking water treatment: A review. *Chemical Engineering Journal, 405*, 126673.

Li, Y., Du, L., & Wei, D., (2021b). Multiscale CNN based on component analysis for SAR ATR. *IEEE Transactions on Geoscience and Remote Sensing, 60*, 1–12.

Lin, K., Zhao, H., Lv, J., Li, C., Liu, X., Chen, R., & Zhao, R., (2020). Face detection and segmentation based on improved mask R-CNN. *Discrete Dynamics in Nature and Society, 2020.*

Lin, T. C., Ishizaka, M., & Ishii, H., (2009). Acibenzolar-S-methyl-induced systemic resistance against anthracnose and powdery mildew diseases on cucumber plants without accumulation of phytoalexins. *Journal of Phytopathology, 157*, 40–50.

Liu, B., Ding, Z., Tian, L., He, D., Li, S., & Wang, H., (2020). Grape leaf disease identification using improved deep convolutional neural networks. *Frontiers in Plant Science, 11*, 1082.

Liu, J., & Wang, X., (2021). Plant diseases and pests detection based on deep learning: A review. *Plant Methods, 17*, 1–18.

Liu, J., Wang, X., & Tai, X. C., (2022). Deep convolutional neural networks with spatial regularization, volume, and star-shape priors for image segmentation. *Journal of Mathematical Imaging and Vision*, 1–21.

Lu, J., Hu, J., Zhao, G., Mei, F., & Zhang, C., (2017). An in-field automatic wheat disease diagnosis system. *Computers and Electronics in Agriculture, 142*, 369–379.

Lu, J., Tan, L., & Jiang, H., (2021). Review on convolutional neural network (CNN) applied to plant leaf disease classification. *Agriculture, 11*, 707.

Ma, J., Du, K., Zheng, F., Zhang, L., Gong, Z., & Sun, Z., (2018). A recognition method for cucumber diseases using leaf symptom images based on deep convolutional neural network. *Computers and Electronics in Agriculture, 154*, 18–24.

Madhiarasan, M., & Deepa, S., (2017). Comparative analysis on hidden neurons estimation in multi layer perceptron neural networks for wind speed forecasting. *Artificial Intelligence Review, 48*, 449–471.

Mahmud, M., Kaiser, M. S., McGinnity, T. M., & Hussain, A., (2021). Deep learning in mining biological data. *Cognitive Computation, 13*, 1–33.

Mane, D., & Kulkarni, U. V., (2020). A survey on supervised convolutional neural network and its major applications. *Deep Learning and Neural Networks: Concepts, Methodologies, Tools, and Applications* (pp. 1058–1071). IGI Global.

Manirambona, E., Uwizeyimana, T., Uwiringiyimana, E., & Reddy, H., (2021). Impact of the COVID-19 pandemic on the food rations of refugees in Rwanda. *International Journal for Equity in Health, 20*, 1–4.

Mansour, R. F., (2018). Deep-learning-based automatic computer-aided diagnosis system for diabetic retinopathy. *Biomedical Engineering Letters, 8*, 41–57.

Mathews, S. M., (2019). Explainable artificial intelligence applications in NLP, biomedical, and malware classification: A literature review. In: *Intelligent Computing-Proceedings of the Computing Conference* (pp. 1269–1292). Springer.

Mesterházy, Á., Oláh, J., & Popp, J., (2020). Losses in the grain supply chain: Causes and solutions. *Sustainability, 12*, 2342.

Mia, M. J., Maria, S. K., Taki, S. S., & Biswas, A. A., (2021). Cucumber disease recognition using machine learning and transfer learning. *Bulletin of Electrical Engineering and Informatics, 10*, 3432–3443.

Nagaraju, M., & Chawla, P., (2020). Systematic review of deep learning techniques in plant disease detection. *International Journal of System Assurance Engineering and Management, 11*, 547–560.

Narayanan, K. L., Krishnan, R. S., Robinson, Y. H., Julie, E. G., Vimal, S., Saravanan, V., & Kaliappan, M., (2022). Banana plant disease classification using hybrid convolutional neural network. *Computational Intelligence and Neuroscience, 2022.*

Nasirahmadi, A., Wilczek, U., & Hensel, O., (2021). Sugar beet damage detection during harvesting using different convolutional neural network models. *Agriculture, 11*, 1111.

Nazir, T., Nawaz, M., Rashid, J., Mahum, R., Masood, M., Mehmood, A., Ali, F., et al., (2021). Detection of diabetic eye disease from retinal images using a deep learning based CenterNet model. *Sensors, 21*, 5283.

Ozguven, M. M., & Adem, K., (2019). Automatic detection and classification of leaf spot disease in sugar beet using deep learning algorithms. *Physica A: Statistical Mechanics and its Applications, 535*, 122537.

Pacal, I., Karaboga, D., Basturk, A., Akay, B., & Nalbantoglu, U., (2020). A comprehensive review of deep learning in colon cancer. *Computers in Biology and Medicine, 126*, 104003.

Pandey, A. K., Iqbal, R., Maniak, T., Karyotis, C., Akuma, S., & Palade, V., (2022). Convolution neural networks for pothole detection of critical road infrastructure. *Computers and Electrical Engineering, 99*, 107725.

Pandey, S. K., & Janghel, R. R., (2019). Recent deep learning techniques, challenges, and its applications for medical healthcare system: A review. *Neural Processing Letters, 50*, 1907–1935.

Pattnaik, G., & Parvathy, K., (2022). Machine learning-based approaches for tomato pest classification. *TELKOMNIKA (Telecommunication Computing Electronics and Control), 20*, 321–328.

Picon, A., Alvarez-Gila, A., Seitz, M., Ortiz-Barredo, A., Echazarra, J., & Johannes, A., (2019). Deep convolutional neural networks for mobile capture device-based crop disease classification in the wild. *Computers and Electronics in Agriculture, 161*, 280–290.

Polder, G., Blok, P. M., De Villiers, H. A., Van, D. W. J. M., & Kamp, J., (2019). Potato virus Y detection in seed potatoes using deep learning on hyperspectral images. *Frontiers in Plant Science, 10*, 209.

Qadir, H. A., Shin, Y., Solhusvik, J., Bergsland, J., Aabakken, L., & Balasingham, I., (2019). Polyp detection and segmentation using mask R-CNN: Does a deeper feature extractor CNN always perform better? In: *2019 13th International Symposium on Medical Information and Communication Technology (ISMICT)* (pp. 1–6). IEEE.

Qin, Z., & Zhang, M., (2005). Detection of rice sheath blight for in-season disease management using multispectral remote sensing. *International Journal of Applied Earth Observation and Geoinformation, 7*, 115–128.

Ramcharan, A., McCloskey, P., Baranowski, K., Mbilinyi, N., Mrisho, L., Ndalahwa, M., Legg, J., & Hughes, D. P., (2019). A mobile-based deep learning model for cassava disease diagnosis. *Frontiers in Plant Science*, 272.

Razzak, M. I., Imran, M., & Xu, G., (2020). Big data analytics for preventive medicine. *Neural Computing and Applications, 32*, 4417–4451.

Rehman, A., Saba, T., Kashif, M., Fati, S. M., Bahaj, S. A., & Chaudhry, H., (2022). A revisit of internet of things technologies for monitoring and control strategies in smart agriculture. *Agronomy, 12*, 127.

Ren, P., Xiao, Y., Chang, X., Huang, P. Y., Li, Z., Gupta, B. B., Chen, X., & Wang, X., (2021). A survey of deep active learning. *ACM Computing Surveys (CSUR), 54*, 1–40.

Roy, A. M., Bose, R., & Bhaduri, J., (2022). A fast accurate fine-grain object detection model based on YOLOv4 deep neural network. *Neural Computing and Applications, 34*, 3895–3921.

Saleem, M. H., Potgieter, J., & Arif, K. M., (2019). Plant disease detection and classification by deep learning. *Plants, 8*, 468.

Sánchez-Reolid, R., De La Rosa, F. L., López, M. T., & Fernández-Caballero, A., (2022). One-dimensional convolutional neural networks for low/high arousal classification from electrodermal activity. *Biomedical Signal Processing and Control, 71*, 103203.

Sarkar, S., (2021). *Intelligent Energy-Efficient Drones: Path Planning, Real-Time Monitoring and Decision-Making*. In: University of Louisiana at Lafayette.

Searchinger, T., Waite, R., Hanson, C., Ranganathan, J., Dumas, P., Matthews, E., & Klirs, C., (2019). *Creating a Sustainable Food Future: A Menu of Solutions to Feed Nearly 10 Billion People by 2050*. Final report. In: WRI.

Shaikh, T. A., Rasool, T., & Lone, F. R., (2022). Towards leveraging the role of machine learning and artificial intelligence in precision agriculture and smart farming. *Computers and Electronics in Agriculture, 198*, 107119.

Sharif, M. I., Khan, M. A., Alhussein, M., Aurangzeb, K., & Raza, M., (2022). A decision support system for multimodal brain tumor classification using deep learning. *Complex & Intelligent Systems, 8*, 3007–3020.

Sharma, M., Nath, K., Sharma, R. K., Kumar, C. J., & Chaudhary, A., (2022). Ensemble averaging of transfer learning models for identification of nutritional deficiency in rice plant. *Electronics, 11*, 148.

Sharma, S., & Guleria, K., (2022). Deep learning models for image classification: Comparison and applications. In: *2022 2nd International Conference on Advance Computing and Innovative Technologies in Engineering (ICACITE)* (pp. 1733–1738). IEEE.

Shi, J., Peng, D., Peng, Z., Zhang, Z., Goebel, K., & Wu, D., (2022). Planetary gearbox fault diagnosis using bidirectional-convolutional LSTM networks. *Mechanical Systems and Signal Processing, 162*, 107996.

Shi, X., An, X., Zhao, Q., Liu, H., Xia, L., Sun, X., & Guo, Y., (2019). State-of-the-art internet of things in protected agriculture. *Sensors, 19*, 1833.

Singh, A. K., Ganapathysubramanian, B., Sarkar, S., & Singh, A., (2018). Deep learning for plant stress phenotyping: Trends and future perspectives. *Trends in Plant Science, 23*, 883–898.

Singh, D., Jain, N., Jain, P., Kayal, P., Kumawat, S., & Batra, N., (2020). PlantDoc: A dataset for visual plant disease detection. *Proceedings of the 7th ACM IKDD CoDS and 25th COMAD* (pp. 249–253).

Singh, V., & Misra, A. K., (2017). Detection of plant leaf diseases using image segmentation and soft computing techniques. *Information Processing in Agriculture, 4*, 41–49.

Sinha, B. B., & Dhanalakshmi, R., (2022). Recent advancements and challenges of internet of things in smart agriculture: A survey. *Future Generation Computer Systems, 126*, 169–184.

Sodjinou, S. G., Mohammadi, V., Mahama, A. T. S., & Gouton, P., (2022). A deep semantic segmentation-based algorithm to segment crops and weeds in agronomic color images. *Information Processing in Agriculture, 9*, 355–364.

Sundin, G. W., Castiblanco, L. F., Yuan, X., Zeng, Q., & Yang, C. H., (2016). Bacterial disease management: Challenges, experience, innovation, and future prospects: Challenges in bacterial molecular plant pathology. *Molecular Plant Pathology, 17*, 1506–1518.

Swain, S., Nayak, S. K., & Barik, S. S., (2020). A review on plant leaf diseases detection and classification based on machine learning models. *Mukt Shabd, 9*, 5195–5205.

Thakur, R., & Rane, D., (2021). Machine learning and deep learning for intelligent and smart applications. *Future Trends in 5G and 6G* (pp. 95–113). CRC Press.

Ukwuoma, C. C., Zhiguang, Q., Bin, H. M. B., Ali, L., Almaspoor, Z., & Monday, H. N., (2022). Recent advancements in fruit detection and classification using deep learning techniques. *Mathematical Problems in Engineering, 2022.*

Valicharla, S. K., (2021). *Weed Recognition in Agriculture: A Mask R-CNN Approach.* West Virginia University.

Vamathevan, J., Clark, D., Czodrowski, P., Dunham, I., Ferran, E., Lee, G., Li, B., et al., (2019). Applications of machine learning in drug discovery and development. *Nature Reviews Drug Discovery, 18*, 463–477.

Varshney, P. K., & Arora, M. K., (2004). *Advanced Image Processing Techniques for Remotely Sensed Hyperspectral Data.* Springer Science & Business Media.

Vecvanags, A., Aktas, K., Pavlovs, I., Avots, E., Filipovs, J., Brauns, A., Done, G., et al., (2022). Ungulate detection and species classification from camera trap images using RetinaNet and faster R-CNN. *Entropy, 24*, 353.

Viola, P., & Jones, M., (2001). Rapid object detection using a boosted cascade of simple features. In: *Proceedings of the 2001 IEEE Computer Society Conference on Computer Vision and Pattern Recognition (CVPR)* (Vol. 1, pp. I–I). IEEE. https://ieeexplore.ieee.org/abstract/document/990517/ (accessed on 30 November 2023).

Wang, C., Gong, L., Yu, Q., Li, X., Xie, Y., & Zhou, X., (2016). DLAU: A scalable deep learning accelerator unit on FPGA. *IEEE Transactions on Computer-Aided Design of Integrated Circuits and Systems, 36*, 513–517.

Wang, D., Cao, W., Zhang, F., Li, Z., Xu, S., & Wu, X., (2022a). A review of deep learning in multiscale agricultural sensing. *Remote Sensing, 14*, 559.

Wang, D., Li, Z., Dey, N., Ashour, A. S., Moraru, L., Biswas, A., & Shi, F., (2019a). Optical pressure sensors based plantar image segmenting using an improved fully convolutional network. *Optik, 179*, 99–114.

Wang, D., Vinson, R., Holmes, M., Seibel, G., Bechar, A., Nof, S., & Tao, Y., (2019b). Early detection of tomato spotted wilt virus by hyperspectral imaging and outlier removal auxiliary classifier generative adversarial nets (OR-AC-GAN). *Scientific Reports, 9*, 1–14.

Wang, N., Wang, Y., & Er, M. J., (2022b). Review on deep learning techniques for marine object recognition: Architectures and algorithms. *Control Engineering Practice, 118*, 104458.

Wang, P., Meng, F., Donaldson, P., Horan, S., Panchy, N. L., Vischulis, E., Winship, E., et al., (2022c). High-throughput measurement of plant fitness traits with an object detection method using faster R-CNN. *New Phytologist, 234*, 1521–1533.

Wang, Q., Qi, F., Sun, M., Qu, J., & Xue, J., (2019c). Identification of tomato disease types and detection of infected areas based on deep convolutional neural networks and object detection techniques. *Computational Intelligence and Neuroscience, 2019.*

Wong, J., Lambert, R., Wurtzel, E., & Rocheford, T., (2004). QTL and candidate genes phytoene synthase and ζ-carotene desaturase associated with the accumulation of carotenoids in maize. *Theoretical and Applied Genetics, 108*, 349–359.

Wu, Z., Chen, Y., Zhao, B., Kang, X., & Ding, Y., (2021). Review of weed detection methods based on computer vision. *Sensors, 21*, 3647.

Xie, X., Ma, Y., Liu, B., He, J., Li, S., & Wang, H., (2020). A deep-learning-based real-time detector for grape leaf diseases using improved convolutional neural networks. *Frontiers in Plant Science, 11*, 751.

Xiong, Y., & Zuo, R., (2022). Robust feature extraction for geochemical anomaly recognition using a stacked convolutional denoising autoencoder. *Mathematical Geosciences, 54*, 623–644.

Xu, G., Zhang, F., Shah, S. G., Ye, Y., & Mao, H., (2011). Use of leaf color images to identify nitrogen and potassium deficient tomatoes. *Pattern Recognition Letters, 32*, 1584–1590.

Xu, X. B., Zhao, X. L., & Wang, G. Q., (2022). KL divergence adaptive weight combination multi-classifier emotion recognition method fused with bimodality. In: *2022 3rd Asia Service Sciences and Software Engineering Conference* (pp. 95–101).

Xu, Y., Liu, X., Cao, X., Huang, C., Liu, E., Qian, S., Liu, X., et al., (2021). Artificial intelligence: A powerful paradigm for scientific research. *The Innovation, 2*, 100179.

Yalcin, H., & Razavi, S., (2016). Plant classification using convolutional neural networks. In, *2016 Fifth International Conference on Agro-Geoinformatics (Agro-Geoinformatics)* (pp. 1–5). IEEE.

Yang, Z., Yang, D., Dyer, C., He, X., Smola, A., & Hovy, E., (2016). Hierarchical attention networks for document classification. In: *Proceedings of the 2016 Conference of the North American Chapter of the Association for Computational Linguistics: Human Language Technologies* (pp. 1480–1489).

Yin, W., Li, L., & Wu, F. X., (2022). Deep learning for brain disorder diagnosis based on fMRI images. *Neurocomputing, 469*, 332–345.

Yu, X., Zhou, Q., Wang, S., & Zhang, Y. D., (2022). A systematic survey of deep learning in breast cancer. *International Journal of Intelligent Systems, 37*, 152–216.

Zhang, C., Sun, G., Fang, Z., Zhou, P., Pan, P., & Cong, J., (2018). Caffeine: Toward uniformed representation and acceleration for deep convolutional neural networks. *IEEE Transactions on Computer-Aided Design of Integrated Circuits and Systems, 38*, 2072–2085.

Zhang, H., Goodfellow, I., Metaxas, D., & Odena, A., (2019). Self-attention generative adversarial networks. In: *International Conference on Machine Learning* (pp. 7354–7363). PMLR.

Zhang, X., Wang, W., Zhao, Y., & Xie, H., (2021). An improved YOLOv3 model based on skipping connections and spatial pyramid pooling. *Systems Science & Control Engineering, 9*, 142–149.

Zhao, C., Shuai, R., Ma, L., Liu, W., Hu, D., & Wu, M., (2021). Dermoscopy image classification based on StyleGAN and DenseNet201. *IEEE Access, 9*, 8659–8679.

Zhao, L., (2022). Image semantic segmentation method based on GAN network and FCN model. *The Journal of Engineering, 2022*, 1–9.

Zheng, H., Liu, J., & Ren, X., (2022). Dim target detection method based on deep learning in complex traffic environment. *Journal of Grid Computing, 20*, 1–12.

Zhou, Q., Wu, X., Zhang, S., Kang, B., Ge, Z., & Latecki, L. J., (2022). Contextual ensemble network for semantic segmentation. *Pattern Recognition, 122*, 108290.

Zihni, E., Madai, V. I., Livne, M., Galinovic, I., Khalil, A. A., Fiebach, J. B., & Frey, D., (2020). Opening the black box of artificial intelligence for clinical decision support: A study predicting stroke outcome. *Plos One, 15*, e0231166.

CHAPTER 6

Deep Learning Approach for Detecting, Classifying, and Managing Insect Pests of Various Crops

OWAIS BASHIR,[1] SHABIR AHMAD BANGROO,[1] NASIR BASHIR NAIKOO,[1] NAZIR AHMAD GANIE,[2] S. S. KUKAL,[3] DEELAK AMIN,[4] SHEIKH ASIF,[4] ABDUL ROUF MALIK,[5] REHANA RASOOL,[1] SHAFEEQ UR REHMAN,[6] AND KHALID REHMAN HAKEEM[7]

[1]*Division of Soil Science, Sher-e-Kashmir University of Agricultural Sciences and Technology of Kashmir, Kashmir, India*

[2]*Division of Animal Genetics and Breeding, Sher-e-Kashmir University of Agricultural Sciences and Technology of Kashmir, Kashmir, India*

[3]*Punjab Water Regulation and Development Authority, Chandigarh, Punjab, India*

[4]*Division of Entomology, Sher-e-Kashmir University of Agricultural Sciences and Technology of Kashmir, Kashmir, India*

[5]*Division of Fruit Science, Sher-e-Kashmir University of Agricultural Sciences and Technology of Kashmir, Kashmir, India*

[6]*School of Environment and Civil Engineering, Dongguan University of Technology, Dongguan, Guangdong, China*

[7]*Department of Biological Sciences, Faculty of Science, King Abdulaziz University, Jeddah, Saudi Arabia*

Soil, Agriculture, and Ecosystem Modelling: Smart Technologies for Sustainable Solutions.
Owais Bashir, Khalid Rehman Hakeem, Surinder S. Kukal, and Razeef Mohd (Eds.)

ABSTRACT

Insect pests pose a considerable threat to the quality and quantity of the crop; hence, pest management is one of the most important activities among all farm activities. Identification and counting (quantifying) are two important aspects done manually in the traditional method for pest monitoring. However, visual or manual monitoring cannot be done in large fields, especially when infested with different insect pest species. Also, manual procedures are time-consuming, cumbersome, and have an increased error rate. Substantial investigations have been done to develop procedures and ways to monitor and detect insect pests in the field remotely. AI is a novel approach used for the early detection and elimination of pests, resulting in reduced damage to crops, vegetables, and fruit. One of the methods used is a combination of proximal digital images with machine learning (**ML**) techniques; however, a single model cannot be used for all insect species and orders. The present chapter will highlight – (i) a brief description of the most commonly used AI-based tools for insect pest detection using proximal digital images and ML; (ii) automated monitoring of some common insect pest orders based on deep learning (DL) methods; and (iii) automated pest management strategies using drone technology.

6.1 INTRODUCTION

An agricultural pest is an insect population that feeds on agricultural plant tissue. Agricultural pests are considered a nuisance influenced by local weather, insect pressure, and crop management approach (Baiano, 2020; French et al., 2021; Rasche, 2021; Shahzad et al., 2021; Singh et al., 2022; Skendžić et al., 2021). Agricultural pests include frugivores, seed-eating insects, and rodents (e.g., viruses, parasitic organisms, fungus, and microbes), which can lower crop output or, under the worst situation, cause major yield losses (Baiano, 2020; Bryson, 2022). Globally, pests may severely harm crops, producing economic, psychological, and social losses (productivity, revenues, and finances) (Aaen et al., 2015; Abbas et al., 2022). Agricultural pests not only threaten the farmer who owns the property but also threaten neighboring and perhaps distant holdings (Aaen et al., 2015). Agricultural pests have long threatened crop development and product storage. A man started protecting the crop from time immemorial. Manual identification is slow, inefficient, and creates significant labor expenses (Aaen et al.,

2015). Agricultural networking has evolved fast in recent years, making access to photographs of agricultural pests easier. Computer vision-based image recognition technology may dramatically cut recognition costs while increasing recognition speed and efficiency (Arunnehru et al., 2022; Dong et al., 2022; Huang et al., 2022; Javaid et al., 2022; Kumar et al., 2022; Liu et al., 2022; Lu et al., 2022; Wang et al., 2020b; Zhang et al., 2022a; Zhao et al., 2022). With no feature pre-processing, classic machine learning (*ML*) classification algorithms like SVMs and back propagation neural networks fail. In such a case, certain new technologies exist for pest recognition and classification.

Recently, *ML* has come a long way in reducing the pest population. *ML* and deep learning (DL) advances allow much-improved pest identification and detection (Høye et al., 2021; Khan et al., 2021). A recent development in *ML* is DL which enables us to work directly with pictures without manually extracting characteristics (Gharaibeh et al., 2022; Tan et al., 2021). With excellent results, DL networks like CNN have been used for picture identification, vehicle detection, pest recognition, pest classification, and human motion recognition (Lilhore et al., 2022; Mohidem et al., 2021; Rakhmatulin et al., 2021; Shanthakumari et al., 2022). As shown in various applications, DL may increase computer vision accuracy. Instead of utilizing input data to make intelligent decisions, DL may learn and create smart decisions using structured algorithms in layers (Nurcahyani & Lee, 2021; Tran et al., 2021). DL and computer vision technology can automate repeated inspection activity, increasing reliability and efficiency (Fang et al., 2020). It will also enable the creation of novel and effective solutions for integrated pest management (IPM) in crops, including robots that can roam within crops and conduct early pest detection and control duties autonomously. The DL-based pest detection requirements vary by species, plant location, and developmental stage (Fuentes et al., 2017; Wang et al., 2022a). Pests acquire various features and esthetic qualities from the egg to the adult stage. The DL inspection of pests in fruits and vegetables consists of four key steps: acquisition, segmentation, feature extraction, and classification.

Structures for multi-level classifications have spontaneously been created to classify crop pests. The use of image processing for insect identification has been described in recent research owing to low computing costs and simple differentiation of insects with similar colors and forms (Böckmann et al., 2021; Che'Ya et al., 2022; Fuentes et al., 2017; Sabanci et al., 2022). Descriptor-based clustering segmentation is used to identify pests in crops having various orientations and illumination conditions. For recognizing individual moths, region, and contour-base segmentation are coupled. DL algorithms have recently excelled in pest categorization and detection (Darwin et al., 2021; Mirzazadeh et al., 2021; Singh & Sabrol, 2021). Based on collected insect features, several models were developed, in which photographs of insects were sorted into distinct groups. For this chapter, we searched Google Scholar, Scopus, Science Direct, and IEEE Xplore for relevant works on pest monitoring using digital photos. The search terms were "pest," "insect," "image," and "detection." These are wide phrases that produced many irrelevant results. Still, it was critical to keep the search broad to prevent missing relevant matches, and only peer-reviewed journal articles were chosen.

6.1.1 PEST DETECTION METHODS

Detection approaches seek to differentiate a target pest from the surroundings. This corresponds to a binary categorization with "target present" and "target-absent." The quantity of discovered specimens is typically counted to assess the extent of the infestation. Earlier work examined image processing without **ML** and had poor results. Simple implementation and low processing cost were the key advantages of this technique. However, they depend largely on handmade parameters and thresholds, making them very vulnerable to changes in picture properties. Barbedo (2020) started an investigation in which two knowledge base systems (KBS) were coupled to identify and count whiteflies in traps. The initial KBS used image processing to choose a few potential specimen locations and for evaluation; the second. KBS employed a set of numerical descriptors to thoroughly define the specimens of interest. A boundary tracking method (series of rules outlining each item of interest) and basic thresholding were employed to identify whiteflies in traps in the same year. Al-Saqer and Hassan described the pest's morphologies using Zernike moments and regional features (Red Palm Weevil). The traits computed by the two approaches were utilized to develop a library of pest templates that could detect the Red Palm Weevil, among other insect species. Maharlooei et al. (2016) detected aphids in soybean leaves using color modifications and image processing. Manual leaf segmentation was followed by pest segmentation utilizing the HSI color space. Aphids were picked by size, and the remaining linked components were counted. Xia et al. (2012) used multifractal analysis to measure an object's irregularity and complexity. This method is suitable for pest identification since it can handle changes in size and rotation while retaining most of the image's information. Certain research used this method to identify candidate blobs indicating probable whitefly trap sites using two thresholding approaches. Then multifractal analysis was used to retrieve the objects' characteristics and used a Mahalanobis distance to eliminate picture background before doing multifractal analysis to pick candidate blobs. The final whitefly detection in the photos used blob features paired with size and shape criteria. Wang et al. (2020) used the K-Clustering method to identify whiteflies in crops. A picture is first divided into 1,00,100 blocks. Then, using both the RGB and L*a*b* color spaces, k-means clustering is utilized to categorize each pixel. The ellipse eccentricity rules remove unwanted objects. Yao et al. (2013) evaluated the efficacy of normalized cuts (NCuts), watershed, and K-means clustering on pest separation in traps. This was done by using the optical flow angle as a weight function. The K-means clustering algorithm was utilized in to improve the outcomes of DL models.

Artificial neural networks (ANN) are models inspired by the way neurons are structured and linked in the brain. ANNs are often used for image categorization. A shallow architecture was almost universal until 2010. Deep architectures rose in popularity as processing power increased and GPUs became the primary training hardware. DL has been heavily used in picture categorization research since 2015. However, several investigations still use shallow ANNs. Asefpour Vakilian & Massah (2013) used image processing and ANNs to identify beet armyworms in controlled settings. The three-layer neural networks were fed with seven morphological and texture characteristics retrieved from the potential pests. To identify thrips and whiteflies in traps (Espinoza et al., 2016), a threshold was used to identify possible item locations from which 14 color and morphological characteristics were retrieved. These characteristics were fed into a multilayer feed-forward neural network, which detected the target. Roldán-Serrato et al. (2018) used ANNs to identify bean and potato pests. The RSC and LRA neural network designs were utilized, with the former being faster and more accurate. For inferring spatial correlations without explicitly extracting features, deep architectures are favored. Convolutional neural networks (CNNs) are especially popular because of their close layer-spatial coupling. The method used a sliding window, and the CNN predicted the existence of a moth for each picture patch. NMS was utilized to keep only patches with probability locally maximal and over a threshold, exposing the specimen locations. Barbedo & Castro (2019) investigated CNNs for psyllid identification in trap photographs. The research focused on how sensor quality, picture resolution, and training/test dataset distribution affect the model's capacity to correctly distinguish psyllids among insects and detritus gathered by the trap. An anchor-free area proposal network and decision networks have also been used to examine contextual information.

6.2 INDIRECT SENSING: DETECTION OF INSECT-DAMAGED CROPS

Remember that soil, nutrient, and water conditions are not always evenly distributed in a field, and that crop vegetation might vary in heterogeneous areas. A field's biomass and crop surface (leaf area index (LAI, yield, vigor) might vary geographically. Specific DL algorithms are used for the indirect inspection of pests. The CROP-Meter sensor does not directly detect insects but indirectly assesses LAI and biomass (Bieganowski et al., 2021; Salam & Raza, 2020; Shi et al., 2020). In low biomass regions, there has been a more significant pest population. To detect changes in plant structure, transpiration rate, and biochemical function, some monitoring systems utilize multi- and hyperspectral cameras, true-color cameras and spectrometers (Barbedo, 2019; Beckschäfer et al., 2014; Hornero, 2021; Kasampalis et al., 2018; Yao et al., 2017). Larger advanced camera and spectrometer systems are too heavy for low-cost UAVs to carry. Optical technologies may quickly analyze huge fields. Devices using stereo cameras can provide 3D data about a crop standing (Paulus, 2019). Using temporary structure from motion techniques, 3D data could be determined by calculating from UAV camera flight (Díaz-Varela et al., 2015). 3D crop data may reveal more about the canopy structure and identify crop damage (Li et al., 2021; Pádua et al., 2017). In optical technology, changing light conditions is the most significant constraint. Otherwise, periodic spectrometer recalibration is required to compare crop canopy reflectance to a white standard reference value. A more complex hardware and technology solution, such as an online measuring system, is needed to simplify manual field measurement (Aslan et al., 2022; Boursianis et al., 2022; Sinha & Dhanalakshmi, 2022). The optical signal is also connected to other variables such as nutrition (fertilizer), water deficit, illness, virus, and mammalian infestation. In crop canopies, these components have the same wavelength making it more complex to distinguish the source of crop loss. Sophisticated atmospheric correction algorithms often cannot altogether remove atmospheric elements that might obstruct the ground for optical signals (Table 6.1).

TABLE 6.1 Various Machine Learning Models for Pest Detection

Pest	Image	Problem	Input	Classifier/Detector	Accuracy	References
Palm weevil	Trap	Detection	Zernike moments and region properties	Template matching.	0.88–0.97	Al-Saqer & Hassan (2011)
Whiteflies	Controlled	Detection	Binary image	Connected objects counting.	0.68–0.95	Barbedo (2020)
Psyllids	Trap	Detection	Image patches.	CNN	0.69–0.92	Barbedo & Castro (2019)
Whitefly	Controlled	Detection	Spatial, color, and texture features.	Knowledge-based system.	0.70–0.97	Boissard et al. (2008)
10 species	Field	Classification	Original images.	Deep residual learning.	0.98	Cheng et al. (2017)
Whiteflies, aphids and thrips	Trap	Classification	Size, shape and color features.	Feature similarity.	0.59–1.00	Cho et al. (2007)
10 species	Field	Classification	Resized images.	AlexNet (CNN)	0.94	Dawei et al. (2019)
10 species	Field	Classification	SIFT, NNSC, LCP features.	SVM	0.85	Deng et al. (2018)
8 species	Field	Classification	Segmented images.	ANN	0.93	Dimililer & Zarrouk (2017)
Moth	Trap	Detection	Patches from original image.	CNN	0.93	Ding & Taylor (2016)
Thrips	Field	Detection	HSI color channels	SVM	0.98	Ebrahimi et al. (2017)
Western flower thrips, whitefly.	Trap	Detection	Patches from original image.	Multi-layer ANN	0.92–0.96	Espinoza et al. (2016)
10 species	Field	Classification	L*a*b* images	K-Means clustering.	N/A	Fina et al. (2013)
6 species	Trap	Classification	Morphological and color features.	ANN	0.82	Han et al. (2012)
24 species	Trap	Detection	Original images.	CNN+AFRPN	0.55–0.99	Jiao et al. (2020)
Whiteflies	Field	Detection	Image with background removed.	Multifractal analysis.	0.88	Li et al. (2015)
Aphids	Field	Detection	HOG features	SVM	0.87	Liu et al. (2016a)

TABLE 6.1 *(Continued)*

Pest	Image	Problem	Input	Classifier/Detector	Accuracy	References
12 species	Field	Classification	Cropped regions.	CNN	0.83–0.95	Liu et al. (2016b)
N/A	N/A	Survey	N/A	N/A	N/A	Liu et al. (2017)
16 butterfly species.	Trap	Classification	Original images.	CNN + RPN + PSSM	0.75	Liu et al. (2019)
Aphids	Controlled	Detection	Binary image.	Object counting.	0.81–0.96	Maharlooei et al. (2017)
N/A	N/A	Survey	N/A	N/A	N/A	Martineau et al. (2017)
Whiteflies	Trap	Detection	Binary image.	Object counting.	0.81–0.99	Qiao et al. (2008)
Mexican bean beetle, Colorado potato beetle	Field	Detection	Original images.	RSC and LRA	0.76–0.89	Roldán-Serrato et al. (2018)
Flies, gnats, moth flies, thrips and whiteflies.	Trap	Classification	Original images.	Tiny YOLOv3 + multistage CNN	0.92–0.94	Rustia et al. (2019)
6 species	Trap	Classification	LOSSV2 + SIFT	Feature similarity	0.96–0.99	Solis-Sánchez et al. (2011)
Red turpentine beetle	Trap	Detection	Original images.	Downsized retina net.	0.75	Sun et al. (2018)
Beet armyworm	Controlled	Detection	Morphological and texture features.	ANN	0.89	Asefpour Vakilian & Massah (2013)
20 insect species	Varied	Classification	SURF + HOG features	SVM	0.90	Venugoban & Ramanan (2014)
Whiteflies	Field	Detection	Thresholding + morphology	Connected objects counting.	0.82–0.97	Wang et al. (2013)
Whiteflies	Field	Detection	Original images	K-means clustering	0.92–0.98	Wang et al. (2018)
3 species	Field	Detection	Original images	CNN + Decision net	0.62–0.91	Wang et al. (2020)

TABLE 6.1 *(Continued)*

Pest	Image	Problem	Input	Classifier/Detector	Accuracy	References
5 species	Trap	Classification	SIFT features	MLSLC, KNNC, PDLC, PCALC, NMC, SVM	0.77–0.89	Wen et al. (2009)
8 species	Trap	Classification	Several features	MLSLC, KNNC, NMC, NDLC, DT	0.60–0.87	Wen & Guyer (2012)
9 moth species	Trap	Classification	154 features	IpSDAE	0.94	Wen et al. (2015)
Whiteflies	Trap	Detection	Original images	Multifractal analysis.	0.96	Xia et al. (2012)
Whiteflies, aphids, thrips	Trap	Classification	Segmented insects.	Mahalanobis distance	0.70–0.91	Xia et al. (2015)
24 species	Field	Classification	Resized images	VGG19 (CNN)	0.89	Xia et al. (2018)
4 species of lepidoptera	Trap	Classification	156 color, shape, and texture features.	SVM	0.97	Qing et al. (2012)
Several	Trap	Detection	Original images	Normalized cuts, watershed, K-means.	0.96	Yao et al. (2013)
Atractomorpha, Erthesina, Pieris	Field	Detection	Original images	DSRNLP	0.70–0.92	Yue et al. (2018)
6 species	Trap	Classification	Original images, variety of features.	YOLO, SVM	0.90–0.92	Zhong et al. (2018)

6.3 DIRECT SENSING: DETECTION OF INSECT SPECIES

Sensor-based insect classification might replace labor-intensive traditional procedures. This section briefly describes field sensor technology at the ground and aerial levels. As of now, satellite-based approaches are unsuitable due to low ground resolution and the inability to detect things smaller than a few millimeters.

6.3.1 REMOTE SENSING

Light, color, and traps can track insect movement throughout a field. Stationary traps for automated insect detection use high-speed cameras (image tracking) and identify flying insects from various perspectives (Souza et al., 2020; Sun & Gaydecki, 2021; Wang et al., 2022a). A sufficient disparity between pests and the background is needed for identification. Despite attempts to design selective traps, beneficial insects may fall into them. In such cases, various image-processing algorithms evolved to recognize single individuals and classify distinct pest species flying with the camera view or collected in traps (Bjerge et al., 2021; Gonzalez et al., 2016; Mamdouh et al., 2022; Preti et al., 2021). The fundamental issue with image-processing algorithms is their varying efficiency. Also, Radar inspects insect monitoring by producing certain energy levels (Perez et al., 2022; van Klink et al., 2022). Radar antennas send out electromagnetic pulses and calculate the reflected wavelengths. Two types of radars are utilized in insect migration research: weather radar and entomological Radar. An entomological radar recognizes pests' motion, orientation, height, and flying action (scanning Radar) (Drake et al., 2020; Maggiora et al., 2019; Noskov et al., 2021). It thus can identify species based on wing beat intensity, insect weight, and body structure (vertical-beam Radar) (Drake et al., 2020). Insect-detecting Radar uses millimeter wavelengths or complete polarization. Radar, LIDAR, and LADAR devices operate using light or laser beams rather than microwaves. These apps could indeed record pests traveling just under 1 meter in height. Besides spectroscopy and image sensors, LIDAR utilizes its own illumination sources; thereby, outcomes are less affected by modifying light sources (Shemesh et al., 2022). The various kinds of LIDAR used to classify insects are high spectral resolution LIDAR, Scheimpflug LIDAR, Continuous-wave LIDAR, and fluorescence LIDAR (Rydhmer et al., 2022b). Recent advances in LIDAR technology allow the estimation of insect body contributions and wing-beat frequencies, which may help discriminate between species and hazardous and benign insects (Rydhmer et al., 2022b; Tauc et al., 2019). Time multiplexing might also provide information regarding insect body cross-section, wing thickness, and biological tissue (Brydegaard et al., 2017; Tsai, 2022). The absorption and dispersion of illumination pulses by airborne particles such as plumes, haze, and moisture may restrict their use. Acoustic technologies can identify dangerous insects in real time through sound analysis of their sounds (Baumgartner et al., 2019; Burfoot et al., 2021; Saleh et al., 2018). Depending on the insect's field conditions and specific characteristics, many kinds of sound sensors may be used to classify nuisance insects (Rydhmer et al., 2022a). Limitations include sound changes across insect life stages (young larvae transmit sound weakly), ambient circumstances, and sensor-insect distances.

No single approach for identifying insect infestations can be developed. Combining several sensor systems may be a highly successful approach (sensor fusion). This includes developing multi-sensors systems on the same scale (ground-based sensors) as well as distinct scales (e.g., fusing aerial and ground-based sensors). Also, combining direct and indirect approaches shows promise. Indirect techniques can pinpoint the affected region, whereas direct methods can determine whether or not insects, or a specific bug species, caused the damage. Finally, farmers need maps of damaged crop regions or insect dispersion to selectively spray damaged areas. They may select which pesticides and fields need to be treated (offline spraying approach). Big field sprayers are inefficient in regions infested by fewer insects (less than one spray boom section). Ground-based and aerial radar technologies are not suitable for detecting microscopic insects or those on the crop surface.

6.3.2 FLUORESCENCE AND THERMAL SYSTEMS

Thermal infrared and fluorescence RS systems can follow plant respiration and photosynthetic activities, allowing pre-symptomatic pest detection. Laser-induced fluorescence (LIF) has been the most viable idea for obtaining plant fluorescence. Induced LED light components with rectified wavelength range (i.e., 650 nm and 730 nm) are used in certain commercial fluorometers to provide essential fluorescent indicators (e.g., maximum possible quantum yield, non-photochemical fluorescent quenching (NPQ), operational efficiencies of PSII (PSII), etc.) (Mohammed et al., 2019). The continuous fluorescence spectra may also be determined by combining a laser emitter source of light (e.g., He–Ne light source) with a quantitative spectroscope. Römer et al. (2011) detected wheat leaf rust accurately (93%) using fluorescence spectra 370–800 nm and a robust polynomial fitting technique. Aside from fluorescence, pests may alter plant evapotranspiration (ET) and respiration rates, causing observable alterations in thermal emission levels (Jud et al., 2018). A fungal infection (*Plasmopara viticola*) assault on grapevines was detected using thermal photography (Stoll et al., 2008). Falkenberg et al. (2007) Using thermal imaging, distinguished biotic (root rot) and abiotic (drought) stressors in cotton. Other investigations have validated the efficacy of fluorescence and thermal approaches in detecting plant pests. However, employing fluorescence and thermal RS systems to monitor plant pests has certain drawbacks. First, key critical fluorescence parameters (e.g., Fv/Fm, NPQ, PSII) need dark adaptation to the plant, limiting the technique's effectiveness and scalability (Medina, 2016). Because fluorescence signals are weak and easily mistaken with natural light, most modern fluorometers were intended for leaf or canopy measurements, limiting their applicability in vast regions. Combining fluorescence and thermal approaches with other RS systems, like hyperspectral systems, may effectively monitor plant diseases and pests.

6.3.3 SAR AND LIDAR SYSTEMS

Few studies currently use SAR and Lidar RS to monitor plant pests, perhaps because of the limited correlation between SAR and Lidar characteristics in pest characters. However, SAR and Lidar can identify plant traits and their surroundings, which may be useful in pest monitoring (Jafarbiglu & Pourreza, 2022; Khosrokhani & Nasr, 2022; Rhodes et al., 2022; Sharma et al., 2022; Taddeo, 2022). Cloudy and wet environments are conducive to many plant pests; hence, clouds prevent using VIS-NIR RS pictures. SAR's ability to penetrate clouds makes it an active remote sensing tool (Singh & Kathpalia, 2007). SAR was utilized to fetch plant moisture content, soil attributes, as well as certain structural properties of a tree canopy, noticeably for tall plants (e.g., trees, maize, sugarcane, etc.). These findings may help identify plant disease and pest habitats. These findings imply the use of microwave RS data in pest monitoring might be useful in identifying sorghum pest infestations (Singh & Kathpalia, 2007).

Lidar can obtain extensive vegetation cover morphometric data. Rebuilding the 3D canopy architecture is possible using the Lidar dataset for tree canopy, which may reveal signs of disease, pest infestation, or biomass reduction. It could be possible to separate their existence and effect using Lidar data gathered during a cropping period. SAR and Lidar observations may be useful tools for monitoring pest infestation.

6.4 PEST CLASSIFICATION METHODS

The difficulty in classifying pests must distinguish between targeted and non-targeted species, which might be many. DL has recently experienced substantial advancements in newly developed solutions for this pest classification. The important aspect of DL is that no human oversight or intervention is needed. Cho et al. (2007) employed image processing to identify whiteflies, aphids, and thrips in traps. The first strategy employed size, shape, and color traits to identify aphids; the second used YUV color space properties to distinguish whiteflies from thrips. The first method, dubbed LOSS, extracts shape characteristics to describe objects. The second method, SIFT, uses a well-known algorithm to detect visual elements. These two features were then used for the pest classification. Also, Han et al. proposed a hardware platform for pest classification. The method collects morphological and color characteristics for classifying three-layer 29-neuron ANN (Han et al., 2012). Pre-processing photos having gray scale alterations and average filtering removes the pest's background and canny edge detection segments before rescaling and feeding them to the two neural networks liable for pest classification.

DL is considered the current state of the art for pest detection and classification. DL uses certain features for pest classification, saliency maps using global contrast regions for pest detection and classification and target bounding boxes with Alex-Net-inspired CNN for training and testing data. DL is an effective tool for pest classification with high-end accuracy. In a study by (Gobinath & Gopinath, 2021), the deep residual learning approach improved two CNN architectures, ResNet101 and ResNet50, and showed better results than simple AlexNet CNN SVM and shallow ANN. Dawei et al. used DL to classify 10 species in field photos collected by an AlexNet CNN (Dawei et al., 2019). The model surpassed a CNN built from scratch, owing to the tiny dataset's lack of information for complete model training. Also, Dawei et al. (2019) employed a DL strategy to classify 16 butterfly species in trap photos. First, the CNN backbone was merged with CSA for feature extraction and improvement. An RPN was used to locate possible pests. A PSSM was employed instead of fully connected (FC) layers for pest categorization and bounding box regression. Finally, contextual areas of interest aided identification using a two-step DL strategy to categorize and count five pest categories in trap photos. The pests on the wax paper were found using the Tiny YOLO v3 object detector. Three levels of hierarchy were employed in the classification process, and CNN models were trained to handle the challenges presented by each level.

6.4.1 AUTOMATIC MONITORING OF LEPIDOPTERA PEST SPECIES

Numerous moth and butterfly species are reported to create a considerable economic loss in many crops globally. These insects may lay many eggs, and the larvae eat voraciously, producing natural defoliation and substantial losses when populations are established (Chandel et al., 2022; Leather, 2022; Lemay & Scott-Dupree, 2022). Currently, most monitoring is done using delta traps and pheromone lures. Developing automated detection and identification models for these insects is difficult owing to their numerous positions when connected to sticky traps (Barbedo & Castro, 2020; Rosado et al., 2022; Ullah et al., 2022). Silveira & Monteiro (2009) used a **ML** system using characteristics based on circularity and symmetry to recognize eyespot trends of the nymphalid dragonfly *Bicyclus anynana*. The program could also distinguish patterns from other butterfly species. Kasinathan & Uyyala (2021) Employed a mix of form, color, texture, and numerical data retrieved for moth description to identify species using photographs. Later, an IPSDAE was developed to construct a deep neural network (DNN) for moth identification independent of insect position. This approach identified 98.13% of moth genera without categorizing species. Guarnieri et al. (2011) introduced an automated monitoring system for codling moth detection, attaining a 100% effectiveness level when compared to local visual assessment by employing a customized trap in the field. Numerous techniques for species identification utilize artificial neural networks (ANNs). ANNs are computer simulations of biological neural networks that may be taught to recognize patterns in photos. A higher proficiency platform (93%) for recognition of above 150 species from 64 insect orders families, including Lepidoptera, Megaloptera, Coleoptera, Hemiptera, Odonata, Hymenoptera, Neuroptera, and Orthoptera, has indeed been created by (Dettner, 2015). Wang et al. (2022) offered an approach based on color and texture traits, whereas (Kayci & Kaya, 2014) proposed a method based on form. In **ML**, texture descriptors, particularly the gray level co-occurrence matrix (GLCM), have proven effective in Lepidoptera identification and monitoring, correctly identifying 19 Pieridae species with 96.3% accuracy. Their accuracy ranged from 95% to 97% when using deep CNNs and DL methods on three publicly accessible bug datasets (Li & Xiong, 2018). HoMSC histograms, Gray-level co-occurrence matrix, and a weight-based k-nearest neighbor (KNN) classifier was used to identify 50 lepidopteran species. Two researchers built a mobile robot vehicle with a camera for field identification of Pyralidae species. Zhao et al. (2019) used the gaussian mixture model (GMM), aggregation dispersion variance (ADV), and distance regularization level set evolution (DRLSE) to accurately identify the target species. The model recognized Pyralidae species with 94.3% accuracy, superior to the support vector machine (SVM) technique.

6.4.2 *MONITORING OF SUCKING INSECTS*

Sucking bugs are a significant cause of losses in greenhouses. Thrips, aphids, and whiteflies are among the most troublesome pests when crops are grown in confined spaces. Sticky traps are often used in greenhouses to monitor insect numbers. This approach is challenging for non-experts due to its size, complexity, and poor efficiency of exhausted or inexperienced human observers. Several experiments were done on automated greenhouse sucking pest detection (Tureček et al., 2022; Xolmurodov et al., 2022). Several writers presented computer vision methods for identifying the adult greenhouse whitefly *Trialeurodes vaporariorum*. Bodhe & Mukherji (2013) used threshold methods to recognize and count whiteflies in images. An algorithm used by (Ghods & Shojaeddini, 2016) successfully identified whiteflies in plant leaf tissue with 85% efficiency. A prototype was built by (Blasco et al., 2019) to track insect captures in the fields by transmitting pictures to a remote server. The device, modeled on the Raspberry Pi design, features a camera for taking pictures, a control system for controlling photo taking, and a transmitter for uploading photos and information to a remote streaming device. To distinguish between images of pest-infested white flies and healthy leaf tissue, the researchers utilized data analysis feature selection methods like GRLM and GLCM before classification models like binary decision tree classifiers, Bayesian classifiers, SVMs, ANNs, and KNN. These classification models have high sensitivity (93.9%–98.8%) and accuracy (90–98%). Ebrahimi et al. (2017) used a SVM to identify and categorize whitefly stages. Automatic color transformation models for counting and quantifying whiteflies in soybean leaves and yellow sticky traps have been presented. Multiple IPM-based approaches were created. Qiao et al. (2008) suggested an image processing technique for silver leaf whitefly Bemisiatabaci density estimation using yellow sticky trap data from greenhouses. The strategy worked better with medium and high pest populations in this study. Sun et al. (2017) used a unique, innovative vision method employing two-dimensional Fourier-transformed spectra to score and identify Bemisiatabaci in yellow sticky traps and *Frankliniella* occidentalis in blue sticky traps.

Espinoza et al. (2016) suggested a model for identifying and monitoring *Frankinella* occidentalis and *Bemisia tabaci* using image processing (segmentation, morphological, and color property estimates) and ANN (96%). Bauch & Rath (2004) employed to form and color to identify *Bemisia tabaci* and *Trialeurodes vaporarium* with 85% accuracy. Lu et al. (2019) have suggested a CNN classification model using GAN image augmentation. A Raspberry Pi 2 sensor was employed to investigate thrips and whiteflies and create synthesized images utilizing generative adversarial network data mining algorithms to enhance CNN classification with negligible visual information. An accuracy of somewhere between 85% and 95% were achieved using this method. To analyze photos efficiently, Xia et al. (2015) presented an embedded system for detecting aphids, whiteflies, and thrips. This approach was 95.5% accurate in a greenhouse and 92.5% outdoors using yellow sticky traps. Liu et al. (2016a) used a SVM and histograms of oriented gradient characteristics to recognize wheat aphids in the crops. This new technique correctly identifies 86.81%. Li et al. (2019) additionally built a region proposal network (RPN) with non-maximum suppression (NMS) and a CNN of the Zeiler and Fergus model. Maharlooei et al. (2017) developed an image processing model for identifying soybean aphids (*Aphis glycines*) using leaf samples and compared the results with human counting. It was best with low lighting and a Sony camera with 96% accuracy. Chen et al. (2018) suggest a methodology for segmenting and counting aphid nymphs using pak choi leaf samples. Unlike human counting, this technique is non-selective and may be utilized for different pests. This technique, like others mentioned, seems to count rather than differentiate between whiteflies and aphids. Rustia et al. (2020) operated a camera, sticky traps, a temperature probe, and an optical sensor to retrieve the V-channel color aspect from RGB images. Also, Gutierrez et al. (2019) investigated whiteflies utilizing a camera module, a data source generating system, and two different microchips have been utilized with a multilayer perceptron (MLP) and KNN to accomplish a 66% to 80% accuracy.

6.4.3 MONITORING OF THE INSECT PEST POPULATION

6.4.3.1 SENSOR-BASED INSECT MONITORING

In an ecosystem, detectors are frequently employed to collect ancillary information like temperatures, rainfall, and illuminance. However, approaches for camera surveillance of invertebrates in their natural environment, such as insects and other invertebrates, are still in their infancy. The advancement and development are driven mainly by agricultural research to anticipate the incidence and availability of advantageous and pest insect species of economically important to provide more effective testing of natural ingredients for intrusive species of insects or to supervise disease vectors such as mosquitos (Early et al., 2016; Fand et al., 2012; Panzavolta et al., 2021). Cameras, radars, and microphones are the most commonly utilized sensors. Such sensor-based surveillance is projected to produce enormous data—orders of huge datasets greater than those typically analyzed in ecology—which calls for effective methods for obtaining pertinent biological material (Al-Naji & Zagrouba, 2020; Høye et al., 2021; Kamminga et al., 2018; Manogaran et al., 2022; Qiu et al., 2022; Shafi et al., 2019; Wang et al., 2020b). **ML** may be an essential technique in this situation.

6.5 DRONES IN PEST CONTROL

Drones are unmanned aerial vehicles (UAVs) used in many fields, including military, surveillance, and disaster management, but are new to agriculture. Drones have three main varieties: fixed-wing, multi-rotor, and hybrid. Other factors are automation, size, weight, and power supply. Drones' operating characteristics, including flying speed, height, and durability, must be improved for agricultural and related industries. When utilizing drone-based measures to mitigate droplet size, dispersion, specific gravity, uniformity, accumulation, and permeability should be presumed. Despite low acceptance, the drone technique is highly appropriate and suitable for the control of pests. Globally, regulatory rules have been established to execute site-specific agricultural management more accurately (Delavarpour et al., 2021; Esposito et al., 2021; Hasanaliyeva et al., 2022; Tsouros et al., 2019). Drones are outstanding instruments for speedy, dependable, and non-destructive field issue identification. Weather variables like wind, temperature, and relative humidity may also impact pesticide performance in the field (Wang et al., 2018a). Because controlling the weather is challenging, scientists have tried to investigate drones in controlled environments. Liu et al. (2021) investigated the usefulness of drones in a fabulous setting and the effects of drone flight speed on particle diameter, form, diffusion, and homogeneity. Successful pesticide spraying using drones may be optimized utilizing the study's criteria. Droplet size is an essential factor in pesticide or other input delivery. Utilizing infra-red thermography, Zhengguang et al. (2019) evaluated particle accumulation. Using photonic technologies, infrared thermography identifies the infrared-specific spectrum output of an entity's thermal radiation. This technique has been used in addition to particle precipitation assessment.

6.5.1 ACTUATION DRONES FOR PRECISION APPLICATION OF PESTICIDES

Sensing drones might assist in discovering insect hotspots, while actuation drones could help manage pests. Variable rate pesticide application might help control insect hotspots. For years, airplanes were used to apply pesticides, but the chemicals are dispersed over large areas and disappear. It has an impact on the local terrestrial ecosystem and marine environments as well as people's health. Drones might be used to apply pesticides and miticides precisely. Drones have advantages over human crop dusters in terms of deployment convenience, possible spray drift reduction and operator exposure to pesticides. Drones are frequently employed in Japan, where they have been since the 1980s, for pesticide spray in rice (Oryza sativa L.) (Iost Filho et al., 2020). These drones are generally above 25 kg; however, they are among the most extensively utilized in pest control. A quarter of Taiwan's paddy fields are sprayed with unmanned drones that are authorized for usage as of March 2016. Additional crops, such as maize, wheat, and soybeans, are increasingly being treated with unmanned crop dusters. In California grape wine, automated Japanese agricultural dusters are currently being tested for pesticide application. In China, automated crop dusters are used sparingly but increasingly for fruits and vegetables, mangos, and rice (Lusas et al., 2012). The countries of the United States, India, South Korea, Ukraine, and Portugal are all working on building special automated crop dusters or spraying rigs for drones that may be bought commercially. Precision pesticide application and precision monitoring might minimize total spraying, reducing pesticide consumption and resistance development while increasing natural enemy presence.

6.5.2 ACTUATION DRONES FOR PRECISION RELEASES OF NATURAL ENEMIES

Drones might benefit from augmentative biological control, which depends on mass releases of natural enemies to manage pests. They might place natural enemies where they are required, increasing biocontrol agent effectiveness and lowering distribution costs. A few natural enemies, such as aphids and nematodes, may be splashed using conventional spray equipment (Woltz et al., 2015). So, like pesticides, these biocontrol chemicals might be sprayed by drones. European corn borer (Lepidoptera: Crambidae) parasitoids are key biocontrol agents of *Trichogramma* spp. Companies and research institutions in Germany, France, Austria, Italy, and Canada have begun using *Trichogramma* drones (IVEZIĆ et al., 2022; Zang et al., 2021). China uses drone-delivered *Trichogramma* parasitoids to manage pests in sugarcane (*Saccharum* spp.). Drones carrying *Trichogramma* spp. and *Cotesia flavipes* Cameron (Hymenoptera: Braconidae) are used in Brazil to control the sugarcane borer. It is also used to control other lepidopteran pests in different crops.

6.6 ADVANTAGES OF MACHINE LEARNING (ML) IN ENTOMOLOGY

The benefits of **ML** in terms of reduced effort in feature entomology were proved in most publications, aside from increased efficiency of the classification/prediction tasks in the studies reviewed (Bhatnagar et al., 2022; Susheel & RajKumar, 2022). Entomological modeling consumes much effort, but **ML** processes these tasks automatically. Additionally, searching manually for effective feature extraction methods is not always straightforward. For instance, manually removing traits that significantly impacted crop development was impossible in the case of insect detection. Additionally, the model is expressly trained to count in the case of pest counting. The model in the challenge of classifying plant pest leaves held up well under difficult circumstances like lighting, complex backgrounds, various image resolutions, sizes, and orientations (Bereciartua-Pérez et al., 2022; Thapa et al., 2020). The models in the publications on pest classification also were resistant to occlusion, variability, brightness, and size (Akintayo et al., 2018; Bereciartua-Pérez et al., 2022; Geissmann et al., 2022; Tetila et al., 2019). As another example, a primary aspect of the Deep Anomaly model was the capability to sense unidentified pests and not just a set of predetermined pests, trying to exploit the homogeneous characteristics of crop fields to identify faraway, heavy occluded, and unknown pests (Xia et al., 2021).

Moreover, in the research articles in which various datasets were used for checking, the model performance was relatively high, with small reductions in performance compared with the version using the same dataset to train and test. Although DL requires more time to train than conventional algorithms (such as SVM and RF), it tests very quickly. For instance, the model's testing time was shorter than that of SVM and KNN, even though it took much longer to train the model to detect pests. Another benefit of DL is the ability to create replicated datasets to train a model, which could be appropriately constructed to resolve actual issues (Alzubaidi et al., 2021; Sarker, 2022). For instance, the authors solved the problem of crop foliage overlaying in the challenge of detecting pests in maize and weed fields by simulating top-down images of overlapping plants on a dirt background. The network could therefore differentiate between pests on weeds and maize even in overlapping situations.

6.7 DISADVANTAGES AND LIMITATIONS OF DEEP LEARNING (DL)

The requirement of large datasets, which would act as the input during the training method, is a significant drawback and hurdle in the application of ML. Despite data augmentation approaches that add label-preserving modifications to a dataset, depending on the complexity of the subject being studied, at least a few 100 photos are still necessary (i.e., number of classes, the precision required, etc.) (Alzubaidi et al., 2021; Arikan et al., 2019; Bayer et al., 2021; Hu et al., 2019; Kamilaris & Prenafeta-Boldú, 2018b; Simões et al., 2016). The bulk of the time, data annotation is essential; nevertheless, some activities are more sophisticated and require the annotation of input pest pictures by experts, who may be challenging to include. When data labeling is complex, such as counting pests or determining whether photographs contain weeds, professionals or volunteers might sometimes make mistakes. The DL models' exceptional ability to learn and in some cases, generalize specific problems is another drawback. For instance, the classification of a single pest facing up on a uniform background is carried out. A practical application should be able to categorize pictures of a pest as it appears on the plant. Not all pests only manifest on the upper side of leaves (Sottocornola et al., 2022). The model failed to recognize pests if their size in a testing image was noticeably smaller than in the training set. Data pre-processing is occasionally a necessary and time-consuming procedure in computer vision, not just in DL, especially when satellite or aerial photographs are involved (Guirado et al., 2019). Hyperspectral data have a high degree of dimensionality and few training examples, which is problematic.

Additionally, there are situations when the current datasets fail to describe the issue they aim to address fully. For instance, it was essential to consider external elements besides the weather while estimating maize production (Kuwata & Shibasaki, 2016) by entering agriculture data such as fertilization and irrigation. Finally, there are not many publicly accessible datasets in pest detection, so researchers frequently have to create their collections of photos to work with. It can take many hours or even days to do this.

6.8 FUTURE OF DEEP LEARNING (DL) IN AGRICULTURE AND ENTOMOLOGY

ML has only been used to approximate the problems of classifying land cover, estimating crop kind, crop phenology, detecting pests, and fruit grading in agricultural computer vision applications. It will be fascinating to see how ML responds to challenges in other areas of agriculture, such as seed identification, water management, plant moisture stress detection, water erosivity evaluation, herbicide use, micropollutants identification, crop hail damage, and greenhouse monitoring. Since many of the research mentioned above areas use data analysis methods (such as linear as well as logistic regression, K means clustering, SVM, KNN, Wavelet-based filtering, and Fourier transform) that share concepts with and undertake similarly to DL, it would seem worthwhile to investigate the applicability of DL to these issues as well. Other potential application areas include the use of satellite photographs (i.e., utilizing drones) to supervise the efficiency of the seed production process, to improve the quality of wine production by harvesting grapes at the right time for the best levels of maturity, to supervise pest movement to consider their overall wellbeing and identify potential diseases, as well as many other situations where computer vision is involved. In addition to these datasets, satellites like MERIS, MODIS, AVHRR, RapidEye, Sentinel, Landsat, and others provide remote sensing data with multi-temporal, multi-spectral, and multi-source images that can be utilized to solve challenges relating to the classification of land and crop cover. More methods using LSTM or other RNN models are anticipated in the future, utilizing the time dimension to provide more accurate pest classification or prediction. More advanced architectures would also emerge, fusing different DL models and classifiers or hand-crafted features with automatic features retrieved using different methodologies to enhance the final result, similar to what was achieved. Researchers should test their models against more realistic and generic datasets to show how well they can generalize to other real-world scenarios. It would be preferable if researchers made their datasets accessible to the public so that other researchers may use them. Finally, some of the solutions presented in the publications surveyed may soon be used commercially. For autonomous robots that harvest crops, detect pests, or predict the predicted yields of diverse crops, the methods incorporating Faster Region-based CNN and DetectNet CNN would be very helpful (Esmaeilzadeh Khorasani, 2022; Kamilaris & Prenafeta-Boldú, 2018a; Saleem et al., 2021; Zhang et al., 2022). This method might be used in microbiology to count the cells of insects or pests in the future.

6.9 CONCLUSION

Artificial intelligence (AI), **ML**, and DL have remarkable potential in insect detection and monitoring both physically in the field and using good quality images. Although a very convenient technique, it is not being used on a large scale due to a lack of robustness in varied practical situations. This limitation can also be attributed to the limited dataset that is used to teach these classification models. To bridge this gap, there is a dire need to construct a large comprehensive database with many HD images of different pests to improve the precision of image recognition. The more effectively we feed our data, the more reliable system will be developed. Efforts can be made to develop an effective image representative database by encouraging the involvement of technical people in quality image collection and labeling. This approach has already been used for plant disease identification (Barbedo, 2018).

Pest monitoring enables us to respond to possible insect pest infestations timely, which can save our crops from severe damage. However, monitoring entire large fields is very challenging. It may need a large network of scattered sensors to collect the data from the entire area, but installing sensors in an inhospitable environment would be extremely difficult and costly. As a substitute, we can fix the sensors in farm tools and machinery used daily on the farm, and in this way, the farm will be monitored for the presence of pests much more effectively daily. Micro-flying robots can be used as a future approach, but they will need much more research and innovations. For fully automated monitoring models, algorithms should be light so that small and cost-effective units can handle the image data and give precise and timely information to the decision-making unit.

However, despite all the advancements, there is still a research gap that needs to be bridged by future research and innovations.

KEYWORDS

- **deep learning**
- **drones**
- **entomology**
- **Lepidoptera**
- **lidar systems**
- **machine learning**
- **pest control**
- **sensor-based insect monitoring**

REFERENCES

Al-Saqer, S. M., & Hassan, G. M., (2011). Red palm weevil (*Rynchophorus ferrugineous*, Olivier) recognition by image processing techniques. *Am. J. Agric. Biol. Sci., 6*, 365–376.

Asefpour, V. K., & Massah, J., (2013). Performance evaluation of a machine vision system for insect pests identification of field crops using artificial neural networks. *Archives of Phytopathology and Plant Protection, 46*, 1262–1269.

Barbedo, J. G. A., (2020). Detecting and classifying pests in crops using proximal images and machine learning: A review. *AI, 1*, 312–328.

Barbedo, J. G., & Castro, G. B., (2019). Influence of image quality on the identification of psyllids using convolutional neural networks. *Biosystems Engineering, 182*, 151–158.

Boissard, P., Martin, V., & Moisan, S., (2008). A cognitive vision approach to early pest detection in greenhouse crops. *Computers and Electronics in Agriculture, 62*, 81–93.

Cheng, X., Zhang, Y., Chen, Y., Wu, Y., & Yue, Y., (2017). Pest identification via deep residual learning in complex background. *Computers and Electronics in Agriculture, 141*, 351–356.

Cho, J., Choi, J., Qiao, M., Ji, C., Kim, H., Uhm, K., & Chon, T., (2007). Automatic identification of whiteflies, aphids and thrips in greenhouse based on image analysis. *Red, 346*, 244.

Dawei, W., Limiao, D., Jiangong, N., Jiyue, G., Hongfei, Z., & Zhongzhi, H., (2019). Recognition pest by image-based transfer learning. *Journal of the Science of Food and Agriculture, 99*, 4524–4531.

Deng, L., Wang, Y., Han, Z., & Yu, R., (2018). Research on insect pest image detection and recognition based on bio-inspired methods. *Biosystems Engineering, 169*, 139–148.

Dimililer, K., & Zarrouk, S., (2017). ICSPI: Intelligent classification system of pest insects based on image processing and neural arbitration. *Applied Engineering in Agriculture, 33*, 453.

Ding, W., & Taylor, G., (2016). Automatic moth detection from trap images for pest management. *Computers and Electronics in Agriculture, 123*, 17–28.

Ebrahimi, M., Khoshtaghaza, M. H., Minaei, S., & Jamshidi, B., (2017). Vision-based pest detection based on SVM classification method. *Computers and Electronics in Agriculture, 137*, 52–58.

Espinoza, K., Valera, D. L., Torres, J. A., López, A., & Molina-Aiz, F. D., (2016). Combination of image processing and artificial neural networks as a novel approach for the identification of *Bemisia tabaci* and *Frankliniella occidentalis* on sticky traps in greenhouse agriculture. *Computers and Electronics in Agriculture, 127*, 495–505.

Fina, F., Birch, P., Young, R., Obu, J., Faithpraise, B., & Chatwin, C., (2013). Automatic plant pest detection and recognition using k-means clustering algorithm and correspondence filters. *International Journal of Advanced Biotechnology and Research, 4*, 189–199.

Han, R., He, Y., & Liu, F., (2012). Feasibility study on a portable field pest classification system design based on DSP and 3G wireless communication technology. *Sensors, 12*, 3118–3130.

Jiao, L., Dong, S., Zhang, S., Xie, C., & Wang, H., (2020). AF-RCNN: An anchor-free convolutional neural network for multi-categories agricultural pest detection. *Computers and Electronics in Agriculture, 174*, 105522.

Li, Y., Xia, C., & Lee, J., (2015). Detection of small-sized insect pest in greenhouses based on multifractal analysis. *Optik-International Journal for Light and Electron Optics, 126*, 2138–2143.

Liu, H., Lee, S. H., & Chahl, J. S., (2017). A review of recent sensing technologies to detect invertebrates on crops. *Precision Agriculture, 18*, 635–666.

Liu, L., Wang, R., Xie, C., Yang, P., Wang, F., Sudirman, S., & Liu, W., (2019). PestNet: An end-to-end deep learning approach for large-scale multi-class pest detection and classification. *IEEE Access, 7*, 45301–45312.

Liu, T., Chen, W., Wu, W., Sun, C., Guo, W., & Zhu, X., (2016a). Detection of aphids in wheat fields using a computer vision technique. *Biosystems Engineering, 141*, 82–93.

Liu, Z., Gao, J., Yang, G., Zhang, H., & He, Y., (2016b). Localization and classification of paddy field pests using a saliency map and deep convolutional neural network. *Scientific Reports, 6*, 1–12.

Maharlooei, M., Sivarajan, S., Bajwa, S. G., Harmon, J. P., & Nowatzki, J., (2017). Detection of soybean aphids in a greenhouse using an image processing technique. *Computers and Electronics in Agriculture, 132*, 63–70.

Maharlooei, M., Sivarajan, S., Shirzadifdar, A., Bajwa, S. G., Nowatzki, J. F., & Harmon, J. P., (2016). Developing an image processing based algorithm to detect and count soybean aphids under field conditions. In: *2016 ASABE Annual International Meeting* (p. 1). American Society of Agricultural and Biological Engineers.

Martineau, M., Conte, D., Raveaux, R., Arnault, I., Munier, D., & Venturini, G., (2017). A survey on image-based insect classification. *Pattern Recognition, 65*, 273–284.

Qiao, M., Lim, J., Ji, C. W., Chung, B. K., Kim, H. Y., Uhm, K. B., Myung, C. S., et al., (2008). Density estimation of *Bemisia tabaci* (*Hemiptera: Aleyrodidae*) in a greenhouse using sticky traps in conjunction with an image processing system. *Journal of Asia-Pacific Entomology, 11*, 25–29.

Qing, Y., Jun, L., Liu, Q. J., Diao, G. Q., Yang, B. J., Chen, H. M., & Jian, T., (2012). An insect imaging system to automate rice light-trap pest identification. *Journal of Integrative Agriculture, 11*, 978–985.

Roldán-Serrato, K. L., Escalante-Estrada, J., & Rodríguez-González, M., (2018). Automatic pest detection on bean and potato crops by applying neural classifiers. *Engineering in Agriculture, Environment and Food, 11*, 245–255.

Rustia, D. J. A., Chao, J. J., Chung, J. Y., & Lin, T. T., (2019). *An Online Unsupervised Deep Learning Approach for an Automated Pest Insect Monitoring System* (p. 1). 2019 ASABE Annual International Meeting. American Society of Agricultural and Biological Engineers. https://elibrary.asabe.org/abstract.asp?aid=47205 (accessed on 30 November 2023).

Solis-Sánchez, L. O., Castañeda-Miranda, R., García-Escalante, J. J., Torres-Pacheco, I., Guevara-González, R. G., Castañeda-Miranda, C. L., & Alaniz-Lumbreras, P. D., (2011). Scale invariant feature approach for insect monitoring. *Computers and Electronics in Agriculture, 75*, 92–99.

Sun, Y., Liu, X., Yuan, M., Ren, L., Wang, J., & Chen, Z., (2018). Automatic in-trap pest detection using deep learning for pheromone-based *Dendroctonus valens* monitoring. *Biosystems Engineering, 176*, 140–150.

Venugoban, K., & Ramanan, A., (2014). Image classification of paddy field insect pests using gradient-based features. *International Journal of Machine Learning and Computing, 4*, 1.

Wang, F., Wang, R., Xie, C., Yang, P., & Liu, L., (2020). Fusing multi-scale context-aware information representation for automatic in-field pest detection and recognition. *Computers and Electronics in Agriculture, 169*, 105222.

Wang, K., Zhang, S., Wang, Z., Liu, Z., & Yang, F., (2013). Mobile smart device-based vegetable disease and insect pest recognition method. *Intelligent Automation & Soft Computing, 19*, 263–273.

Wang, Z., Wang, K., Liu, Z., Wang, X., & Pan, S., (2018). A cognitive vision method for insect pest image segmentation. *IFAC-PapersOnLine, 51*, 85–89.

Wen, C., & Guyer, D., (2012). Image-based orchard insect automated identification and classification method. *Computers and Electronics in Agriculture, 89*, 110–115.

Wen, C., Guyer, D. E., & Li, W., (2009). Local feature-based identification and classification for orchard insects. *Biosystems Engineering, 104*, 299–307.

Wen, C., Wu, D., Hu, H., & Pan, W., (2015). Pose estimation-dependent identification method for field moth images using deep learning architecture. *Biosystems Engineering, 136*, 117–128.

Xia, C., Chon, T. S., Ren, Z., & Lee, J. M., (2015). Automatic identification and counting of small size pests in greenhouse conditions with low computational cost. *Ecological Informatics, 29*, 139–146.

Xia, C., Lee, J. M., Li, Y., Chung, B. K., & Chon, T. S., (2012). In situ detection of small-size insect pests sampled on traps using multifractal analysis. *Optical Engineering, 51*, 027001.

Xia, D., Chen, P., Wang, B., Zhang, J., & Xie, C., (2018). Insect detection and classification based on an improved convolutional neural network. *Sensors, 18*, 4169.

Yao, Q., Liu, Q., Dietterich, T. G., Todorovic, S., Lin, J., Diao, G., Yang, B., & Tang, J., (2013). Segmentation of touching insects based on optical flow and NCuts. *Biosystems Engineering, 114*, 67–77.

Yue, Y., Cheng, X., Zhang, D., Wu, Y., Zhao, Y., Chen, Y., Fan, G., & Zhang, Y., (2018). Deep recursive super resolution network with Laplacian pyramid for better agricultural pest surveillance and detection. *Computers and Electronics in Agriculture, 150*, 26–32.

Zhong, Y., Gao, J., Lei, Q., & Zhou, Y., (2018). A vision-based counting and recognition system for flying insects in intelligent agriculture. *Sensors, 18*, 1489.

CHAPTER 7

A Systemic Machine Learning Approach for Fruit Yield Estimation Under Changing Global Climatic Conditions

OWAIS BASHIR,[1] SHABIR AHMAD BANGROO,[1] NASIR BASHIR NAIKOO,[1] NAZIR AHMAD GANIE,[2] S. S. KUKAL,[3] ASMAT ARA,[4] RANA AFREEN,[1] REHANA JAVID,[5] ABDUL ROUF MALIK,[5] GHULAM HASSAN,[5] HASEEB AHMED,[5] and KHALID REHMAN HAKEEM[6]

[1]*Division of Soil Science, Sher-e-Kashmir University of Agricultural Sciences and Technology of Kashmir, Kashmir, India*

[2]*Division of Animal Genetics and Breeding, Sher-e-Kashmir University of Agricultural Sciences and Technology of Kashmir, Kashmir, India*

[3]*Punjab Water Regulation and Development Authority, Chandigarh, Punjab, India*

[4]*Division of Plant Breeding and Genetics, Sher-e-Kashmir University of Agricultural Sciences and Technology of Kashmir, Kashmir, India*

[5]*Division of Fruit Science, Sher-e-Kashmir University of Agricultural Sciences and Technology of Kashmir, Kashmir, India*

[6]*Department of Biological Sciences, Faculty of Science, King Abdulaziz University, Jeddah, Saudi Arabia*

ABSTRACT

With cutting-edge computer technology and data analysis techniques, machine learning (**ML**) (Heidari et al.), a crucial element of artificial intelligence (AI), enables farmers to produce high-quality horticultural fruits free from

Soil, Agriculture, and Ecosystem Modelling: Smart Technologies for Sustainable Solutions.
Owais Bashir, Khalid Rehman Hakeem, Surinder S. Kukal, and Razeef Mohd (Eds.)

insects, pests, and illnesses. A new technology in the horticulture industry is **ML**. By offering guidance at each step of crop growth and development, from planting the crop to the stage of harvesting, ML utilizes technologies that help the agricultural community improve the overall cultivation and management of horticultural crops. ML assists with planting, weeding, and spraying tasks. It makes it easier to apply fertilizers precisely where they are needed, increasing fertilizer effectiveness while preserving soil fertility. ML-based systems stop the overuse of insecticides, herbicides, and irrigation water. Compared to older, more conventional ways of crop production, these technologies require less time and labor. The primary objective of intelligent horticulture is to increase the efficiency of these techniques and the tools employed. Scientists have developed several instruments through the use of this technology that assist farmers in preventing crop losses. To acquire nutrient-dense and cost-effective food, devices have been created to identify pests and illnesses, locate trees, identify fruit based on size, color, texture, and maturity, find fruit between leaves, approach fruit during harvesting, etc. These technologies support the farming community in making timely management decisions, improving yield estimation accuracy. Therefore, ML-based solutions boost production and quality, which are necessary for a country to experience economic success. This review study is intended to inform both researchers and farmers on AI, promote their use of these tools in the field for effective input management, and examine the potential applications of **ML** in horticulture.

7.1 INTRODUCTION

Due to increasing population pressure, the loss of natural resources, soil degradation, and changes in climatic conditions, various solutions are needed in agricultural production for its service delivery and expansion (Ferreira et al., 2022; Foley et al., 2011). These degradation challenges underscore the need for a land appraisal, crop protection, and yield prediction for global food supply. Accurate forecasting requires collaboration between governmental and non-governmental organizations (NGOs). Therefore, policymakers need accurate agricultural output forecasting to simplify export and import evaluations and enhance national food security (Abdelkader & Elshorbagy, 2021; Giller et al., 2021; Nosratabadi et al., 2021). Due to several complex variables, policymakers and other organizations make it difficult to anticipate fruit productivity. These complex factors, which all have an impact on fruit productivity, include soil quality, landscape features, insect damage, genotype, water quality, weather conditions, harvest scheduling, etc. (Romero et al., 2022; Thakur & Uphoff, 2017). The ultimate goal of all agricultural operations is cropping output. In crop yield, mechanisms and techniques are time-dependent and nonlinear. As a result of the interdependence between external factors and non-arbitration, these solutions are typically challenging. To predict crop yields and make important production-related decisions in the past, farmers primarily relied on their own knowledge and historical (Shook et al., 2021). However, recent technological developments have increased yield prediction and the capacity to use powerful computers to analyze vast quantities of data (Heidari et al., 2022). Examples include crop model simulation and machine learning (**ML**). Today, a growing body of research demonstrates that **ML** algorithms have more promising results than traditional statistics. The fundamental procedure is explained in Figure 7.1.

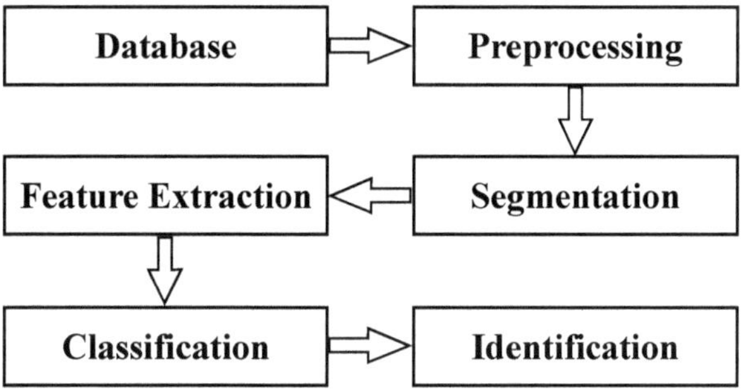

FIGURE 7.1 Block diagram of identification in image processing.

Fruit production prediction has recently been transformed by **ML** technologies, including decision trees (**DT**), DNNs, artificial neural networks (ANNs), and multivariate regression (Çetin et al., 2022; Chen et al., 2022b; Hong Khai et al., 2022; Kaur et al., 2022). In order to obtain an initial estimate of yield losses and gains, it is quite typical to predict fruit production based on several regulating characteristics (Silvestri et al., 2021). **ML** is used for scheduling of future processes, such as intercultural activities, worker participation, harvest timing, and the decision of operating strategies for a particular production process (Zhang et al., 2021). According to the necessary outputs and inputs for various crops, some **ML** algorithms have recently been tested and shown to produce accurate fruit yield predictions (Nosratabadi et al., 2021). These techniques include M5-Prime Regression Tree, KNN, support vector regression (SVR), ANNs, and DNNs (PS, 2019).

ML may be utilized from genetic and environmental data to predict fruit production, check yield, and yield variation (He et al., 2022a). A subtype of **ML** model called DNNs can ascertain the underlying structure of data without having to explicitly enter its features. DNNs also have a large number of constructed non-linear components that, at each stacked layer, transform the basic data input into a much more meaningful and conceptual interpretation (Li et al., 2021; Tripathi, 2021). DNNs are regarded as global approximator characteristics, which suggests that they can estimate virtually any function, even if it may be challenging to find the right parameters. Because of its promising potential, several academics have actually worked on fruit yield prediction using Ml techniques during the preceding decades.

The current study focuses on – (i) fruit yield prediction and the features employed in fruit prediction. (ii) Recognition and classification of horticulture crops using **ML** approach. And (iii) yield prediction of various crops, including apple, mango, almond, grapes, and banana.

7.2 FRUIT YIELD PREDICTION PROCESS

Data collection, preprocessing, segmentation, and assessment are among the stages of the **ML** method for forecasting fruit production. Utilizing **ML**-based regression or classification approaches, data analysis is performed to predict fruit yield systems (Maheswari et al., 2021). Numerous factors affect the method for calculating fruit yield. Recent studies indicate that leaf and fruit information, cultivation practices, soil characteristics, documentation on cropland as well as vegetation indexes, water management, weather data, satellite measurements, information on yield history, characteristics of underground water, fertilizer application specifics, and phenology information are all factors that positively influenced fruit yield (Panda et al., 2010) (Figure 7.2).

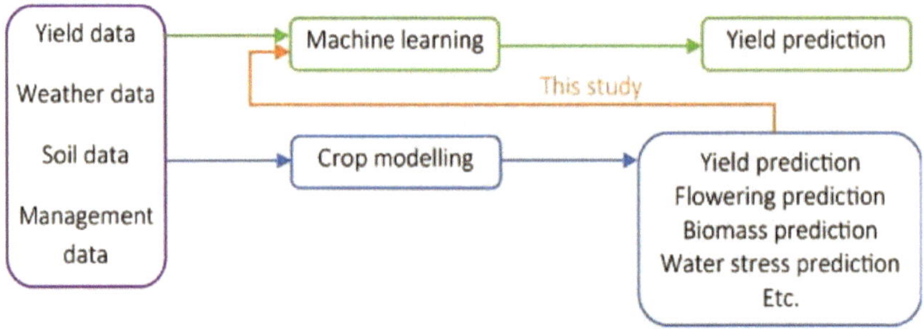

FIGURE 7.2 Various governing actors involved in the fruit yield prediction.

7.3 ALGORITHMS OF MACHINE LEARNING (ML) THAT FORECAST FRUIT HARVEST

To forecast fruit yields, a variety of regression- and classification-based prediction techniques are being used. DNNs, convolution neural networks (CNN), ANNs, random forests (Cherfi et al.), extremely randomized trees (ERT), support vector machines (SVM), KNN, and SVR, as well as linear regression (LR) and multiple linear regression (MLR), multivariate adaptive regression splines (MARS), DNN, and convolution neural networks (CNN), have all been used (Alibabaei et al., 2021; Darwin et al., 2021; Lombardo et al., 2021; Sabry et al., 2022; Singaravel et al., 2017).

In the **ML** one or more independent and dependent variables may be correlated using the LR model. Relationships can be found in **ML** classification algorithms by using the data and reducing the error (RMSE or MSE) associated with regression techniques. Numerous MLR analysis applications have shown that using numerous independent factors is more accurate and efficient than using a single independent factor (Khorramifar et al., 2022; Sabzi-Nojadeh et al., 2021). In MLR models, the estimation of regression is usually done using the least-squares method (LSM). A MARS model may be assessed using the generalized cross-validation (GCV) index, which typically uses MLR techniques to account for variable nonlinearity and the correlations methodology. Additionally, the learning input has been split up into a variety of customizable splines with various slopes. The K-NN approach is frequently used for both classification and regression, and it gives near neighbors more weight when making predictions than distant neighbors since they are closer to the average (Ali et al., 2019). Different distance formulae, such as the Euclidean, Manhattan, and Minkowski formulas, can be used to determine the distance between two neighbors.

An SVM, or multimodal classification, creates a collection of features that are linearly differentiating for classifying the input. SVM is different from SVR, which may be applied to resolve regression-related concerns. Decision tree (DT) is a supervised type of classification model which may be adapted to both regression and classification (Charbuty & Abdulazeez, 2021). It contains three nodes in total: the root node, which lacks any incoming edges, the decision node, which has two edges—incoming and outgoing—and the leaf node, which has both incoming and outgoing edges (but has not any outgoing edges). In the case of a decision tree, each outgoing node's attribute is split into two or more branches based on the splitting criterion. Breiman et al.'s classifications and regression trees

(Koirala et al.), a nonparametric method that constructs binary trees from discrete and continuous data features, is one of the most efficient methods for DT induction (Cherfi et al., 2018; Ozsagir et al., 2022). The information gain, Gini Diversity Index (GDI), and gain ratio are also used to differentiate the attributes in CART. Random forest (Cherfi et al.) is a highly useful tool in agricultural research for predicting production (Ghosh & Maiti, 2021). Due to the extensive usage of **DT** in the regression computation, it offers a varied variety of regression trees. The bias generated by the single decision tree is offset by the randomization, and the RF beats all other **DT** in terms of precision. ERT (extra tree) is a technique that uses unpruned **DT**, although it is more like RF in that it is an ensemble model (Abbas et al., 2021; Chen et al., 2022a). Actually, it develops the trees using the entire learning sample and separates the nodes into randomly chosen cut-points (without bootstrap copying). Similar to the RF, the number of variables and the trees used to partition nodes are present. Given that it enables the modeling of the intricate nonlinear relationship between input and output, an ANN technique is really the methodology of **ML** that is employed the most frequently for predicting agricultural production (Yin et al., 2021). An input layer, a hidden layer, and an output layer are the three layers that make it up. The number of hidden layer nodes, learning rate, and training tolerance are only a few of the variables that affect how effectively ANNs work. To put the value predicted within the range of acceptable compared to the value observed, the learning rate regulates weight change over a number of repetitions.

7.3.1 DEEP LEARNING (DL) ALGORITHMS

In contrast to the globally optimal process, the classical ANN is a problem of local minima, where the optimization process often ends in a locally optimized state. Additionally, the overfitting problem in the generic **ML** models shows the complexity process. DNNs may address local minima and overfitting issues through a thorough optimization procedure inside a deep network structure (Pan et al., 2022; Sun & Liang, 2022). To improve performance, the back-propagation process also goes through forward and backward tweaking. The adoption of efficient activation functions, such as the sigmoid and rectified linear units, would solve the issue of depleting the gradients of loss functions during the back-propagation process (ReLU). A series of stacked fundamental units are present between the input and output layers of CNN (convolutional, pooling, and activation layers). While a pooling layer may utilize multiple procedures to extract low-dimensional data from the input data, a convolution layer employs a number of local filters to perform the convolution on data provided by the input (max-pooling and average pooling (Pandey, 2019). The performance of CNN's nonlinear fitting can be improved by using the nonlinear operations in the activation layer. CNN updates weight through BP, much like BPNN. Since an LSTM network has forgetting, input, and output gates that control the filtering of the previous state, it is a helpful tool for classifying and forecasting sequence and time series data. Instead of aiming for the most recent statuses, this structure prefers to obtain more significant historical ones. Rashid et al. go into great detail on the network structure and underlying methods of the LSTM models.

7.3.2 MULTIPLE ALGORITHMS

In addition to these, a range of algorithms for classification and regression, including XGBoost, MARS, elastic net, gradient boosting, TOMGRO, and Vanthoor, ridge regression (RIDGE), SNN, DRL, and others, have been effectively employed for the prediction of crop production. Agricultural yield predictions have also been made using hybrid approaches like CNN-RNN, CNN-LSTM, and MLR-ANN.

7.4 GROWTH MONITORING IN HORTICULTURAL CROPS USING MACHINE LEARNING (ML) TECHNIQUES

Fruit farmers and scientists use fruit attributes to control fruit production and choose genotypes based on qualities that will increase output or improve quality (de Jesus et al., 2022; Mirheidari et al., 2022; Rahimi & Abdolinasab, 2022). Fruit scientists may employ autonomous intelligent collecting of data on horticulture crop growth to track yield and plan the timing of fruit and vegetable harvests (Felföldi et al., 2022; Pereira Miranda et al., 2022). Lu et al. (2019) came up with strategies for finding and observing mushrooms. In order to find the mushrooms in the photos, Lu et al. employed the YOLO technique and recommended a placement correction method. He developed a method for measuring images in order to measure and record the mushroom caps' diameter during the course of their growth. To calculate mushroom circles from constantly captured images and thereafter collect data on mushroom cap growth, the recommended method (YOLO-V3 + SP) was used; it performed better than the current circle. Orchard management thus necessitates automated detection and recognition of fruits and flowers at various developmental phases. A MASU R-CNN instance segmentation model, for instance, was created by (Tian et al., 2020) to recognize and separate apple blossoms at their three distinct stages—bud, semiopen, and fully open. He has suggested a better YOLO-V3 model to identify immature, emerging, and ripe apples in farms with complicated backdrops. The suggested YOLO-V3-dense model outperformed both the conventional YOLO-V3 and the Faster R-CNN in terms of detection. An automated method for tracking apple development in orchards was created by (Wang et al., 2020) to monitor fruit ripening and fruit thinning. They used the FCF model to divide up the images of apples. The method produced a horizontal diameter of an apple with an absolute error of 0.90 mm. Tu et al. (2018) used outdoor natural RGB-D pictures with the DSIFT (dense scale invariant features transform) approach and the LLC (locality-constrained linear coding) technique to detect and grade the quality of passion fruits. The fruit maturity at the following four stages was then assessed using the SVM classifier: young (Y), near-young (NY), near-mature (NM), and aftermature (M) (AM). The proposed technique successfully achieves 92.7% recognition and 91.5% maturation accuracy. In order to identify and categorize blueberry fruit and track its development, Ni et al. (2020) used an iterative annotation method and a Mask R-CNN model. The proposed model performed reasonably well, with an R2 of 88.6% for the determined berry number and an RMSE of 1.484. The diameter of the panicle-associated picture region and the number of panicles are significant indicators of blooming. To recognize the evolution of green to light yellow, light pink, and finally brown-red flower panicles on mango trees at various phases of maturity, Wang et al. (2020) developed a machine vision assessment system. In a distinct investigation (Koirala et al., 2019) suggested four deep-learning architectures for categorizing the phases of mango panicles at three distinct stages: not entirely opened (white) flowers, opened flowers, and flower drop and fruit set. In terms of the total panicle count, the YOLO-V3 rotational model outperformed the R2 CNN upright model. Bauer et al. (2019) created the AirSurfLettuce platform, which uses a CNN-based learning model to extract growth and quantity information from huge aerial photographs captured in a lettuce field. With a focus on yield-related characteristics including field size distribution, the AirSurf Lettuce can independently assess in-field iceberg lettuces to support knowledgeable and accurate crop monitoring. In horticulture crop growth monitoring, deep-learning algorithms have enhanced prediction and classification impacts, according to the earlier research.

7.5 RECOGNITION AND CLASSIFICATION OF HORTICULTURAL CROPS USING MACHINE LEARNING (ML) TECHNIQUES

Automatic crop identification and classification are essential domain in horticultural science. The diversity of horticultural crops makes accurate labeling challenging. There are at approximately 91 types of fruiting plants, 200 types of vegetable crops, and over 6,000 types of ornamental plants, with new varieties sprouting every year (Gebreeyessus, 2022; Pandey et al., 2019). The same is true for horticultural crops; they can undergo mutations that lead to intraclass variation. For classifying horticultural plants, a large expertise and skill are needed. Although it is not feasible to manually classify big datasets, doing so is laborious and error-prone. **ML** methods are going to be a huge step forward in horticultural study because of how quickly and accurately they can identify objects (Hasan et al., 2021; Oppenheim et al., 2019; Yang & Xu, 2021). CNNs have improved the precision of image classification and event forecasting in horticultural crops. Using a two-stage deep learning (DL) model (Hiary et al., 2018) achieved excellent results in identifying different kinds of flowers. The robust classifier consisted of two phases: first, a fully convolutional network (FCN) technique was used to segment the floral regions, and then, a CNN technique was utilized to classify the various types of flowers. Researchers have built horticultural image databases like ImageNet and OUFD (Yang & Xu, 2021). Large datasets are utilized to train DNN models for crop identification. Due to the lack of horticultural crop identification algorithms, most researchers collect their own photos. DCNNs have been used to classify seven olive varieties (Arbequina, Changlot Real, Arbosana, Picual, Lechn de Sevilla, Ocal, and Touriga Franca) and three plum varieties (Angelino, BlackSplendor, and Owent) (Salvucci et al., 2022). The proposed approach accurately classified 91%–97% of plum varieties. Inception-ResNet-V2 identified olive varietals from 2,800 fruits with 95.91% accuracy, whereas AlexNet got just 77.30%. Fernandes et al. (2019) discriminated TFvar and TN using spectral and RGB pictures (TNvar). Around 35,833 spectra from 626 plant species and 64 species were utilized to build the categorization model. The SVM model outperformed the CNN, correctly recognizing 81.9% of TNvar spectra and 63.3% of non-TNvar spectra. CNN's accuracy rose to 93.8% from 91.6% when tested using TFvar data.

DL is used for recognizing fruit kinds in orchards. Developing a dependable fruit identification system is the first step in building a future auto harvester. Numerous academics have tested ways to deal with real-world problems including illumination, leaf shadows, and diverse persons. Kang & Chen (2020) suggested LedNet, ResNet (a lightweight backbone), and DaSNet to recognize and segment apple fruits and branches (a network that used a GFPN combined with ASPA to improve detection performance). LedNet trained with ResNet-101 segmented branches and apples with 77.2% accuracy, 0.849% F1 score, and 84.16% recall. Computer vision's semantic segmentation breaks a picture into pixels. Label every pixel that corresponds to an image element. If fruits overlap or are too dark to view clearly, these traits can be used to deduce their placements. Next, they offered DaSNet-V247. DaSNet-V2 with ResNet-101 achieves 86.8% recall, 88.8% precision, 79.4% branch and 87.3% apple segmentation accuracy. DaSNet-V2 quickly detects apple orchards. Gené-Mola et al. (2019) researched two-dimensional apple identification using Mask R-CNN and three-dimensional placement with SfM photogrammetry. The system performed well with 11 typical Fuji apple trees and 1,455 fruit, scoring 0.816 for 2D detection and 0.881 for 3D detection and placement. Scientists have sought to generalize techniques for recognizing and classifying fruits and vegetables due to their diverse forms and colors. Mao et al. (2020) presented a multipath convolutional neural network (CNN) using SVMs and color component selection (MPCNN). The accurate categorized rate for cucumber photos was above 90%, whereas the false classified rate was as follows. Around 88% recall, 86% accuracy, and 0.86 F1 Mask R-CNN helped a strawberry-picking robot recognize strawberries more accurately. A collection of 100 strawberry pictures earned 95.41% recall, 95.78% accuracy, and 88.55% intersection over union (MIoU). Giefer et al. (2019) constructed SSD model Deep-6DPose (an end-to-end deep learning system) for automated fruit grading and packing. Le & Lin (2019) used DL to categorize grouped bananas without physical sampling. When trained with just one side of a banana picture, the proposed deep-learning model scored 96.5% classification accuracy using Mask R-CNN. Horticultural crops are grown differently than other crops. Regular pruning and fruit thinning maintain healthy trees and high-quality harvests. To automate horticulture cultivation, plants must be split. Majeed et al. (2020) automated apple orchard tree cutting using a CNN-based SegNet.

Many researchers have spent a lot of time researching into the challenge of detecting and categorizing horticulture items, although the current popular DNNs require high-performance computer equipment for reasoning, making them unsuitable for most farms and orchards. Shi et al. (2022) suggested a generalized attribution approach for trimming unneeded connections to efficiently determine mangos in a well-designed large-scale network. The suggested pruning strategy has the ability to fine-tune the network while keeping critical pretrained parameters, and it may also be used to compute attributions for convolutional kernels. With just a 2.4% loss in accuracy, the suggested network lowered computational cost by 83.4% compared to the YOLO-based network without pruning. DL has been applied in horticulture studies to identify sites, plant types, and organs (Gracelin & Raimond, 2022). DL has yielded noteworthy benefits in manufacturing, especially in speed and performance. As people's lives improve, so do their demands for fresh fruit. After harvest, problems may include disease transmission, insect infestations, mechanical damage, and inappropriate storage. Horticultural items that are damaged might lose marketability and quality. Therefore, quality assurance for horticulture items, especially fresh produce like fruits and vegetables, is a hot topic at present. Recent research have employed DL using RGB pictures or spectrographic approaches to predict post-harvest grade, maturity, bruises, and nutritional content in horticultural goods (Eissa et al., 2022; Fracarolli et al., 2021; He et al., 2022b; Kaur et al., 2021; Sanchez et al., 2020). DL outperforms classic **ML** algorithms in image, spectral, and audio classification accuracy. Nasiri et al. (2019) used VGG-16-trained RGB photos to identify bad dates. New CNN model improves classification to 96.98%. Similarly, Ucat & Cruz (2019) applied DL and image processing to categorize bananas with an accuracy of above 90%. Dates and bananas are analyzed using RGB image data, which can identify obvious surface flaws. The quality of fruits and vegetables, as well as any internal or subdermal damage, are not easily evident to the human eye. Using near-infrared spectrographs, hyperspectral imagers, and thermal imagers, the inside status of some things may be detected without damaging them. It's tough to appropriately evaluate vast volumes of hyperspectral picture data. A lot of difficult data has been processed using DL, a prominent machine-learning approach. Using DL and hyperspectral transmittance data, Wang et al. (2021) explored internal and hidden mechanical damage in blueberries. CNN uses ResNet. LR, sequential minimum optimization, bagging and multilayer perceptron (BMP), and RF were also compared to serve as examples of more standard **ML** approaches. The average accuracy of the revised ResNet is 88%, and it has an F1-score of 0.90. DL can identify grape damage. Using hyperspectral transmittance photos, Zheng et al. (2020) constructed a deep-learning-based FCN model to identify internal blueberry bruises (HSTIs). Predictions of unbruised and wounded fruit were more accurate using the FCN approach than the SVM classifier. The novel whole-wavelength method with random initialization obtained 81.2% accuracy on the whole test set of blueberry data. Feng et al. (2019) employed hyperspectral imaging to detect winter jujube deterioration. Researchers employed LR, SVM, and CNN models. CNN claims the fastest prediction speed and detection accuracies above 85% (Table 7.1).

TABLE 7.1 Application of Deep Learning (DL) for the Detection of Stress in Fruit Crops

SL. No.	Disease Name	Crop	Sensor	Model Used	Performance Metric and Value
1.	Anthracnose disease	Mango	RGB images	MCNN	CA: 97.13% PSO: 88.39%
2.	Three diseases (healthy, black sigatoka and black speckle).	Banana	RGB images	LeNet	CA: 99.72% Precision: 99.70% Recall: 99.72% F1: 0.9971
3.	Six diseases (healthy apple, general apple scab, serious apple scab, apple gray spot, general cedar apple rust and serious cedar apple rust)	Apple	RGB images	Based on DenseNet-121	CA: 93.31% F1: 0.9339
4.	Six diseases (anthracnose, canker, lepra fruit rot, peacock spot, Parlatoriaoleae, *Aspidiotus nerii*)	Olive	RGB images	The modified AlexNet	CA: 99.11% Precision: 99.49% Recall: 99.11% F1: 0.9929
5.	The fungal diseased peaches in the slight, moderate or severe level.	Peach	Hyperspectral images	DBN based on the combined features.	CA: 93.3%, 100% and 100% for slightly, moderately and severely decayed samples, respectively

7.6 YIELD PREDICTION OF VARIOUS CROPS USING MACHINE LEARNING (ML)

7.6.1 APPLE YIELD PREDICTION USING ML

For better crop management and marketing of apples and other fruits, it is essential to anticipate yields from fruit trees. Site-specific yield estimates also allow variable-rate input (such as water, fertilizer, and herbicide)

and activity application in the field (e.g., fruit thinning). As a result of the inability of most crop models to foresee yield and geographical variation, supplementary features were used to make predictions during the growing season. For optimal orchard management, a prediction should be made as soon as possible. Apple crop estimation with multispectral fruit images (Safren et al., 2007). Yield variability was estimated using digital photos of flowering trees by (Aggelopoulou et al., 2011). Precision agriculture predicts yields using crop modeling, algorithms, and statistical methods. Without polynomial and interaction factors, correlation and MLR results are unconvincing. In 2003, MLR may have trouble expressing nonlinear connections between crop attributes and site factors, leading to incorrect findings (Ayoubi & Sahrawat, 2011; Feng et al., 2018). Multiple regressions and PCA are more options. These strategies decreased data dimensionality, analyzed complicated connections, and identified key variables. Later attempts at precision agriculture included AI and softer computing for geographical analysis and crop management (Ampatzidis et al., 2020; Raj et al., 2021). Also, NHL FCMs produce more accurate forecasts than traditional FCMs and **ML**. Fuzzy Cognitive Maps were utilized to forecast apple production categories for precision farming (Jayashree et al., 2015; Papageorgiou et al., 2013). This work aimed to construct an AI system based on FCM learning to predict apple fruit production. The results reveal that FCM learning predicts phenomena, gives a forward-looking option on apple fruit development, and is congruent with horticulturists' opinions. This model represents how horticulturists think and operate. The proposed model can be completely understood, and its parameters are all important.

7.6.2 YIELD PREDICTING OF ALMOND

LR and **ML** are only two examples of the various statistical approaches that have been developed for almond yield prediction. Due to their reliance on a single regression model, these approaches might overfit if the available training data is insufficient. Integrating many separate learning models into a single prediction model is becoming increasingly popular in the **ML** field as a means of improving model generalization (Mosavi et al., 2018). By combining many learners, flaws in one model may be offset by others, resulting in more robust and precise predictions. The optimal hypothesis may be outside the domain of any given model under some conditions. The search space can be increased and the fit to the data enhanced by combining multiple models.

Bagging and boosting are common practices here. Using bootstrap sampling, bagging collects data subsets for use in training and developing initialized learners. Zhang et al. (2019) predicted almond output using a representative bagging model called Random Forest (Cherfi et al.). Boosting generates base models in a sequential fashion, capitalizing on learner dependencies at the base level. As a representative boosting approach, SGB uses the decision tree as its primary learner. In contrast to RF's output of many distinct trees, SGB constructs the next tree by incorporating corrections made to earlier trees.

An almond orchard's output is very variable, both between years and between individual trees, as a result of a myriad of circumstances. Both statistical "machine learning models" and process-based "crop models" are used to make predictions about crop yields. Daily simulations of the crop's evolution in response to changes in the nitrogen cycle, water availability, and energy expenditure allow the "crop model" to provide accurate predictions of harvest success. Crop models are estimated based on plant physiology features. Some well-known crop models are APSIM and SALUS. For these mechanistic models to work, though, researchers require a lot of information on things like cultivars, management styles, and soil types. There are a lot of unknown input components, a complex procedure, and a paucity of field data across a wide range of environmental gradients, all of which make calibration challenging (Chong et al., 2021). However, **ML** models attempt to build prediction algorithms by utilizing data from a wide variety of sources. It is possible to forecast yield without considering specific crop characteristics, as the approach is independent of established physiological systems that regulate plant growth (Chang et al., 2021; Hara et al., 2021). Using process models and statistical **ML**, scientists have investigated how weather impacts almond yields (Hara et al., 2021; Kang et al., 2020).

7.6.3 GRAPE YIELD FORECASTING

About 7 million hectares of grapevines are cultivated between 50°N and 43°S on every continent except Antarctica, making them the most significant perennial crop. Extremely erratic weather has affected winemakers in several regions throughout the world. In spite of progress in vineyard technology, climate change scenarios may make this problem much more severe. Grape yields are also affected by the farmer's decisions on irrigation, fertilization, and grape variety selection. The wine industry is quite concerned about the inherent instability of its annual production. Therefore, improved grapevine yield models are in great demand to aid commercial endeavors and boost the productivity of vineyards and wineries. There is still a need for practical methodologies despite the economic importance of grapevines and the international work done in this area of research, particularly to predict grape output at farm level in tiny patches with specific grape varieties, agronomic practices, and ecological circumstances. Crop simulation and data-driven models provide the basis of the most popular approaches of estimating future crop yields. Software for simulating crops uses an abstract mathematical model to represent the dynamic mechanics of the physiological stages of a plant (Park et al., 2003; Prusinkiewicz, 2004; Yan et al., 2004). Prior agroclimatic conditions and the effects of nutrients and carbon (mediated by the grapevine permanent structure) on crop yield are major factors in these process-based models. This makes it difficult and expensive to collect the biophysical data necessary for crop simulation models (CSMs), which in turn hinders their use and portability (Kraft et al., 2021). On the other hand, data-driven yield prediction models are developed empirically without a deep familiarity with the biophysical systems that provided the raw data. These techniques may be used with little to no expense and do not require a predetermined model framework to predict crop yields. Using regression analysis and time-frequency methods, data-driven models have become increasingly popular in recent years for estimating agricultural yields (Kumar et al., 2021). A **ML** approach to forecasting agricultural yields with a focus on both nonparametric and semiparametric models, with validation based on the precision with which forecasts are made.

Consequently, in recent years, ML techniques such as regression trees, random forests, ANNs, and SVMs have been employed to foretell agricultural output (Rodriguez-Galiano et al., 2015). For the most part, these models are used for annual crops, however, there is little work in applying ML to estimate grape yields. Previous studies on estimating grape yields were examined. By way of illustration, Fraga & Santos (2017) analyzed daily data on the three vineyards' minimum and maximum air temperatures and precipitation totals. It appears that the optimal conditions for growing berries are a cool start to the blooming process and a warm end. High yield requires above-average rainfall before to blooming. We used the phenological data of bunches per vine and pollen per anther (Fernández-González et al., 2011) and the phytopathological variables of *B. cinerea* and *P. viticola* for the same Ribeiro of origin vineyard. The purpose of (Nendel, 2010) predictive bud-break simulation was to determine the optimum temperature and the earliest date to begin accumulating heat units for northern European vines. Variability in grape yield was forecasted using environmental factors including temperature, sunshine, and carbon dioxide (CO_2) concentration (Bindi et al., 1996). Visual inputs (Aquino et al., 2018) are used to predict yield in various varieties of grapevine by counting the number of identifiable berries in photographs taken between the berry-set and cluster-closure phenostages.

7.7 DIFFICULTIES AND FUTURE PATHS FOR THE FRUIT PREDICTION

7.7.1 DIFFERENT FEATURES ARE NOT STANDARDIZED

Many studies rely on satellite imagery and vegetation indices. In addition to these, information on farmland, soil, irrigation, and crop management is also used to make predictions about harvest yields. The most widely used and successful feature group for crop yield prediction may be easily identified; however, the most efficient sub-feature list within each feature group is less easily identified. Precipitation, vapor pressure deficit, humidity, solar radiation, temperature, rainfall, and vapor pressure are only few of the meteorological variables that have been heavily employed by scientists (Oogathoo et al., 2020). Previous studies do not clearly indicate the ideal set of climatic sub-features for any crop. Thus, further investigation is required to establish the best combination of climate variables for each crop. As an additional step, research on the best climate attributes to have at each site. As climate change affects agricultural productivity, it is important to record the yearly fluctuations in preexisting components including weather, temperature, chemistry, and physics. Although environmental data is not as crucial in predicting crop yields, the genotype is represented by a wide range of traits. The most important characteristics must be determined, and the less important ones eliminated, since else the prediction models might be damaged. It is possible that satellite-based attributes can be utilized to make estimates of agricultural yields. Certain study discovered that compared to utilizing coarse-resolution satellite data, employing high-resolution features improved the accuracy of yield forecasts (Peng et al., 2020). Satellite data may be used in a number of ways to estimate harvests. Examples include the potential for enhanced crop production prediction as a result of a stronger correlation between satellite data converged into a whole canopy and total canopy photosynthetic emission.

Second, it's possible that better crop production prediction results might be achieved with more training data. And last, future satellite missions like Fluorescence Explorer may increase the accuracy of yield forecasts by employing satellite data products with larger geographic resolutions than existing ones (FLEX) (Ustin & Middleton, 2021). Earlier methods mostly concerned themselves with increasing the scale of low-resolution end results. The fPAR allowed the most recently identified SIF to get a somewhat better score than EVI. Fluorescence output was also regulated by the amount of moisture in the root zone soil, providing additional insight into the effects of drought and heat stress. Research has shown that LST may be a useful predictor of agricultural water stress, and our study confirms that LST-based KDD and GDD are more susceptible to plant water issues than Tair across all AEZs. More accurate yield predictions might be made by combining satellite data and static environmental factors than by using only static weather data (i.e., climate information, soil information, and irrigation properties). There is a possibility that the model's precision might be enhanced by using peak or late climatic information. Despite evidence from satellites demonstrating the usefulness of crop biomass tracking, yield weights eventually affect precise measurements of individual grains at the time of filling. Multiple agronomy studies have demonstrated the negative effects of drought and heat on crop yields. The findings of this study shed light on why the incorporation of meteorological indicators like silking or maturation stage had such a significant effect on crop yield forecasting. Predicting regional agricultural production requires other factors, such as soil quality, to complement satellite data. It was interesting to see what they discovered about how environmental stress affects crop growth. Thus, it is important to include satellite data from several spectral bands and a wide range of climatic parameters into predictions of agricultural output on a global scale (Cai et al., 2019; Forkuor et al., 2017; Jeyaseelan, 2003). It's possible that the graphic doesn't accurately depict the nonlinearity of the connection between crop production and its environs. By integrating multi/hyperspectral data with temporal image data, soil, and environmental factors, the function matrix may be effectively constrained. There is typically a preponderance of VIs that are visible or near-infrared in nature. Potentially more insight about plant development can be gleaned from other satellite data. Abiotic elements, such as weather, soil, cropland, irrigation, and crop management, can also have an effect on crop yield but are not caught by satellite data. To better predict agricultural output using multi-band satellite data and weather conditions, more study is required.

7.8 DISADVANTAGES AND BENEFITS OF MACHINE LEARNING (ML)

Autonomous feature extraction, classification, and prediction are areas where **ML** systems really shine. When features can be extracted automatically, less time is spent manually extracting them and writing feature descriptions. As the number of model layers increases, so does the accuracy of machine-learning models' predictions. It's not easy to make a good feature extractor, either. These studies relied on preexisting models or made only minor modifications to them, and their primary authors were mostly computer scientists and image processing researchers. Because of this, more time and effort was spent on strategies for general image categorization and feature extraction than on developing methods specifically tailored to horticultural products. For instance, in fruit yield estimation applications, manually isolating elements that affect crop development was nearly impossible (Campobenedetto et al., 2021; Navas et al., 2021; O'Sullivan et al., 2021; Riera et al., 2021; Wu et al., 2021). **ML** models should be developed with a focus on feature extraction activities in order to be efficiently applied to the horticultural sciences (Liu & Wang, 2021; Sun et al., 2019). Thankfully, transfer learning is a viable option within the realm of **ML**. Using this strategy, scientists may apply pre-existing models prepared with massive quantities of data to similar problems. In order to fine-tune the trained model, they must adjust certain layers and use only the goal data (the data they will learn from). The effectiveness and performance of subsequent modeling assignments are enhanced by fine-tuning. Big datasets (like ImageNet) served as the basis for experiments like VGGNet and AlexNet, which were subsequently fine-tuned to achieve optimal performance. Since the model architecture and computational requirements of **ML** are higher, only highly trained computer scientists should attempt to create or apply it. To help non-expert researchers quickly and easily build machine-learning technologies, computer scientists and machine-learning enthusiasts have developed a number of software and hardware solutions (Wu et al., 2022). In order to facilitate network construction by researchers without specialist knowledge, software has adopted a number of standard models and adjustments. AlexNet, CNN-SSAE, DaSNet, LedNet, and VGGNet are just a few examples. As an alternative to these pre-built models, horticulture researchers can find resources online for creating their own machine-learning models. The recommended network frameworks and hardware acceleration tools can be used together to dramatically reduce computation time and successfully carry out prediction and/or classification in real-time data processing applications. DL isn't perfect, though. First, optimization tasks are hard and time intensive because of the massive datasets and the high number of weights to modify. Discussions, conclusions, and a look ahead.

7.9 FUTURE PERSPECTIVES OF DEEP LEARNING (DL) IN THE HORTICULTURE DOMAIN

Many issues (such as those concerning productivity, environmental changes, food security, and sustainability) in the agriculture-based economy might be addressed by employing machine-learning techniques and applications. The deep-learning methods are used for species classification, quality detection, yield predictions, and pest management. For instance, deep-learning technology provides accurate detection of fruits, flowers, and vegetables, and may generate yield maps that provide real-time information on geographical variance, enabling efficient and precise harvesting practices that enhance the marketable output. This technology has the potential to lessen the reliance on the unreliable and inaccurate human experience now used by farmers and gardeners. Detection of stress in plants is another interesting example. The use of **ML** in conjunction with digital images or spectral data has shown great potential for improving the early identification, categorization, and quantification of plant diseases and stressors. More information on horticulture crops may be gathered with the use of inexpensive digital imaging equipment, IoT capabilities, computing, and data storage capabilities, all of which can be used to train deep-learning models and address specific horticultural issues. Despite the fact that DL often beats competing models in trials, picking the right model for any particular task can be challenging. Setting up the necessary gear and software to implement a machine-learning model is just as challenging. Since it is more of an art than a science, **ML** is. The detection accuracy may suffer if a tried and true model is used widely to typical farms and orchards. With DL becoming the standard in research for image object recognition, a CNN was utilized to identify plants and calculate yield. LeNet, AlexNet, VGGNet43,91, GoogLeNet, and ResNet are only few of the popular CNN-based object identification algorithms. Included in this category of methods are R-CNN, Faster R-CNN, and Mask R-CNN. End-to-end, single-stage deep-learning object recognition methods include the YOLO and SSD techniques. These models are able to recognize early signs of plant stress, perform flower and fruitlet thinning, and make accurate predictions about fruit output. When it comes to horticultural plants, visual inspection is less useful for spotting quality indications or early bruises. Authors have utilized SAE-FNN27, CNN-SSAE86, and DBN100, among others, to extract features from hyperspectral pictures in order to determine fruit quality (SSC and hardness) and plant stress (such as early bruising detection and disease identification). Deep-learning algorithms may be helpful in this context since Vis/NIR hyperspectral imaging can detect plant stress/internal bruising of fruit before the onset of symptoms. Integrating deep-learning technology with these rapid nondestructive testing methods is necessary for feature extraction for immediate flaw diagnosis. RNN or LSTM models coupled with high-performance regression or classifiers are promising options for future horticulture study. The purpose of this analysis was to apply LSTM to foresee banana harvest output for the sake of improved decision making and long-term planning. Effective uses of DL in plant science include image recognition and quality assessment of horticulture products, early detection systems for plant stress, and yield prediction. Several authors have developed fruit, branch, and tree trunk identification algorithms for mobile devices, after first training the model. There are a number of obstacles that need to be overcome before DL can be used on mobile devices effectively. These include things like embedding a model into a mobile device and miniaturizing sensing equipment. Therefore, more research is necessary. Intelligent horticulture for demanding gardening activities is now feasible because to advancements in DL technology. These detailed phenotypic measurements provide accurate tracking of key stages in crop growth and factors that affect output. Gradually, DL has advanced via the use of picture data, and these models have been put to good use in the field of agriculture. However, as was previously said, it is difficult to choose the most applicable model among the several techniques now available for a given application in horticultural research. Large numbers of plant photos are desperately needed to develop reliable models. When creating training datasets for plant phenotypic data, we strongly recommend utilizing in-field real-condition imaging data (varying shadow, light, and mutual occlusion circumstances). Locating and making use of suitable open-source datasets may help decrease the time spent on data collection. We are confident that deep-learning technology has a great deal of potential to serve the horticulture industry in a more efficient and correct manner, resulting in increased output and improved quality of the products, and improved ability to recognize plant illnesses.

7.10 CONCLUSION

The horticultural industry stands to benefit greatly from ML and DL-based artificial intelligence (AI) technology. Productivity and food quality might both benefit from these technological advancements. Horticultural crop issues and illnesses are investigated using image-based, model-specific approaches that yield very reliable findings. By empowering farmers to make informed choices at crucial junctures, it helps alleviate shortages of materials, labor, and produce. The vast potential uses in agriculture and horticulture have recently piqued the interest of scientists and researchers. Thanks to ML and DL, intelligent farming and gardening have been furthered at a rapid pace. Fruits might be graded with the use of **ML** techniques like ANNs, SVMs, and convolution neural networks (CNNs) based on 3D photographs of the fruits. Because of this, **ML** is a reliable technique for identifying and labeling various types of fruit. Improving and modifying categorization in **ML** is crucial to achieving process automation. To keep up with the ever-increasing need for food from an expanding population, the agricultural industry must use AI technology. To protect farmers and the economy as a whole from the severe effects of agricultural diseases, these technologies must be implemented at the field level. The use of AI in agriculture and horticulture has helped farmers automate their methods and make the transition to more accurate cultivation procedures, which have increased yield, productivity, and product quality while making better use of existing resources.

KEYWORDS

- **deep learning algorithms**
- **fruit prediction**
- **grape yield forecasting**
- **horticultural crops**
- **horticulture domain**
- **machine learning**
- **yield prediction**

REFERENCES

Abbas, S., Jalil, Z., Javed, A. R., Batool, I., Khan, M. Z., Noorwali, A., Gadekallu, T. R., & Akbar, A., (2021). BCD-WERT: A novel approach for breast cancer detection using whale optimization based efficient features and extremely randomized tree algorithm. *PeerJ Computer Science, 7*, e390.

Abdelkader, A., & Elshorbagy, A., (2021). ACPAR: A framework for linking national water and food security management with global conditions. *Advances in Water Resources, 147*, 103809.

Aggelopoulou, A., Bochtis, D., Fountas, S., Swain, K. C., Gemtos, T., & Nanos, G., (2011). Yield prediction in apple orchards based on image processing. *Precision Agriculture, 12*, 448–456.

Ali, N., Neagu, D., & Trundle, P., (2019). Evaluation of k-nearest neighbor classifier performance for heterogeneous data sets. *SN Applied Sciences, 1*, 1–15.

Alibabaei, K., Gaspar, P. D., & Lima, T. M., (2021). Crop yield estimation using deep learning based on climate big data and irrigation scheduling. *Energies, 14*, 3004.

Ampatzidis, Y., Partel, V., & Costa, L., (2020). Agroview: Cloud-based application to process, analyze and visualize UAV-collected data for precision agriculture applications utilizing artificial intelligence. *Computers and Electronics in Agriculture, 174*, 105457.

Aquino, A., Millan, B., Diago, M. P., & Tardaguila, J., (2018). Automated early yield prediction in vineyards from on-the-go image acquisition. *Computers and Electronics in Agriculture, 144*, 26–36.

Ayoubi, S., & Sahrawat, K. L., (2011). Comparing multivariate regression and artificial neural network to predict barley production from soil characteristics in northern Iran. *Archives of Agronomy and Soil Science, 57,* 549–565.

Bauer, A., Bostrom, A. G., Ball, J., Applegate, C., Cheng, T., Laycock, S., Rojas, S. M., et al., (2019). Combining computer vision and deep learning to enable ultra-scale aerial phenotyping and precision agriculture: A case study of lettuce production. *Horticulture Research, 6.*

Bindi, M., Fibbi, L., Gozzini, B., Orlandini, S., & Miglietta, F., (1996). Modeling the impact of future climate scenarios on yield and yield variability of grapevine. *Climate Research, 7*, 213–224.

Cai, Y., Guan, K., Lobell, D., Potgieter, A. B., Wang, S., Peng, J., Xu, T., et al., (2019). Integrating satellite and climate data to predict wheat yield in Australia using machine learning approaches. *Agricultural and Forest Meteorology, 274*, 144–159.

Campobenedetto, C., Mannino, G., Beekwilder, J., Contartese, V., Karlova, R., & Bertea, C. M., (2021). The application of a biostimulant based on tannins affects root architecture and improves tolerance to salinity in tomato plants. *Scientific Reports, 11*, 1–15.

Çetin, N., Karaman, K., Kavuncuoğlu, E., Yıldırım, B., & Jahanbakhshi, A., (2022). Using hyperspectral imaging technology and machine learning algorithms for assessing internal quality parameters of apple fruits. *Chemometrics and Intelligent Laboratory Systems*, 104650.

Chang, C. L., Chung, S. C., Fu, W. L., & Huang, C. C., (2021). Artificial intelligence approaches to predict growth, harvest day, and quality of lettuce (*Lactuca sativa* L.) in a IoT-enabled greenhouse system. *Biosystems Engineering, 212*, 77–105.

Charbuty, B., & Abdulazeez, A., (2021). Classification based on decision tree algorithm for machine learning. *Journal of Applied Science and Technology Trends, 2*, 20–28.

Chen, L., Huang, H., Thangavelu, L., Abdelbasset, W. K., Bokov, D. O., Algarni, M., Ghazali, S., & Alashwal, M., (2022a). Optimization and comparison of machine learning methods in estimation of carbon dioxide loading in chemical solvents for environmental applications. *Journal of Molecular Liquids, 349*, 118513.

Chen, W. H., Lo, H. J., Aniza, R., Lin, B. J., Park, Y. K., Kwon, E. E., Sheen, H. K., & Grafilo, L. A. D. R., (2022b). Forecast of glucose production from biomass wet torrefaction using statistical approach along with multivariate adaptive regression splines, neural network, and decision tree. *Applied Energy, 324*, 119775.

Cherfi, A., Nouira, K., & Ferchichi, A., (2018). Very fast C4. 5 decision tree algorithm. *Applied Artificial Intelligence, 32*, 119–137.

Chong, A., Gu, Y., & Jia, H., (2021). Calibrating building energy simulation models: A review of the basics to guide future work. *Energy and Buildings, 253*, 111533.

Darwin, B., Dharmaraj, P., Prince, S., Popescu, D. E., & Hemanth, D. J., (2021). Recognition of bloom/yield in crop images using deep learning models for smart agriculture: A review. *Agronomy, 11*, 646.

De Jesus, O. N., Lima, L. K. S., Soares, T. L., Da Silva, L. N., Dos Santos, I. S., Sampaio, S. R., & De Oliveira, E. J., (2022). Phenotypic diversity and alternative methods for characterization and prediction of pulp yield in passion fruit (Passiflora spp.) germplasm. *Scientia Horticulturae, 292*, 110573.

Eissa, A., Helyes, L., Romano, E., Albandary, A., & Ibrahim, A., (2022). Thoughts for foods: Imaging technology opportunities for monitoring and measuring food quality. *A Glance at Food Processing Applications*, 157.

Felföldi, Z., Ranga, F., Roman, I. A., Sestras, A. F., Vodnar, D. C., Prohens, J., & Sestras, R. E., (2022). Analysis of physico-chemical and organoleptic fruit parameters relevant for tomato quality. *Agronomy, 12*, 1232.

Feng, L., Zhu, S., Zhou, L., Zhao, Y., Bao, Y., Zhang, C., & He, Y., (2019). Detection of subtle bruises on winter jujube using hyperspectral imaging with pixel-wise deep learning method. *IEEE Access, 7*, 64494–64505.

Feng, P., Wang, B., Liu, D. L., Xing, H., Ji, F., Macadam, I., Ruan, H., & Yu, Q., (2018). Impacts of rainfall extremes on wheat yield in semi-arid cropping systems in eastern Australia. *Climatic Change, 147*, 555–569.

Fernandes, A. M., Utkin, A. B., Eiras-Dias, J., Cunha, J., Silvestre, J., & Melo-Pinto, P., (2019). Grapevine variety identification using "big data" collected with miniaturized spectrometer combined with support vector machines and convolutional neural networks. *Computers and Electronics in Agriculture, 163*, 104855.

Fernández-González, M., Rodríguez-Rajo, F. J., Jato, V., Escuredo, O., & Aira, M. J., (2011). Estimation of yield 'loureira' variety with an aerobiological and phenological model. *Grana, 50*, 63–72.

Ferreira, C. S., Seifollahi-Aghmiuni, S., Destouni, G., Ghajarnia, N., & Kalantari, Z., (2022). Soil degradation in the European Mediterranean region: Processes, status, and consequences. *Science of the Total Environment, 805*, 150106.

Foley, J. A., Ramankutty, N., Brauman, K. A., Cassidy, E. S., Gerber, J. S., Johnston, M., Mueller, N. D., et al., (2011). Solutions for a cultivated planet. *Nature, 478*, 337–342.

Forkuor, G., Hounkpatin, O. K., Welp, G., & Thiel, M., (2017). High resolution mapping of soil properties using remote sensing variables in south-western Burkina Faso: A comparison of machine learning and multiple linear regression models. *PloS One, 12*, e0170478.

Fracarolli, J. A., Pavarin, F. F. A., Castro, W., & Blasco, J., (2021). Computer vision applied to food and agricultural products. *Revista Ciência Agronômica, 51.*

Fraga, H., & Santos, J. A., (2017). Daily prediction of seasonal grapevine production in the Douro wine region based on favorable meteorological conditions. *Australian Journal of Grape and Wine Research, 23*, 296–304.

Gebreeyessus, G. D., (2022). Towards the sustainable and circular bioeconomy: Insights on spent coffee grounds valorization. *Science of The Total Environment,* 155113.

Gené-Mola, J., Vilaplana, V., Rosell-Polo, J. R., Morros, J. R., Ruiz-Hidalgo, J., & Gregorio, E., (2019). Multi-modal deep learning for Fuji apple detection using RGB-D cameras and their radiometric capabilities. *Computers and Electronics in Agriculture, 162*, 689–698.

Ghosh, A., & Maiti, R., (2021). Application of SWAT, random forest and artificial neural network models for sediment yield estimation and prediction of gully erosion susceptible zones: Study on Mayurakshi River Basin of Eastern India. *Geocarto International*, 1–25.

Giefer, L. A., Arango, C. J. D., Babr, M. M., & Freitag, M., (2019). Deep learning-based pose estimation of apples for inspection in logistic centers using single-perspective imaging. *Processes, 7*, 424.

Giller, K. E., Delaune, T., Silva, J. V., Descheemaeker, K., Van De, V. G., Schut, A. G., Van, W. M., et al., (2021). The future of farming: Who will produce our food? *Food Security, 13*, 1073–1099.

Gracelin, S., & Raimond, K., (2022). Deep learning based indigenous herbal medicinal plants recognition: A comprehensive review. In: *2022 6th International Conference on Computing Methodologies and Communication (ICCMC)*. IEEE.

Hara, P., Piekutowska, M., & Niedbała, G., (2021). Selection of independent variables for crop yield prediction using artificial neural network models with remote sensing data. *Land, 10*, 609.

Hasan, A. M., Sohel, F., Diepeveen, D., Laga, H., & Jones, M. G., (2021). A survey of deep learning techniques for weed detection from images. *Computers and Electronics in Agriculture, 184*, 106067.

He, L., Fang, W., Zhao, G., Wu, Z., Fu, L., Li, R., Majeed, Y., & Dhupia, J., (2022a). Fruit yield prediction and estimation in orchards: A state-of-the-art comprehensive review for both direct and indirect methods. *Computers and Electronics in Agriculture, 195*, 106812.

He, Y., Xiao, Q., Bai, X., Zhou, L., Liu, F., & Zhang, C., (2022b). Recent progress of nondestructive techniques for fruits damage inspection: A review. *Critical Reviews in Food Science and Nutrition, 62*, 5476–5494.

Heidari, A., Navimipour, N. J., & Unal, M., (2022). Applications of ML/DL in the management of smart cities and societies based on new trends in information technologies: A systematic literature review. *Sustainable Cities and Society*, 104089.

Hiary, H., Saadeh, H., Saadeh, M., & Yaqub, M., (2018). Flower classification using deep convolutional neural networks. *IET Computer Vision, 12*, 855–862.

Hong, K. T., Abdullah, S. N. H. S., Hasan, M. K., & Tarmizi, A., (2022). Underwater fish detection and counting using mask regional convolutional neural network. *Water, 14*, 222.

Jayashree, L., Palakkal, N., Papageorgiou, E. I., & Papageorgiou, K., (2015). Application of fuzzy cognitive maps in precision agriculture: A case study on coconut yield management of southern India's Malabar region. *Neural Computing and Applications, 26*, 1963–1978.

Jeyaseelan, A., (2003). Droughts & floods assessment and monitoring using remote sensing and GIS. *Satellite Remote Sensing and GIS Applications in Agricultural Meteorology*, 291.

Kang, H., & Chen, C., (2020). Fast implementation of real-time fruit detection in apple orchards using deep learning. *Computers and Electronics in Agriculture, 168*, 105108.

Kang, Y., Ozdogan, M., Zhu, X., Ye, Z., Hain, C., & Anderson, M., (2020). Comparative assessment of environmental variables and machine learning algorithms for maize yield prediction in the US midwest. *Environmental Research Letters, 15*, 064005.

Kaur, P., Harnal, S., Tiwari, R., Upadhyay, S., Bhatia, S., Mashat, A., & Alabdali, A. M., (2022). Recognition of leaf disease using hybrid convolutional neural network by applying feature reduction. *Sensors, 22*, 575.

Kaur, S., Randhawa, S., & Malhi, A., (2021). An efficient ANFIS based pre-harvest ripeness estimation technique for fruits. *Multimedia Tools and Applications, 80*, 19459–19489.

Khorramifar, A., Rasekh, M., Karami, H., Covington, J. A., Derakhshani, S. M., Ramos, J., & Gancarz, M., (2022). Application of MOS gas sensors coupled with chemometrics methods to predict the amount of sugar and carbohydrates in potatoes. *Molecules, 27*, 3508.

Koirala, A., Walsh, K. B., Wang, Z., & McCarthy, C., (2019). Deep learning–method overview and review of use for fruit detection and yield estimation. *Computers and Electronics in Agriculture, 162,* 219–234.

Kraft, P., Rezaei, E. E., Breuer, L., Ewert, F., Große-Stoltenberg, A., Kleinebecker, T., Seserman, D. M., & Nendel, C., (2021). Modeling agroforestry's contributions to people—A review of available models. *Agronomy, 11*, 2106.

Kumar, M., Kumari, A., Kumar, D., Al-Ansari, N., Ali, R., Kumar, R., Kumar, A., et al., (2021). The superiority of data-driven techniques for estimation of daily pan evaporation. *Atmosphere, 12*, 701.

Le, T. T., & Lin, C. Y., (2019). Deep learning for noninvasive classification of clustered horticultural crops–A case for banana fruit tiers. *Postharvest Biology and Technology, 156*, 110922.

Li, Z., Tian, L., Jiang, Q., & Yan, X., (2021). Distributed-ensemble stacked autoencoder model for non-linear process monitoring. *Information Sciences, 542*, 302–316.

Liu, J., & Wang, X., (2021). Plant diseases and pests detection based on deep learning: A review. *Plant Methods, 17,* 1–18.

Lombardo, T., Duquesnoy, M., El-Bouysidy, H., Årén, F., Gallo-Bucno, A., Jørgensen, P. B., Bhowmik, A., et al., (2021). Artificial intelligence applied to battery research: Hype or reality? *Chemical Reviews.*

Lu, C. P., Liaw, J. J., Wu, T. C., & Hung, T. F., (2019). Development of a mushroom growth measurement system applying deep learning for image recognition. *Agronomy, 9*, 32.

Maheswari, P., Raja, P., Apolo-Apolo, O. E., & Pérez-Ruiz, M., (2021). Intelligent fruit yield estimation for orchards using deep learning based semantic segmentation techniques—A review. *Frontiers in Plant Science, 12*, 684328.

Majeed, Y., Zhang, J., Zhang, X., Fu, L., Karkee, M., Zhang, Q., & Whiting, M. D., (2020). Deep learning-based segmentation for automated training of apple trees on trellis wires. *Computers and Electronics in Agriculture, 170*, 105277.

Mao, S., Li, Y., Ma, Y., Zhang, B., Zhou, J., & Wang, K., (2020). Automatic cucumber recognition algorithm for harvesting robots in the natural environment using deep learning and multi-feature fusion. *Computers and Electronics in Agriculture, 170*, 105254.

Mirheidari, F., Khadivi, A., Saeidifar, A., & Moradi, Y., (2022). Selection of superior genotypes of Indian jujube (*Ziziphus mauritiana* Lamk.) as revealed by fruit-related traits. *Food Science & Nutrition, 10*, 903–913.

Mosavi, A., Ozturk, P., & Chau, K. W., (2018). Flood prediction using machine learning models: Literature review. *Water, 10*, 1536.

Nasiri, A., Taheri-Garavand, A., & Zhang, Y. D., (2019). Image-based deep learning automated sorting of date fruit. *Postharvest Biology and Technology, 153*, 133–141.

Navas, E., Fernández, R., Sepúlveda, D., Armada, M., & Gonzalez-de-Santos, P., (2021). Soft grippers for automatic crop harvesting: A review. *Sensors, 21*, 2689.

Nendel, C., (2010). Grapevine bud break prediction for cool winter climates. *International Journal of Biometeorology, 54*, 231–241.

Ni, X., Li, C., Jiang, H., & Takeda, F., (2020). Deep learning image segmentation and extraction of blueberry fruit traits associated with harvestability and yield. *Horticulture Research, 7*.

Nosratabadi, S., Ardabili, S., Lakner, Z., Mako, C., & Mosavi, A., (2021). Prediction of food production using machine learning algorithms of multilayer perceptron and ANFIS. *Agriculture, 11*, 408.

O'Sullivan, H., Raumonen, P., Kaitaniemi, P., Perttunen, J., & Sievänen, R., (2021). Integrating terrestrial laser scanning with functional–structural plant models to investigate ecological and evolutionary processes of forest communities. *Annals of Botany, 128*, 663–684.

Oogathoo, S., Houle, D., Duchesne, L., & Kneeshaw, D., (2020). Vapor pressure deficit and solar radiation are the major drivers of transpiration of balsam fir and black spruce tree species in humid boreal regions, even during a short-term drought. *Agricultural and Forest Meteorology, 291*, 108063.

Oppenheim, D., Shani, G., Erlich, O., & Tsror, L., (2019). Using deep learning for image-based potato tuber disease detection. *Phytopathology, 109*, 1083–1087.

Ozsagir, M., Erden, C., Bol, E., Sert, S., & Özocak, A., (2022). Machine learning approaches for prediction of fine-grained soils liquefaction. *Computers and Geotechnics, 152*, 105014.

Pan, X., Chen, M., Zhao, T., & Low, S. H., (2022). DeepOPF: A feasibility-optimized deep neural network approach for AC optimal power flow problems. *IEEE Systems Journal.*

Panda, S. S., Hoogenboom, G., & Paz, J. O., (2010). Remote sensing and geospatial technological applications for site-specific management of fruit and nut crops: A review. *Remote Sensing, 2*, 1973–1997.

Pandey, M., Singh, S., & Verma, S., (2019). Review on cyanobacterial dominance in rice agriculture.

Papageorgiou, E., Aggelopoulou, K., Gemtos, T., & Nanos, G., (2013). Yield prediction in apples using fuzzy cognitive map learning approach. *Computers and Electronics in Agriculture, 91*, 19–29.

Park, S. E., Benjamin, L. R., & Watkinson, A. R., (2003). The theory and application of plant competition models: An agronomic perspective. *Annals of Botany, 92*, 741–748.

Peng, B., Guan, K., Zhou, W., Jiang, C., Frankenberg, C., Sun, Y., He, L., & Köhler, P., (2020). Assessing the benefit of satellite-based solar-induced chlorophyll fluorescence in crop yield prediction. *International Journal of Applied Earth Observation and Geoinformation, 90*, 102126.

Pereira, M. D., Christine, C. R. H., Santa-Catarina, R., Cesar, F. V. J., Santana, J. G. S., Pastana De, S. P. T., Azevedo, V. P. A., et al., (2022). Topcross hybrids in papaya (*Carica papaya* L.): Evaluation of the potential for increasing fruit quality in new cultivars. *Archives of Agronomy and Soil Science, 68*, 1473–1486.

Prusinkiewicz, P., (2004). Modeling plant growth and development. *Current Opinion in Plant Biology, 7*, 79–83.

PS, M. G., (2019). Performance evaluation of best feature subsets for crop yield prediction using machine learning algorithms. *Applied Artificial Intelligence, 33*, 621–642.

Rahimi, M., & Abdolinasab, M., (2022). Examining the inheritance of watermelon fruit traits by Hayman's graphical approach. *BioMed Research International*, 2022.

Raj, E., Appadurai, M., & Athiappan, K., (2021). Precision farming in modern agriculture. *Smart Agriculture Automation Using Advanced Technologies* (pp. 61–87). Springer.

Riera, L. G., Carroll, M. E., Zhang, Z., Shook, J. M., Ghosal, S., Gao, T., Singh, A., et al., (2021). Deep multiview image fusion for soybean yield estimation in breeding applications. *Plant Phenomics, 2021.*

Rodriguez-Galiano, V., Sanchez-Castillo, M., Chica-Olmo, M., & Chica-Rivas, M., (2015). Machine learning predictive models for mineral prospectivity: An evaluation of neural networks, random forest, regression trees and support vector machines. *Ore Geology Reviews, 71*, 804–818.

Romero, P., Navarro, J. M., & Ordaz, P. B., (2022). Towards a sustainable viticulture: The combination of deficit irrigation strategies and agroecological practices in Mediterranean vineyards. A review and update. *Agricultural Water Management, 259*, 107216.

Sabry, F., Eltaras, T., Labda, W., Alzoubi, K., & Malluhi, Q., (2022). Machine learning for healthcare wearable devices: The big picture. *Journal of Healthcare Engineering, 2022.*

Sabzi-Nojadeh, M., Niedbała, G., Younessi-Hamzekhanlu, M., Aharizad, S., Esmaeilpour, M., Abdipour, M., Kujawa, S., & Niazian, M., (2021). Modeling the essential oil and trans-anethole yield of fennel (*Foeniculum vulgare* Mill. var. vulgare) by application artificial neural network and multiple linear regression methods. *Agriculture, 11*, 1191.

Safren, O., Alchanatis, V., Ostrovsky, V., & Levi, O., (2007). Detection of green apples in hyperspectral images of apple-tree foliage using machine vision. *Transactions of the ASABE 50*, 2303–2313.

Salvucci, G., Pallottino, F., De Laurentiis, L., Del Frate, F., Manganiello, R., Tocci, F., Vasta, S., et al., (2022). Fast olive quality assessment through RGB images and advanced convolutional neural network modeling. *European Food Research and Technology, 248*, 1395–1405.

Sanchez, P. D. C., Hashim, N., Shamsudin, R., & Nor, M. Z. M., (2020). Applications of imaging and spectroscopy techniques for non-destructive quality evaluation of potatoes and sweet potatoes: A review. *Trends in Food Science & Technology, 96*, 208–221.

Shi, R., Li, T., & Yamaguchi, Y., (2022). Output-targeted baseline for neuron attribution calculation. *Image and Vision Computing, 124*, 104516.

Shook, J., Gangopadhyay, T., Wu, L., Ganapathysubramanian, B., Sarkar, S., & Singh, A. K., (2021). Crop yield prediction integrating genotype and weather variables using deep learning. *Plos One, 16*, e0252402.

Silvestri, C., Bacchetta, L., Bellincontro, A., & Cristofori, V., (2021). Advances in cultivar choice, hazelnut orchard management, and nut storage to enhance product quality and safety: An overview. *Journal of the Science of Food and Agriculture, 101*, 27–43.

Singaravel, S., Geyer, P., & Suykens, J., (2017). Component-based machine learning modeling approach for design stage building energy prediction: Weather conditions and size. *Proceedings of the 15th IBPSA Conference* (pp. 2617–2626).

Sun, C., Tian, Y., Gao, L., Niu, Y., Zhang, T., Li, H., Zhang, Y., et al., (2019). Machine learning allows calibration models to predict trace element concentration in soils with generalized LIBS spectra. *Scientific Reports, 9*, 1–18.

Sun, Y., & Liang, F., (2022). *A Kernel-Expanded Stochastic Neural Network.* arXiv preprint arXiv:2201.05319.

Thakur, A. K., & Uphoff, N. T., (2017). How the system of rice intensification can contribute to climate-smart agriculture. *Agronomy Journal, 109*, 1163–1182.

Tian, Y., Yang, G., Wang, Z., Li, E., & Liang, Z., (2020). Instance segmentation of apple flowers using the improved mask R–CNN model. *Biosystems Engineering, 193*, 264–278.

Tripathi, M., (2021). Analysis of convolutional neural network-based image classification techniques. *Journal of Innovative Image Processing (JIIP), 3*, 100–117.

Tu, S., Xue, Y., Zheng, C., Qi, Y., Wan, H., & Mao, L., (2018). Detection of passion fruits and maturity classification using red-green-blue depth images. *Biosystems Engineering, 175*, 156–167.

Ucat, R. C., & Cruz, J. C. D., (2019). Postharvest grading classification of cavendish banana using deep learning and TensorFlow. In: *2019 International Symposium on Multimedia and Communication Technology (ISMAC)* (pp. 1–6). IEEE.

Ustin, S. L., & Middleton, E. M., (2021). Current and near-term advances in Earth observation for ecological applications. *Ecological Processes, 10*, 1–57.

Wang, D., Li, C., Song, H., Xiong, H., Liu, C., & He, D., (2020). Deep learning approach for apple edge detection to remotely monitor apple growth in orchards. *IEEE Access, 8*, 26911–26925.

Wang, D., Zhang, M., Mujumdar, A. S., & Yu, D., (2021). Advanced detection techniques using artificial intelligence in processing of berries. *Food Engineering Reviews*, 1–24.

Wu, X., Xiao, L., Sun, Y., Zhang, J., Ma, T., & He, L., (2022). A survey of human-in-the-loop for machine learning. *Future Generation Computer Systems.*

Wu, Z., Chen, Y., Zhao, B., Kang, X., & Ding, Y., (2021). Review of weed detection methods based on computer vision. *Sensors, 21*, 3647.

Yan, H. P., Kang, M. Z., De Reffye, P., & Dingkuhn, M., (2004). A dynamic, architectural plant model simulating resource-dependent growth. *Annals of Botany, 93*, 591–602.

Yang, B., & Xu, Y., (2021). Applications of deep-learning approaches in horticultural research: A review. *Horticulture Research, 8.*

Yin, J., Medellín-Azuara, J., Escriva-Bou, A., & Liu, Z., (2021). Bayesian machine learning ensemble approach to quantify model uncertainty in predicting groundwater storage change. *Science of the Total Environment, 769*, 144715.

Zhang, D., Pee, L., & Cui, L., (2021). Artificial intelligence in E-commerce fulfillment: A case study of resource orchestration at Alibaba's smart warehouse. *International Journal of Information Management, 57*, 102304.

Zhang, Z., Jin, Y., Chen, B., & Brown, P., (2019). California almond yield prediction at the orchard level with a machine learning approach. *Frontiers in Plant Science, 10*, 809.

Zheng, W., Bai, Y., Luo, H., Li, Y., Yang, X., & Zhang, B., (2020). Self-adaptive models for predicting soluble solid content of blueberries with biological variability by using near-infrared spectroscopy and chemometrics. *Postharvest Biology and Technology, 169*, 111286.

CHAPTER 8

Current Trends and Future Perspectives for Modeling Agronomic Crops and Improving Intercropping in Climate Change Scenarios: A Review

OWAIS BASHIR,[1] SHABIR AHMAD BANGROO,[1] NASIR BASHIR NAIKOO,[1] NAZIR AHMAD GANIE,[2] S. S. KUKAL,[3] MIR WASEEMA JEHAAN,[1] REHANA MOHIUDDIN,[4] ZAHIDA RASHID,[4] TANVEER AHMED,[4] SHAFEEQ UR REHMAN,[5] AND KHALID REHMAN HAKEEM[6]

[1]*Division of Soil Science, Sher-e-Kashmir University of Agricultural Sciences and Technology of Kashmir, Kashmir, India*

[2]*Division of Animal Genetics and Breeding, Sher-e-Kashmir University of Agricultural Sciences and Technology of Kashmir, Kashmir, India*

[3]*Punjab Water Regulation and Development Authority, Chandigarh, Punjab, India*

[4]*Division of Agronomy, Sher-e-Kashmir University of Agricultural Sciences and Technology of Kashmir, Kashmir, India*

[5]*School of Environment and Civil Engineering, Dongguan University of Technology, Dongguan, Guangdong, China*

[6]*Department of Biological Sciences, Faculty of Science, King Abdulaziz University, Jeddah, Saudi Arabia*

ABSTRACT

For many purposes, including management choices, adaptation policies, and potential adaptation strategies to control the effects of climate change

Soil, Agriculture, and Ecosystem Modelling: Smart Technologies for Sustainable Solutions.
Owais Bashir, Khalid Rehman Hakeem, Surinder S. Kukal, and Razeef Mohd (Eds.)

on future crop production, a wide range of crop models with different data input requirements are being used. The lack of trustworthy past data for model calibrations is a persistent problem for crop model simulation, particularly for future crop achievement forecasts and impact studies under various conditions. There are times when the quantity and quality of input data required to power the majority of crop models may not be present. Even when a suitable crop simulation model (CSM) is chosen, data shortages limit some of the models' ability to serve as reliable projection tools. A review of the factors preventing efficient use of CSMs and complementary sources of input data has not yet been conducted. This review examined various CSMs, pertinent sources from which crop models can obtain their input data, and a framework for gathering that data. Therefore, using multiple methods of data collection for model input data can lessen the difficulties crop modelers face due to data scarcity. The future of modeling is dependent on the quality and accessibility of the input data, the willingness of modelers to work together on modularity and standardization, and the communication skills of potential user groups.

8.1 INTRODUCTION

Intercropping is a century-old practice on the margins of modern agriculture, which is dominated by monoculture, resource-intensive, and high-yielding crops (Biswas et al., 2022; Li et al., 2021; Raza et al., 2021b). Intercropping can provide efficient and resilient agriculture by addressing some of the primary problems associated with modern agriculture, such as poor yield, insect, and pathogen build-up, soil degradation, and ecological pollution (Biswas et al., 2022; Brooker et al., 2021). Intercropping has attracted the attention of numerous agronomic, environmental, and ecological scientists with various study interests, enabling multidisciplinary syntheses that combine diverse knowledge on intercropping's potential (Giller et al., 2021; Mensah et al., 2021). Intercropping can help stabilize enhanced accumulated yields per unit of input under these circumstances, preserve and improve soil quality, provide financial protection against crop loss and price fluctuations, satisfy food preferences and cultural needs, preserve and enhance soil quality, and raise revenue (Mogaka et al., 2022; Thierfelder et al., 2015). Intercropping has existed and remains the predominant agricultural practice in several regions around the globe. In 70–90% of the region's maize-growing regions, small-scale farmers in Central America intercrop maize with beans, potatoes, and other crops (Wasaki & Dissanayaka, 2021); Intercropping is used by 98% of beans and peas in Africa, and 92% of beans in Colombia (Mofolo, 2021); the percentage of intercropped farmland in the tropical land varies between 17% in India to 94% in Malawi (Malik et al., 2021). China has more than 22% of the planet's population but only 9% of the world's productive land. Intercropping has traditionally been an important factor in Chinese agricultural productivity (Liang & Shi, 2021). Better use of the land, less reliance on agrochemicals, improved soil management, and yield diversity are some advantages of intercropping systems. The utilization of appropriate plant densities, management layout or planting strip widths, and the appropriate times of sowing for the different crops are some of the problems encountered by intercropping (Ren et al., 2021; Yin et al., 2021). Multi-species cropping system modeling helps facilitate and achieve a better intercropping cultivation plan.

Crop simulation models (CSMs) are now universally recognized as valuable tools for examining cause-and-effect interactions in crop production. CSMs evaluate crop ecosystems by establishing boundaries and identifying key elements. They use mathematical formulas to explain variations in the elements and afterwards link the results to create a model of the system. CSMs are statistical equation-based computerized simulations of crop growth, maturation, and output as functions of the environment, soil conditions, field crop genotypes, and management practices (Dhakar et al., 2022; Iqbal et al., 2018; Lamsal et al., 2017; Li et al., 2022; Zhang et al., 2020). CSMs are essentially intended to answer, "what if" and "when" type queries concerning production as the interesting primary perspective (Wenkel et al., 2013). They differ in regard to fundamental structure, application, and core development. Models with various levels of detail have been created based on the information and knowledge accessible in a specific subject. Since preliminary work began utilizing CSMs, two types such as empirical and mechanistic, have emerged as dominant supermodels (Singh et al., 2020b; Wu et al., 2022).

Mechanistic models are created to provide a statistical explanation of a system's physical operations that is fairly complete (Marwedel, 2021). Mechanistic models (explanatory models) are made up of quantifiable data about the processes and mechanisms that control the system's dynamic behavior. The models individually characterize the system's methodologies in order to describe it. The model would then combine such different parts to create a simulation. Fleisher (2009) asserts that the reductionistic nature of mechanistic models results from deploying mechanisms that descend from high levels (the canopy) to low levels (cellular processes). A solid grasp of the temporal and spatial notions of sharing resources and how these affect processes and mechanisms at the plant level is required to recreate relationships in intercrop systems. Less reliable models exist, though; they frequently fall under descriptive or empirical models. Empirical models often form regression equations with one or more components and are straightforward representations of measured data used to predict eventual yield.

The technology and benefits of intercropping must be thoroughly understood to increase interest in intercropping studies and potential farmer adoption. This chapter aims to discuss intercropping resource capture principles while highlighting its benefits and potential downsides. It is necessary to produce a significant amount of detailed data; hence an outline of crop simulation for intercropping as a technique for doing so was made. Since models in intercropping systems are currently in their initial stage, selected crop models which simulate resource use in multidrop mechanisms are herewith reviewed, and constraints of the modeling techniques are highlighted. The chapter discusses the problems with intercropping and resource utilization and explains the numerous intercropping model concepts.

8.2 NEED FOR CROP SIMULATION MODELS (CSMS)

Field studies and other traditional agronomic research methods have been and continue to be trusted sources of information for establishing connections between agricultural land use practices and observable consequences in the real world (Kephe et al., 2021; Lindblom et al., 2017). On the other hand, conventional agronomic research approaches cannot keep up with the rate at which data is required to inform policy and decision-making procedures. Additionally, traditional agronomic studies occasionally only present the findings of experiments conducted at particular times and places, leading to season- and location-specific results. Moreover, these investigations are frequently expensive, time-consuming, and laborious, which delays the data gathering period. Once more, it is improbable that research on recommendations for many different qualities could be undertaken over several years and in many regions.

Additionally, these trials were unable to offer sufficient data in terms of time and geography to identify effective management approaches. Due to the limitations of conventional field experiments, a tool or approach that enables the quick collection of new information and research findings and the quick dissemination of information to end-users in many industries that depend on such data and outcomes for decision-making is urgently needed (Keim et al., 2010; Longo et al., 2017; Sutanto et al., 2013). The intended outputs and effects will be available sooner owing to the acceleration of these plant-environment interactions, which will also lessen the requirement for data collection during farm experiments. To accurately forecast the production of food crops under various conditions and periods, techniques like crop modeling for the intercrops must be created (Benami et al., 2021; Divya et al., 2021). Instead, there is a growing understanding, particularly among experts in intercrop modeling, that high-quality field trials, notwithstanding their rarity, are essential for developing and validating intercrop models (Keating & Thorburn, 2018; Trivedi et al., 2021). In light of this, we investigate the relationship between formal experimental studies and crop models using various crop models as a bridge. Before investing in expensive and time-consuming field trials, researchers can examine cultivar potential in new locations using intercrop simulation modeling. Experiments on a laptop or desktop computer may be employed to pre-evaluate costly and time-consuming agricultural and modeling field trials with many treatments.

To support more efficient and environmentally friendly agriculture, intercrop models are crucial decision-making techniques for evaluating the impact of climate change/variability and determining control measures on alternative cropping systems' production and natural environments. They offer an easier and more cost-effective way to examine how agricultural land management practices affect crop productivity and the environment. They also serve as a tool to demonstrate how much management is required to produce economically viable yields. To assess farm management approaches' risk and financial implications, models for intercrop simulation may be utilized as systems to aid decision-making. Zhao claims that the modeling approach may result in fairly reliable conclusions when constructing agricultural land management policies if reliable, observable field data is used to check and verify the models (Zhao et al., 2020). Intercrop models, for instance, have been employed to improve management tactics, including agricultural chemical application, agricultural water use, and plot sizes, as well as to evaluate the performance of various agricultural land management strategies under various climate change scenarios (Chimonyo et al., 2016; Kansiime et al., 2019). Because they can incorporate the understanding of a given crop's physiology accumulated over many years of laboratory and field studies, intercrop models are essential for forecasts. They also provide a useful way of examining how crops respond to climate change and alternative management scenarios. According to data from local and regional studies and worldwide meta-analyzes of modeling and simulation, adaptation tactics are essential in preventing any negative effects or maximizing any positive outcomes that may result from climate change (Akinyi et al., 2021; Flouris et al., 2018; Gouel & Laborde, 2021; Naulleau et al., 2021; Sampaio et al., 2021). To meet future demands for food security, adaptation techniques are most likely the only means to maintain or increase the availability and stability of food. Consequently, crop models could only facilitate land executives and decision-makers recognizing management alternatives for maximizing sustainability goals across time and space if the necessary soil maintenance and weather data are accessible.

8.2.1 THE COMPONENTS OF CROP SIMULATION MODELS (CSMS)

Crop models, which are collections of mathematical formulas that depict the various processes that occur inside the plants and the linkages between the plant and its environment, are now an essential tool for forecasting how climate change may affect crop productivity in the future. CSMs anticipate plant growth and development in connection to production-influencing elements, including environmental conditions and crop management techniques, by using explicit descriptions (as model input data) of ecophysiological processes (Bartram, 2017; Lawrence & Friel, 2019). The crop models are often built around four essential elements (Figure 8.1).

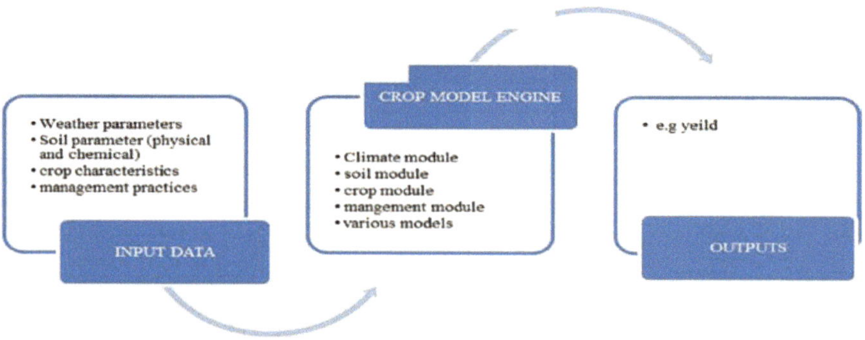

FIGURE 8.1 A generic crop model layout and modeling procedures.

A collection of input data known as a "Minimum Data Set" must be needed to execute a cropping model and complete a simulation. These data are necessary for model evaluation, implementation, construction, and improvement.

Crop management data from the experiment, such as information on at least two significant morphological and physiological phases, namely inflorescence or flower initiation and physiological or harvest maturity, yield, and yield components, are required in order to calibrate the model (Martiniello & Teixeira da Silva, 2011; Siakou et al., 2021). Additionally required is site-specific weather data throughout the growing season, ideally for the entire year. Data on the growth of plants, carbon sequestration, moisture retention, and fertilizer should all be included, according to Craufurd et al. (1991). This is because the timing and length of significant developmental events like flowering are controlled by plant growth or phenology (anthesis). As a result, it offers the foundation for procedures that gather and utilize nutrients, water, and carbon. The following information is the very minimum needed: The weather station's latitude and longitude, the sun's daily input radiation, the lowest and maximum daily air temperatures (in degrees Celsius), and the daily rainfall totals (mm). To use more reliable methods, the wind velocity, dry and wet bulb temperature, and evapotranspiration (ET) may all be simulated. In order to allow for a "what-if" type analysis, climate data for evaluation should preferably start a few days before sowing and last a few days after harvest. This would give enough time for the experiment as a whole. The soil data includes information on the lower and upper horizon depths (cm), bulk density, organic carbon, sand, silt, clay content, pH in water, aluminum concentration, and root abundance. The management data includes information on planting time and days when soil conditions were evaluated before planting, plant population, plant spacing, planting depth, crop type, irrigation schedules, and fertilizer application methods (Barison & Uphoff, 2011; Mishra & Salokhe, 2010). Test data are required for the study of both models and strategies. According to (Hoogenboom et al., 2019), observable data, which are necessary for model evaluation, should be included in experimental data along with the site, soil, and weather information. Examples of observable data include crop growth metrics, soil water content, and fertility metrics. It is important to note that this input set of data works includes a wide range of crops and models. These crop- and site-specific parameters should be calibrated to adjust model simulation runs to local conditions. While other parameters should be calibrated using useful field data, some values are conservative and require little modification during calibration. The models also include experimentally derived characteristics that replicate the performance of various types from the various crop modules that are included in the models; the values for these characteristics should be modified for specific crops.

Empirical models, geographic suitability models, physiological models, meta-models, and decision models are a few of the modeling approaches utilized in simulation (Kephe et al., 2021; Klabunde & Willekens, 2016). According to Lobell et al. (2006), agricultural climate impact evaluations are primarily carried out using the empirical crop model technique. The significance of the technology is that it can apply yield response algorithms to already available data, even if it is scant or only available in an integrative form, such as climate data monitored monthly (Ransan-Cooper et al., 2021; Yu et al., 2021). It can also be employed to identify the main climatic influences on yield and yield variations in certain geographical areas. Validated model predictions are considered valid within the data range used to fit the empirical models. However, the potential for accurate forecasts beyond observed circumstances may be constrained by the possibility that causal ties inferred from observable data do not reflect process relationships beyond certain observations.

On the contrary, regional adaptation models are widely used to assess the local biophysical land use options under the current and projected climatic conditions (Berger & Troost, 2014; Tufféry et al., 2021; Van Asselen & Verburg, 2013). By quantifying the land potential for specific crops, this method also pinpoints places that are more or less suitable for different land uses and variations in cropping zones (Sandström et al., 2018). Regional adaptation models combine data from numerous sources to provide a broad foundation for multi-criteria decision-making in meta-models. The advantage of this model is that having a shorter run time, which makes it easier to use for exploratory research, like examining various possible explanations and their capacity for integration. A process-based empirical production or biophysical model and a farm optimization model are frequently combined to create decision models, commonly referred to as bio-economic models (Huber et al., 2022; Schreinemachers & Berger, 2011).

On the contrary hand, biophysical models mimic biophysical processes, including plant growth, nitrogen & carbon flows, and rainfall patterns using statistically organized mechanistic process knowledge (Paul et al., 2021). These models, which incorporate multiple biophysical processes, provide an excellent basis for analyzing the effects of climate change on numerous agroecosystems. To simulate agricultural processes, numerous static and dynamic models are still being developed (Guse et al., 2015; Kephe et al., 2021; Muflikh et al., 2021). They differ from models that primarily focus on characterizing farmland innovation process of various levels of complexity, designed to mimic the effects of water restrictions on potential crop yields, designed to simulate the responses of crops to the changes of carbon, soil water and nutrients and integrating management & climate impacts (Kephe et al., 2021). Crop models such as CROPGRO for major grain legumes (Antolin et al., 2021), CERES for cereal crops (Yang et al., 2021), and SUBSTOR for crops with belowground storage organs are part of the decision support system for agro-technology transfer (DSSAT) (Divya et al., 2021). Other models include agricultural production systems simulator model (APSIM), cropping systems simulation model (CropSyst), and CROPWAT/AquaCrop, simulator (Kephe et al., 2021). On the other hand, spatially distributed models such as SWAT/SWIM & MIKE are frequently used at a regional or even global level (Kephe et al., 2021). Many models, including APSIM, CROPSYST, EPIC, STICS, system approach to land use sustainability (SALUS), environment resource synthesis (CERES), aqua crop and InVEST, have been developed to assist decision-makers in evaluating trade-offs between ecosystem services (ES) and forecasting changes in biodiversity under various demographic, land use, and climate scenarios (Kersebaum & Stöckle, 2022). Numerous studies have employed these crop development models to aid farmers worldwide make management choices concerning the frequency and timing of irrigation, plant population density, and sowing time.

8.3 VARIOUS MODELS FOR MODELING AGRONOMIC CROPS

8.3.1 ALMANAC MODEL

ALMANAC is a crop modeling tool used for crop development, intercropping, fertilizer use efficiency, water productivity, biomass accumulation, which determines plant species competitiveness, and other growth challenges, such as water, temperature, and nutrient stresses (Chimonyo et al., 2015; Luedeling et al., 2016). The ALMANAC model has been extensively used to forecast plant phenology, radiation usage efficiency, agriculture output, biomass assessment, climate change, and their impact on ES (Druille et al., 2020). Temperature influences plant growth and the length of degree days and determine the duration of growth phases. All plant species have a certain ideal and base temperature. Using Beer's law, simulations of species viewing for the light permit a variable extinction coefficient (k) for each species. The light distribution between species is determined by LAI, k-values, and plant heights. LAI, light interference with Beer's law, and potential routine bioenergy increase with species-specific implications for radiation use efficiency (RUE). Based on the current rooting zone and demand of each species, the model can mimic competition for nutrients and water. When available water at the plant's current roots depth is insufficient to meet potential ET, daily gains in and biomass are reduced. Total biomass is modeled with radiation usage efficiency and crop production with a method based on a harvest index (HI) sensitive to water stress. The HI, which measures grain yield like the percentage of all the above-ground dry matter at maturity, is the basis for grain production simulations (Xu et al., 2021).

ALMANAC's Batch Run simulation generates results for multiple scenarios concurrently. For its successful determination, ALMANAC uses inputs such as soil, climate, tillage, and growth parameter (Jacot et al., 2021; Kim et al., 2020; Mitchell et al., 2016). Users frequently utilize both NOAA daily weather information and NRCS soil data. Normal minimum and maximum temperatures, precipitation, and solar radiation are considered weather inputs. The weather generator function in ALMANAC is based on the WGEN model's presumptions (Percival et al., 2021). The generator is utilized if no meteorological information is available, or the operator does not wish to use or prepare the already collected data. Using multiple years of climate data, it is possible to mimic management, plant, and soil qualities rapidly. Tillage necessitates the creation/selection of management data and gives data on water runoff, hydration, fertilizers, furrow diking, and lime. Data from existing plants must be collected to calibrate and verify models for new plants. Once created for a particular plant species or cultivar, plant process parameters can be easily shared among models (EPIC, APEX, SWAT). ALMANAC has proved its capabilities and simulations in various crop forecasting and natural resource issues (Kim & Kim, 2021). ALMANAC can also predict future crop yields, their influence on competing species, and the impact of changing land management on these species (Yoon et al., 2021). ALMANAC is also used to analyze natural resource issues such as soil erosion, climate change, risk assessment, management decisions, plant competition, and conservation results (Kiniry et al., 2008). This model estimates not just plant yield but also the movement of water and nutrients through the system and the influence of management decisions on plant growth. Actual field data are used to validate model plant properties. In addition to national evaluations, the method has been employed to evaluate specific farmer fields.

8.3.2 APSIM MODEL

The Agricultural Production Systems Simulator (APSIM), a tool for simulating physiological mechanisms in crop production, is especially valuable for evaluating the economic and biodiversity implications of management practices under climate-related hazards (Divya et al., 2021; Ewert et al., 2015). It arose due to the need for research methods that addressed long-term issues with natural resource management and provided precise estimates of crop output concerning weather, genotyping, land, and farmer management factors (Panchasara et al., 2021). The modeling of fallow periods, crop rotations, and crop sequencing in connection to common soil and climatic variables is particularly important. The crop, soil, environment, and management methods are all described in the several modules comprising APSIM's modular structure (Raza et al., 2021). It comprised a vast array of crops, grasslands, vegetation, and soil operations, such as the hydrological cycle, phosphate, and nitrogen transformations, soil response, and soil erosion, as well as a highly versatile range of organizational management. Aside from the required clock, output file, and meteorological modules, each extra module can be added or removed depending on the needs of any particular simulation (Leufen et al., 2021).

Consequently, simulations are plug-and-play and can be modified by the user to focus on particular system components. For instance, a study of residue dynamics in wheat cropping systems may include modules reflecting soil moisture, soil fertilizer, surface organic matter (OM), and management approaches. Other models attempting to investigate water efficiency in irrigated rice cropping processes may include components characterizing soil moisture and nitrogen. However, instead of wheat and surface organic material components, they may include modules that model rice, pond operations, water management, and the effect of imposed management. In this way, APSIM is a highly versatile modeling framework that can execute many tasks for a wide range of users (Elliott et al., 2014).

APSIM is a dynamic, constant-time simulator that models crop systems by combining management and physiological modules into a single engine (Nurulhuda et al., 2020). The model, controlled by daily climate data, can mimic soil moisture, Carbon, Nitrogen, and Phosphorus dynamics and their interactions within agricultural systems (min. and max. temperatures, solar energy, precipitation). The phase radiation usage efficiency (RUE) governed by the environment and accessible leaf area is utilized to evaluate the average potential yield of a range of crop species. The availability of soil moisture, nitrogen, and (for some crop modules) phosphate determines the daily average production limits (Keating et al., 2003). The SOILWAT control system (Probert et al., 1998) generates a hybrid, spiraling soil moisture stability strategy. The SURFACE module suggests that above-ground agricultural wastes could be eliminated, assimilated into the soil, or left to deteriorate on the soil's surface (Probert & Dimes, 2004). The SOILN2 module models soil variations in C and N. Fresh organic matter (FOM) decomposition, N immobilization, urea breakdown, ammonification, nitrogen fixation, and volatilization are a few of these processes. FOM consists of crop residues and roots tilled into the soil. SOILN and SOILWAT modules provide APSIM crop modules with information regarding soil moisture and nutrient (N) accessibility to regulate the daily growth of crops.

8.3.3 GROWIT MODEL

Utilizing scientific knowledge and cutting-edge technology, the Growit model has greatly improved pasture grass yield. Enhancing plant breeding initiatives and employing simulation models are hoped to make more advances. These algorithms may be utilized to identify shortages and predict crop growth. Agricultural practice needs "specific qualitative criteria," often obtained through expensive and time-consuming research. Modeling may be used to forecast the outcomes of certain management actions and generate answers to brand-new challenges. GROWIT is a generic model of plant development used to predict daily forage yields (Lowenberg-Deboer et al., 1991). This model was developed by (Loewer et al., 1981) as a component of the University of Kentucky's bigger BEEF production model. BEEF is a simulation model for assessing alternative beef production techniques while considering environmental, energy, and economic restrictions (Loewer et al., 1981). GROWIT has been used to imitate the vegetative development of crops such as Coastal Bermudagrass, Tall Festuca, Kentucky Bluegrass, Red Clover, Alfalfa, and Corn.

GROWIT is a generic model not limited by location, crop, or management techniques (Lowenberg-Deboer et al., 1991). Forecasting the potential forage growth rate requires a function relating the greatest growth rate to the ambient temperature, and a curve represents the relationship between growth rate and temperature. It consists of two parabolas. The first describes the growth rate between the minimum and optimal temperature ranges for growth. In contrast, the second describes the growth rate between the optimal and maximum temperature ranges for growth. Day length is calculated using the latitude of the area where the crop grows and the Julian day. As a representation of air temperature as a function of time, the lowest air temperature is predicted to fluctuate at sunrise, the highest at solar noon, and the mean at sunset. There is no development between nightfall and morning. The growth rate is then calculated hourly. GROWIT adjusts for losses in growth owing to water by connecting rainfall distribution and total cumulative daily rainfall with normal and accumulated daily precipitation (Lowenberg-Deboer et al., 1991). This variable decreases until the second-day length is attained, at which point it is presumed that photoperiod no longer affects plant development. Using these criteria, GROWIT estimates the crop's effective rainfall (Loewer et al., 1981). Multiplying the daily rainfall factor by the photoperiod factor, ideal growth rate, and leaf area factors yields the expected yield in pounds per acre.

8.3.4 INTERCOM MODEL

INTERCOM calculates how temperature, sunshine, precipitation, and hydrologic soil factors impact plant development in wild or farmed situations. The model models competition for light and water intake, morphological and physiological development, dry matter building, distribution of dry mass over plant organs, soil moisture balance, evaporation, and transpiration (Lindgren & Gabrielsson, 2021; Querol, 2021; Requena-Mesa et al., 2021; Ruffin et al., 2021; Townsend, 2021). Latitude, daily meteorological data (incident solar radiation, min and max temperature, precipitation, wind speed, and vapor pressure), soil field capacity, and various physiological and morphological parameter values are required as model inputs (Anstey et al., 2021; Bu et al., 2021). The model's water-balancing component may be deleted when the soil has enough water. The model integrates every day. It can replicate one species' behavior or the growth of two venomous species. This study simulates the biomass and LAI of two monoculture understory plants.

Hybrid-Maize was created by combining the growth and development characteristics of CERES-Maize with the mechanistic formulation of photosynthesis and respiration in common crop models like INTERCOM and WOFOST (Yang et al., 2004). It integrates temperature-driven maize phenological development, vertical canopy photosynthetic integration, and organ-specific growth respiration. Hybrid-Maize might be more responsive to environmental changes than CERES-Maize since it comprises gross absorption, growth respiration, and maintenance respiration. While maintaining precision, hybrid-maize needs fewer genotype-specific parameters. The Intercom model's linear relationship between growing degree-days from emergence to silking and emergence to physiological maturity is used to predict silking when the former is unavailable. We measured the leaf area, dry matter accumulation, final grain and stover yields, and HI at two high-yielding locations. Values predicted and measured agreed nicely (HI). Future model development will emphasize subjects like LAI predictions at high plant densities, biomass partitioning, carbohydrate translocation, and maintenance respiration throughout reproductive development. The Intercom model has not been tested in water and nutrient stress, while the Hybrid-Maise model's water and nitrogen balance components are being constructed and validated (Han, 2016). *Zea mays* L.'s growth and development have been modeled. To replicate the morphological and physiological characteristics of specific crops like maize, rice, or potatoes, generic crop models are modified to explain absorption, respiration, development, and growth without respect to crop species (Boote et al., 2013; Brisson et al., 2003; Obidiegwu et al., 2015; Swain et al., 2021). SUCROS, INTERCOM, and WOFOST are examples of crop models. The genetic connections between environment, genotype, and agricultural management are shown in intercom models. They are essential for scientific research, managing farms, and analyzing policy (Fischer & Sun, 2001; Hammer et al., 2002; Hansen, 2005). The goal for which intercom models have created impacts the depth necessary for a mechanistic description of important processes, environmental and managerial sensitivity, data demands, and model outputs. Effective intercom model simulation requires knowing the basic ecophysiological processes and using that information in the model's mathematical formulations.

For instance, precise modeling of grain and stover yields, assimilating partitioned below-ground to root biomass, and exudation is necessary for predicting carbon (C) sequestration potential in a cropping system. The vegetative and reproductive stages are the only ones recognized by INTERCOM (Kropff et al., 1993). Each action needs GDD. In general models like WOFOST (Van Diepen et al., 1989) and INTERCOM (Colas et al., 2021), dry matter production is explicitly predicted, considering both growth and maintenance of respiration model-driven generic crop assimilation. The explicit respiration and photosynthesis functions from INTERCOM and WOFOST were combined to create "Hybrid-Maise." Then, in field tests, the Hybrid-Maise model was utilized to simulate the increase of maize leaf area, the accumulation of above-ground dry matter, and grain and stover yields. Crop management was used in these investigations to lessen biotic and abiotic stress. In order to compare field experiments, CERES Maize and INTERCOM were used.

8.3.5 STICS CROP MODEL

INRA has been developing STICS alongside other research or professional and educational organizations since 1996. STICS has been used for over 10 years and has improved. STICS focuses on cropping system diversity, although it shares many other models' objectives (Whisler et al., 1986). It is a crop simulator with a daily crop, soil, and climatic inputs. Drainage, nitrate leaching, and yield quality and quantity are output criteria. An agricultural environment with a physical medium and a schedule for crop management is simulated. Crop development and growth, as well as water and nitrogen balances, are all simulated processes. Brisson et al. (2009) explain crop models and their main principles.

Conceptually, STICS contains a few more novel elements (such as crop temperature simulation), while most of the remaining elements are founded on conventional formalizations or borrowed from earlier models. Advantages include:

- STICS advantages extend its use across a diverse range of crops, including grassland, tomato, flax, vineyard, pea, beetroot, sunflower, wheat, maize, soybean, banana, sorghum, sugarcane, rapeseed, carrot, lettuce, among others.
- STICS system prioritizes robustness, aiming for minimal output error in replicating soil-climate conditions by utilizing practical technical and soil inputs. However, heightened robustness might potentially trade-off with reduced local accuracy.
- STICS emphasizes "conceptual" modularity, showcasing adaptability to integrate additional modules or expand the system's description. This modularity encompasses aspects such as plant mulch, stony soils, ammonia volatilization, symbiotic nitrogen fixation, and various organic residues, expediting the system's developmental progress.

STICS simulates a single crop cycle or a series of cycles in one dimension. Radiation, minimum and maximum temperatures, precipitation, reference ET, wind, and humidity all define the upper border of the atmosphere (Dumont et al., 2014; Gobin, 2010). The bottom border is the soil/subsoil interface.

Nitrogen content, above-ground biomass, LAI, number of harvested organs, and organ biomass (and nitrogen content) are used to evaluate crops. Vegetative organs (leaves, stems, branches, roots) act differently as radiation, water, and nutrient sensors or reservoirs. Each horizontal layer's water, mineral and organic nitrogen levels identify it. The distribution of root density in the soil profile makes soil and crop interact. STICS may also be compared to intercropping, the concurrent growth of two crops (perennial or annual) that develop and mature according to their individual rhythms. This scenario's soil-plant-atmosphere system has three subsystems at the canopy level. A dominant canopy and an under-story canopy are split into two sections: a shaded portion and a sunlit section. Each area has a light microclimate that impacts how the subsystems perform.

Each module of the STICS model addresses certain mechanisms. The first set has three lessons on above-ground plant ecophysiology (phenology, shoot growth, and yield formation). Four sections describe how soil responds to below-ground plant components (root growth, water balance, nitrogen balance, and soil transfers). The soil-crop system is examined in the crop management module—microclimate studies how the water balance and the climate affect the temperature and humidity of the canopy.

8.3.6 SWAP CROP MODEL

SWAP models vadose zone water, solute, heat transfer, and vegetation growth (Healy, 2008; Kassaye et al., 2021; van Dam et al., 2008). Using the Richards equation, the model simulates soil moisture transport in variably saturated soils. Macroscopic and microscopic principles can replicate root water extraction. SWAP's macroporous flow module covers clay and peat. SWAP uses convection, dispersion, adsorption, and breakdown for solute transport (Crescimanno & Garofalo, 2005). For chemical transport models such as ORCHESTRA for heavy metals, ANIMO for nutrients, and PEARL for pesticides, SWAP generates soil water fluxes. Heat capacities and thermal conductivities are used in SWAP to represent soil heat flow (Overduin et al., 2006). SWAP model scope extends vertically from canopy to shallow groundwater. SWAP is a one-dimensional, vertical model since most transit in this zone is vertical. Runoff or daily transpiration fluxes can be modeled using minute-by-minute data on ET and rainfall. Bottom line: Head and flux-based conditions promote interactions between soil and surface water in delta zones.

SWAP gives field scale a horizontal priority. Since a field only contains one type of plant, one type of soil, one type of drainage condition, and one microclimate at this scale, most transport mechanisms may be described deterministically. There are many field-scale production methods; hence there are numerous management options. SWAP simulations can be scaled up from the field to the regional level for policy evaluations using geographic information systems (GIS) (Hartkamp et al., 1999; Zahraee et al., 2022). SWAP requires inputs for crop and soil parameters.

SWAP offers three advantages over earlier PMP-based irrigation methods. SWAP's regional exponential PMP land cost functions rectify earlier models' incapacity to manage significant policy shocks. SWAP contains regional CES agricultural production functions that allow modest input substitution. Previous versions included Leontief manufacturing. Regional agricultural prices are established endogenously depending on statewide demand.

8.3.7 FORTRAN CROP MODEL

FORTRAN is still commonly used for growth and yield simulators. It is a procedural language with a scientific numeric base (Ahmed et al., 2022; Romain et al., 2022; Viver, 2022). As applications get more sophisticated, procedural approaches and languages have given way to object-oriented approaches and languages. OO is popular because it helps maintain software (Salminen et al., 2005). Existing FORTRAN procedures may not be as straightforward to maintain and adapt as their OO counterparts, but they are valuable. Therefore, reusing code is cost-effective.

Fortran 90 is a versatile and capable programming environment for crop modelers. A framework for establishing (sub)model processes, integrating rate variables, updating time, entering weather data, reading model parameters from files, and visually displaying results is the simulation environment. Fortran was created as a basic simulation language that allows users to switch to Forton's more sophisticated and flexible environment (Neumann et al., 2021). The FST translator builds FSE data files and compact, extensible Fortran programs from FST source code (Craig et al., 2020). A simple modeling language for teaching is called Forton. Generated Fortran is a nice place to start for individuals who need more flexibility than Fortran provides.

Fortran separates state and rate computations in plant and soil subroutines. Task-controlled execution determines which portion is active. Per major program calls, only one of the rate and state components of the plant and soil water sub-processes is carried out. Differentiation exists between initialization, rate calculation, integration, and terminal computation. Clear error messages are provided, and FSTWin—the Windows version of FST that uses the Compaq Visual Fortran Compiler—is simple. The syntax of FST is CSMP. FST elements include INITIAL, DYNAMIC, and TERMINAL. These terms denote calculations carried out before, during, and following a simulation run. The initial conditions and parameters, time variables, data output, and the model's integration process are all described in the Starting section. Here, you may also compute results for the program's active part. The DYNAMIC section describes model dynamics and associated simulation computations. It is the model's longest part. The TERMINAL section is used for end-of-simulation computations and output. This might be a calculation using variable end values. FST's sorting method allows users to focus on process descriptions rather than statement sequencing. Each section's assertions are sorted and checked individually. FST provides clear error warnings, graphical output, and external weather data reading (Van Kraalingen, 1991). The FST/FSE coding was used for the Wageningen models INTERCOM (Kropff & Van Laar, 1993), ORYZA1 (Kropff et al., 1994), and SUCROS (van der Geest et al., 1997). The International Rice Research Institute in the Philippines developed ORYZA2000, which includes modules for water and nitrogen (Bouman & Tuong, 2001). Modelers in Gainesville, Florida have modified their CROPGRO model to fit the FSE modular architecture.

8.3.8 SIMPLE AND UNIVERSAL CROP GROWTH SIMULATOR SUCROS CROP MODEL

SUCROS (Goudriaan & Van Laar, 1994; van der Geest et al., 1997) is a Wageningen school crop growth model (van Ittersum et al., 2003). SUCROS employs a more advanced photosynthesis model than LINTUL. LINTUL optimizes radiation use (RUE). SUCROS combines photosynthesis during the day, canopy depths, and direct and diffuse radiation using a light response curve (LRC). SUCROS never implemented the biochemically more advanced Farquhar-von Caemmerer-Berry model of leaf photosynthesis. SUCROS 1 and 2 is an environmental crop growth model. In SUCROS2, dry matter accumulation in a pest-, disease-, and weed-free environment under the current meteorological conditions are modeled along with potential and water-constrained crop growth (Liu & Bull, 2001; Soualiou et al., 2021). A buildup of dry matter is influenced by irradiation, temperature, crop characteristics, and water availability. The amount of light and incoming radiation affects how quickly the canopy absorbs CO_2 during photosynthesis. Subtracting maintenance respiration is necessary to simulate the development of leaves, stems, roots, and storage organs. Biomass partitioning is a function of crop stage and temperature. Simulating crops requires crop-specific input parameters.

SUCROS1 is a mechanistic model which describes crop development based on processes like CO_2 uptake and respiration (Biernath et al., 2011). Mechanistic models are not always accurate. Each parameter estimate and process design is inaccurate, and mistakes can compound in ultimate yield predictions. Models' secondary goal is yielding prediction. Their main objective is to increase system knowledge through quantitatively integrating current data into a dynamic simulation model. The model's behavior can be used to understand the whole system's behavior better. Crop development may be hampered by nutrient, pest, and disease deficiencies. Therefore, numerous model variants have been produced.

SUCROS1 simulates crop development under sufficient water and nutrients, a clean environment free of pests, diseases, and weeds, and normal weather patterns. Irradiation, temperature, and crop characteristics affect dry matter accumulation. The calculation is based on the canopy photosynthetic rate. This rate depends on the crop's leaf area and the amount of incoming radiation. The crop's daily gross CO_2 absorption rate is calculated based on absorbed sunlight and leaf photosynthetic characteristics. These calculations are made using model subroutines (Goudriaan, 1986; Spitters et al., 1986; Van Laar et al., 1992). Submodel (MOMASSP.FST) creates diurnal gross assimilation for particular circumstances. Part of the generated carbohydrates (CH_2O) is utilized to sustain biomass. The remaining carbs become dry structural matter (plant organs). Growth respiration loses weight during conversion. The dry matter generated is divided among plant organs utilizing phenological parameters. Integrating growth rates across time yields plant organ dry weights.

8.3.9 CROPSYST CROP MODEL

CropSyst is a multi-year, multi-crop, daily time-step simulation model of cropping systems designed to investigate the effects of climate, soils, and management on crop productivity and the environment (Benli et al., 2007; Donatelli et al., 2015; Jalota et al., 2010; Stockle, 1996). The emphasis is on providing a user-friendly interface, connections to GIS applications, a weather generator, and other utilities. CropSyst mimics salinity, water erosiveness, biomass production, crop yield, soil water, soil-plant nitrogen, crop phenology, canopy, and root development, and crop phenology (Saljnikov et al., 2022). Crop rotation, cultivar selection, irrigation, nitrogen fertilizer, the salinity of the soil and irrigation water, tillage, and residue management are some of the management options that have an impact on these processes depending on the weather, the soil, the crop, and the cropping system (Debaeke & Aboudrare, 2004). In the 1990s, CropSyst got its start. It was created due to the dearth of cropping systems models, particularly those with the ability to support crop rotation. Quick, affordable progress was made possible by excellent communication between academics from different countries, a free distribution policy, active collaboration between model creators and users, and an early focus on software design.

To model crop growth and its interactions with management and the environment, CropSyst makes use of EPIC's conceptual capabilities. Less emphasis on software design was placed on the DSSAT and EPIC approaches. CropSyst has always focused on balancing superior research and effective software design techniques. It has certain goals in common with APSIM, a modeling technique that emphasizes quality software engineering processes (Keating et al., 2003; McCown et al., 1996).

CropSyst is a group of programs created to work together to evaluate crop rotations and crop management systems at various spatial and temporal scales. A parameter editor, a cropping systems simulator, a weather generator, a GIS-CropSyst simulation co-operator, a watershed study tool, and many other useful tools are all included in the CropSyst Suite.

8.3.10 EPIC CROP MODEL

The erosion Productivity Impact Calculator was designed to assess the link between soil erosion and production in the U.S. (Pan & Wen, 2014; Raza et al., 2021). Erosion Productivity Impact Calculator simulates erosion, plant development, and other processes using physical components. It incorporates economic components for evaluating erosion costs, effective management options, etc. Simulated physical processes use easily available inputs. EPIC can simulate hundreds of years of erosion if needed. For the 1985 RCA, the EPIC model was utilized to evaluate the needs for fertilizer, productivity, and erosion control. EPIC examined 5,00,000 different crop, tillage, and conservation strategies in the United States. The model needed to be versatile, effective, and able to calculate the effects of management decisions. Model elements include weather simulation, hydrology, erosion, and sedimentation, cycling of nutrients, crop development, tillage, soil temperature, economics, and control of the plant environment. Several elements have been described in other places (Anache et al., 2018; Gessesse et al., 2015; Himanshu et al., 2019). The only thing discussed is crop growth.

Given that it impacts soil productivity, crop growth is a crucial EPIC activity. To calculate the impact of erosion on crop yield, the model must be responsive to crop features, weather, soil fertility, and other parameters. Crop solar radiation absorption, light conversion to biomass, biomass division into roots, above-ground biomass, economic yield, root growth, water use, and nutrient uptake are all subject to simulation (Trybula et al., 2015). These simulations of plant growth are limited by five stress factors daily (water, nitrogen, phosphorus, temperature, and aeration). Three factors prevent roots from growing fully (soil strength, temperature, and aluminum toxicity). Erosion indirectly influences agricultural output by affecting soil characteristics and plant and root development. EPIC replicates all crops using one model with unique parameter values (Castaño-Sánchez et al., 2021; Trybula et al., 2015). EPIC simulates annual and perennial crop development (Powers et al., 2011; Williams et al., 2021). Annual crops develop from planting to harvest or until the sum of the gathered and potential heat is equal. All year long, the root systems of perennial crops like alfalfa, grasses, and pine trees are maintained (Powers et al., 2011). They begin to grow when the daily air temperature is higher than the plant's base temperature. Soybeans, alfalfa, maize, grain sorghum, wheat, barley, oats, sunflower, cotton, pasture, peanuts, potatoes, Durham wheat, winter peas, faba beans, rapeseed, sugarcane, sorghum hay, range grass, rice, cassava, lentils, and pine trees are among the crops listed in the EPIC crop table.

8.3.11 AQUA CROP MODEL

To reproduce crop canopy development, dry matter, and soil moisture content (SMC) under actual growing conditions, the Aqua crop model was calibrated for canopy cover expansion (development), dry matter accumulation, and SMC in the root zone under optimum growth circumstances (Montoya et al., 2016; Shan et al., 2022; Zeleke, 2019). The model replicates soil moisture balance, crop canopy, dry matter, and grain yield (using normalized water productivity function and dry matter and HI). Model performance is measured using the root mean square error (RMSE), normalized RMSE, index of agreement (d), model efficiency (E), and coefficient of determination (R2). As a decision-support tool, the aqua crop model can be used to analyze how climate change would affect water management strategies (rain-fed, mulching, and bunds), sowing dates, plant spacing, and fertilizer management tactics. According to Hsiao et al. Aqua crop model strikes a compromise between use, robustness, precision, and simplicity. The model is meant for extension experts, water managers, irrigation organization employees, economists, and policy specialists who use fundamental models for planning and scenario analysis (Heng et al., 2009). It has been employed in different soils, climates, and management styles worldwide on forage crops, vegetables, cereals, fruits, and root and tuber crops (Heng et al., 2009). Users cannot easily access information regarding stress coefficients (Ks), conservative, and user-specific factors, input and output parameters, management strategies, and applications. This research examines the use of the Aqua crop model in crop growth simulation models as a decision support tool and its implications for crop yield, food security, and climate change adaptation. For the convenience of the user, it also offers a thorough analysis of crop parameters, crop coefficients, calibration, and validation methods.

The aqua crop model is used for on-farm management, irrigation planning, climate change adaptation strategies, and soil water balance by irrigation and water specialists, extension workers, government, and NGOs, researchers, and others. It mimics the growth and production of herbaceous crops under various water, soil, and management conditions (Hsiao et al., 2009).

Due to its simplicity, adaptability, and accuracy, the Aqua Crop model simulates crop development and production (Heng et al., 2009; Hsiao et al., 2009). It has been used to simulate the growth and production of maize, barley, wheat, rice, sunflower, cotton, potato, hot pepper, cabbage, sugarcane, and faba bean under various management, soil, and climate conditions.

Crop penology, rooting depth, aerial canopy, biomass production, and harvestable yield are the five main parameters for aqua crops (Hsiao et al., 2009). Aqua crop uses conservative parameters for initial canopy cover per seedling (CCo)=plant density X CCo, canopy growth coefficient (CGC), and maximum canopy cover (CCx) to compute crop canopy development with time on emergence (Hsiao et al., 2009). Based on a plant's effective rooting depth and water extraction pattern, the aqua crop model creates a model of the root system. Crop rooting depth varies by species, and model inputs include extraction patterns and characteristics for the soil profile.

This diagram shows how soil dynamics and climate impact dynamic physiological systems. Biomass is created by ET. The biomass HI converts crop yield. For non-stressed conditions, the canopy expansion from emergence to complete canopy development follows an exponential rise during the first half and an exponential decrease during the second half (Heng et al., 2009). In Aqua crop, the canopy is represented as the green canopy ground cover (CC) proportion.

8.3.12 SALUS CROP MODEL

To make memory allocation dynamic and code platform independent, SALUS is written in "C." The SALUS software simulates crop, soil, water, and nutrient conditions under various management strategies over a long period (Lopez-Jimenez et al., 2022; Sarkar & Lupi, 2022; Shuai & Basso, 2022). There may be variations in crop rotations, planting dates, plant populations, irrigation, fertilizer, and tillage. The software replicates soil and plant growth every day for any period, taking into account weather patterns. Numerous management strategies could be used in any simulation at once. We can compare the effects of several strategies on crops and soil by employing them simultaneously. This provides a foundation for contrasting the management styles of various locations.

The SALUS model does not anticipate pest and disease outbreaks or harsh weather (e.g., hail). These factors can affect crop output, yield, and a management system's sustainability. Explicit inclusion of pest dynamics submodels in a general cropping systems model like SALUS limits the range of pest species and disease-causing organisms of important U.S. crop species. Modeling instances of extreme weather is also not practical. SALUS uses linkage points to communicate with external models of pest dynamics or data on the occurrence and harm of pests (e.g., integrated pest management (IPM) models) (Ntawubizi et al., 2020). A recognized, quantified pest effect on a crop is a linking point; as a result, pests can be classified according to their potential agricultural effects (i.e., the type of linkage point).

To connect with a particular pest model, SALUS makes available the necessary data (plant number, leaf or root tissue biomass, potential photosynthesis rates, senescence rates, etc.). It is possible to link SALUS to depleting leaf biomass, which affects photosynthesis, growth, and allocation, N intake, and other processes (Deines et al., 2021; Dzotsi et al., 2015; O'Neill et al., 2022; Shuai & Basso, 2022). Defoliating insect populations and tissue consumption rates. The model may also estimate the implications of variable pest incidence based on past data or expert opinion to provide risk analysis replies. By using connection points, the model may contain a wide variety of pest impacts (or damaging variables like hail) for a given application.

8.3.13 INFOCROP MODEL

InfoCrop was created to satisfy these needs. It replicates how weather, soils, agronomic practices (planting, nitrogen, residues, irrigation), and pests affect agricultural yield and impact the environment. Its structure is based on ORYZA1, MACROS, WTGROWS, and SUCROS (Aggarwal et al., 2006; Akula & Shekh, 2005). The approachable model strives to advance crop model R&D and calls for readily accessible inputs. The specifications of the model's cereal crop are covered in this document. The performance of rice and wheat in tropical agro-environments is described in the companion paper.

Rabbinge et al. (1994) describe how InfoCrop evaluates weather, soil, and pests in that order. Model processes include:

- Exploring phenology, photosynthesis, partitioning, leaf area expansion, storage organ numbers, sink balance, transpiration, nitrogen intake, allocation, and redistribution within crop growth processes.
- Evaluating stresses impacting crop growth, such as water scarcity, nitrogen deficiency, temperature fluctuations, floods, and frost.
- Addressing threats to crops posed by insects and diseases, leading to potential crop destruction.
- Considering aspects related to root absorption, inter-layer movement, drainage, evaporation, runoff, and ponding in the context of crop growth.
- Examining processes including mineralization, uptake, nitrification, volatilization, inter-layer mobility, denitrification, and leaching in the context of nutrient dynamics.
- Analyzing organic carbon mineralization and immobilization within the soil system.
- Assessing emissions of CO_2, CH_4, and NOx (N_2O) as part of the environmental impact related to agricultural activities.

Soil tillage and pest and disease population dynamics are missing from InfoCrop. These processes must be quantified to enhance the model. Our results show that the model functions effectively for rice and wheat crops in most-tropical agro-environments that differ in weather, soils, varieties, insect infestation, and agronomic management, despite these restrictions. InfoCrop's performance for other significant annual cereals, legumes, oilseeds, potatoes, and sugarcane will be provided later.

Errors are possible when building huge systems like InfoCrop, such as pulling out biomass from a plant portion and not adding it elsewhere. Carbon, water, and nitrogen balance checks were integrated into the model to prevent inaccuracies. These daily checks guaranteed that the sum of all incoming and outgoing fluxes matched the simulation's change.

We anticipate that InfoCrop will help analyze and interpret experimental data, estimate potential yields and yield gaps, assess climatic variability and change, optimizing agronomic management, quantify the change in soil organic carbon (SOC) with land use, assess greenhouse gas (GHG) emissions, and yield loss due to numerous pests (Gowindaknshnan et al., 2007; Singh et al., 2020). In most cases, the model's input requirements are restricted to readily accessible or quantifiable variables. Based on these fundamental inputs, the model calculates more sophisticated inputs. This compromise was needed to enhance model use in agricultural research, planning, and development.

8.3.14 GLAM CROP MODEL

A crop model GLAM is made to work on a grid of regional and global climate models (GCMs). As a result, GLAM is process-based yet less complex than field-scale models (Challinor & Wheeler, 2008; Challinor et al., 2015). Biotic stressors are not explicitly modeled, but their impact is implicitly taken into account via a yield gap parameter. It parameterizes how weather and climate affect crops. GLAM simulates the effect of climatic fluctuation and changes on agricultural growth and development, from sowing to harvest, by analyzing daily meteorological data (Challinor et al., 2015; Jennings et al., 2020; Ramirez-Villegas et al., 2017; Watson et al., 2015). The model may be utilized for generating and evaluating genotypic adaptation techniques by simulating various varietal features (Challinor et al., 2015; Watson et al., 2015). This model may be utilized in research that transforms gridded meteorological data into agricultural productivity results. In order to measure uncertainty and produce reliable outcomes, our infrastructure is perfectly suited for performing tens of thousands of simulations. The variety of this model's variations sets it apart GLAM-Parti with the new structure and improved partitioning; the 2020 version incorporating ozone damage; and a nitrogen-focused version in development). While crop breeding under climate change is now being studied, GLAM is mostly designed for use with climate models. The 2004 version is licensed with later modifications for heat stress and elevated CO_2.

8.4 CROP MODEL APPLICATIONS

Simulation models are becoming increasingly popular in fields like research, instruction, resource management, policy analysis, and production forecasting. It is possible to use research, crop management systems, and policy analyzes.

1. **Research Understanding:** Model building promotes the integration of research information achieved from the discrete disciplinary study, enables the identification of the system's driving elements, and can identify knowledge gaps. Adopting a modeling strategy might lead to more targeted and effective research planning.
2. **Integration of Knowledge Across Disciplines:** The adoption of modular architecture permits the integration of fundamental research conducted in many locations, nations, and continents. This assures a decrease in research expenditures and worldwide collaboration amongst researchers.
3. **Improvement in Experiment Documentation and Data Organization:** Development, testing, and deployment of simulation models need the utilization of a vast quantity of technical and observational data delivered in specified units and in precise order. Data management necessitates systematic data structure and database systems for the modeler.
4. **Site-Specific Experimentation:** The model can be used to select a specific site. Crop models can be used to predict crop performance in regions where the crop has never been grown or has not been grown under optimum conditions.
5. **Yield Analysis:** It is feasible to extrapolate to different contexts when a model with a solid physiological foundation is utilized. Various crop yields are climatically calculated using simulation models. The

modeling technique enables the assessment of production losses due to non-climatic factors (for example: retarded sowing, crop spacing, soil quality, pests, and diseases). It has also been noted that simulation models are beneficial for isolating the components of production advances attributable to changing weather patterns, genetic advancements, and technological advancements.

6. **Climate Change Projections:** The unpredictability of our climate and in particular, the accompanying weather extremes are a contemporary concern of both the scientific and public communities. The utilization of the crop model to investigate the potential impacts of climate change are felt across all continents. An increase in global temperature is projected due to the increased concentration of carbon dioxide (CO_2) and other GHGs. Because weather variability has a large impact on crop output, any alteration in the global climate is going to have a big impact on crop yields and agricultural productivity. Temperature and CO_2 levels affect biological processes such as respiration, photosynthesis, plant growth, reproduction, etc. Scientists will be better able to advise farmers on crop management decisions, such as crop selection, cultivar choice, sowing dates, and irrigation scheduling if they have a comprehensive grasp of the consequences of climate change.
7. **Yield Forecasting:** Before the actual harvest, reasonably accurate projections of crop yield across wide regions are of tremendous planning importance to both the researcher and the farmer using real meteorological data for cropping season in the geologically relevant region, the model is run using this manner. The simulations are run till harvest using typical weather years.
8. **Breeding and Introduction of a New Crop Variety:** The development and introduction of the new variety is a difficult process which might last between 5 and 15 years. Since the modeling systems method incorporates many agroecosystem components, it may be utilized to perform multi-location field experiments to comprehend genotype by environment relationships (G × E). These studies might help lower the number of sites/seasons required during field evaluation, enhancing the efficacy of variety creation. Again, by simulating a variety of likely genotypes and settings known to discriminate between the genotypes, it may be feasible to identify the crop characteristic that determines the specific interaction. Using the crop

factors that are most beneficial as a predictor of desirable traits and a breeding goal, hypothetical values might then be predicted. Based on understanding the G x E connection, a modeling technique can also provide yield probability estimates in target conditions.

8.5 CROP MODEL LIMITATIONS

Due to an insufficient grasp of natural processes and a lack of computing capacity, crop models cannot provide precise forecasts (Liu et al., 2022; Lowe et al., 2022; Prodhan et al., 2022). Consequently, the estimates of potential consequences of climate change are based on estimations. Furthermore, most models cannot provide reliable projections of shifts in local climate variability or the frequency of extreme events like storms and droughts (Stott et al., 2016; Trenberth et al., 2014). Despite the possibility that such changes could significantly impact agricultural production, general circulation models (GCMs) have not been able to produce reliable predictions of changes in climatic variabilities, such as increases in the number of droughts and storms (Macias Fauria et al., 2011). Models may be misused because different users have varying levels of modeling expertise. The user must choose the crop model that best suits his objectives because crop models are not all the same. As a result, assessments of the likely impacts of climate change are based on estimates.

Furthermore, most models cannot provide reliable projections of changes in regional climate variability or the frequency of extreme events like storms and droughts (Shewmake et al., 2008). Despite the possibility that such changes could significantly impact agricultural production, GCMs have not been able to produce reliable predictions of variations in climatic variabilities, like increases in droughts and storms. GCMs perform badly at the regional scale but reasonably well when simulating global surface air temperatures and rainfall (Kamruzzaman et al., 2021). Additionally, since the behavior of some components in biological and agricultural systems is not fully understood, differences between model output and actual-world systems cannot be justified. As a result of insufficient understanding of natural processes and limitations in computer capacity, crop models are incapable of providing reliable forecasts. Again, the model validation technique is still in its infancy.

In contrast to traditional/disciplinary experiments, a model concurrently tests many hypotheses. Field data are rarely conclusive enough for model validation to be effective, which hinders validation efforts. This is because model parameters and driving forces are derived from site-specific conditions, which should ideally be measurable and available. However, plant, soil, and climate data are frequently erroneous and may come from nearby sites. Sometimes, arbitrary estimates are made for irregularly measured parameters which turn out to be significant. The variability of recorded metrics is further influenced by inherent soil heterogeneity over shorter distances and changes brought on by agricultural operations' impact on soil properties. Throughout the growing season, crop statistics reflect soil heterogeneity and environmental conditions variance. The quality of input data constrains the performance of a model. In cropping systems, it is customary to have factual data on crop growth and development above ground, while data on root growth and soil parameters are often less thorough (Corbeels et al., 2016; de Vries & Wallenstein, 2017; Freschet et al., 2017). Most simulation models require accurate and exhaustive weather data. In addition, sampling mistakes contribute to the inaccuracy of observed data. A perfect crop model would outline all physical and physiological connections between the variables it reproduces and behavior in the real world. Due to the complexity of the biological system and the inadequate knowledge of many of its functions, such a model cannot be made. Even if the ideal crop model were made, it would be difficult to collect the extremely precise system parameters and input data for such a crop environment. As a result, the volume of information included in a crop model is directly tied to its intended purpose and the required level of precision. Even when a sensible choice is made, it is crucial to consider model limits so that modeling investigations are understood in their proper context.

8.6 CONCLUSION

For years, crop models have been routinely used to postulate strategies to enhance agricultural output under various climatic, soil, and management circumstances. Crop models are increasingly used to analyze how climate change affects crop development and productivity, model soil water balance, model canopy cover (LAI), model dry matter accumulation (d), and model crop yield. Depending on the goals, crop models are used as decision support systems for crop genetics, irrigation, soil moisture conservation, and runoff, as well as to quantify nutrient and water stress at various developmental stages. In order to evaluate crop Water Use Efficiency and Water Productivity, Irrigation Scheduling and Operations, Planting/Sowing Date, and Policy

Implications of Food Security, Crop Models are used. The geographical and temporal effects of temperature and precipitation variability may be investigated using crop modeling.

To provide comprehensive and in-depth research findings in crop modeling, integrated research involving various activities and professionals (computer engineers, agronomists, economists, breeders, soil scientists, climate scientists, etc.) is necessary. Crop modeling may therefore simulate soil water balance, crop development, and production under ideal and unfavorable water and fertilizer levels; as a result, farmers can benefit from the findings of these studies and boost their output and food security.

KEYWORDS

- **agronomic crops**
- **AQUA crop model**
- **crop growth simulator**
- **crop model applications**
- **crop simulation models**
- **EPIC crop model**
- **intercropping**

REFERENCES

Aggarwal, P. K., Kalra, N., Chander, S., & Pathak, H., (2006). InfoCrop: A dynamic simulation model for the assessment of crop yields, losses due to pests, and environmental impact of agro-ecosystems in tropical environments. I. Model description. *Agricultural Systems, 89*, 1–25.

Ahmed, Z., Gui, D., Qi, Z., Liu, Y., Liu, Y., & Azmat, M., (2022). Agricultural system modeling: Current achievements, innovations, and future roadmap. *Arabian Journal of Geosciences, 15*, 1–13.

Akula, B., & Shekh, A., (2005). Sensitivity analysis of InfoCrop to weather and non-weather parameters. *Indian Journal of Plant Physiology, 10*, 236.

Anache, J. A., Flanagan, D. C., Srivastava, A., & Wendland, E. C., (2018). Land use and climate change impacts on runoff and soil erosion at the hillslope scale in the Brazilian Cerrado. *Science of the Total Environment, 622*, 140–151.

Anstey, J., Butchart, N., Hamilton, K., Osprey, S., Bushell, A., Holt, L., Richter, Y., et al., (2021). Improving the QBO in climate models. *SPARC News Letter, 57*, 12–47.

Benli, B., Pala, M., Stockle, C., & Oweis, T., (2007). Assessment of winter wheat production under early sowing with supplemental irrigation in a cold highland environment using CropSyst simulation model. *Agricultural Water Management, 93*, 45–53.

Biernath, C., Gayler, S., Bittner, S., Klein, C., Högy, P., Fangmeier, A., & Priesack, E., (2011). Evaluating the ability of four crop models to predict different environmental impacts on spring wheat grown in open-top chambers. *European Journal of Agronomy, 35*, 71–82.

Boote, K. J., Jones, J. W., White, J. W., Asseng, S., & Lizaso, J. I., (2013). Putting mechanisms into crop production models. *Plant, Cell & Environment, 36*, 1658–1672.

Bouman, B., & Tuong, T. P., (2001). Field water management to save water and increase its productivity in irrigated lowland rice. *Agricultural Water Management, 49*, 11–30.

Brisson, N., Beaudoin, N., Mary, B., & Launay, M., (2009). Conceptual basis, formalizations, and parameterization of the STICS crop model. *Conceptual Basis, Formalizations, and Parameterization of the STICS Crop Model*, 1–298.

Brisson, N., Gary, C., Justes, E., Roche, R., Mary, B., Ripoche, D., Zimmer, D., et al., (2003). An overview of the crop model STICS. *European Journal of Agronomy, 18*, 309–332.

Bu, Q., Li, Q., Liu, Y., & Cai, C., (2021). Performance comparison between the specific and baseline prediction models of ecotoxicity for pharmaceuticals: Is a specific QSAR model inevitable? *Journal of Chemistry*, 2021.

Castaño-Sánchez, J. P., Izaurralde, R. C., & Prince, S. D., (2021). Land-use conversions from managed grasslands to croplands in Uruguay increase medium-term net carbon emissions to the atmosphere. *Journal of Land Use Science, 16*, 240–259.

Challinor, A. J., Parkes, B., & Ramirez-Villegas, J., (2015). Crop yield response to climate change varies with cropping intensity. *Global Change Biology, 21*, 1679–1688.

Challinor, A., & Wheeler, T., (2008). Crop yield reduction in the tropics under climate change: Processes and uncertainties. *Agricultural and Forest Meteorology, 148*, 343–356.

Colas, F., Gauchi, J. P., Villerd, J., & Colbach, N., (2021). Simplifying a complex computer model: Sensitivity analysis and metamodeling of an 3D individual-based crop-weed canopy model. *Ecological Modeling, 454*, 109607.

Corbeels, M., Chirat, G., Messad, S., & Thierfelder, C., (2016). Performance and sensitivity of the DSSAT crop growth model in simulating maize yield under conservation agriculture. *European Journal of Agronomy, 76*, 41–53.

Craig, J. R., Brown, G., Chlumsky, R., Jenkinson, R. W., Jost, G., Lee, K., Mai, J., et al., (2020). Flexible watershed simulation with the Raven hydrological modeling framework. *Environmental Modeling & Software, 129*, 104728.

Crescimanno, G., & Garofalo, P., (2005). Application and evaluation of the SWAP model for simulating water and solute transport in a cracking clay soil. *Soil Science Society of America Journal, 69*, 1943–1954.

De Vries, F. T., & Wallenstein, M. D., (2017). Below-ground connections underlying above-ground food production: A framework for optimizing ecological connections in the rhizosphere. *Journal of Ecology, 105*, 913–920.

Debaeke, P., & Aboudrare, A., (2004). Adaptation of crop management to water-limited environments. *European Journal of Agronomy, 21*, 433–446.

Deines, J. M., Kendall, A. D., Butler, J. J., Basso, B., & Hyndman, D. W., (2021). Combining remote sensing and crop models to assess the sustainability of stakeholder-driven groundwater management in the US high plains aquifer. *Water Resources Research, 57*, e2020WR027756.

Donatelli, M., Srivastava, A. K., Duveiller, G., Niemeyer, S., & Fumagalli, D., (2015). Climate change impact and potential adaptation strategies under alternate realizations of climate scenarios for three major crops in Europe. *Environmental Research Letters, 10*, 075005.

Dumont, B., Leemans, V., Ferrandis, S., Bodson, B., Destain, J. P., & Destain, M. F., (2014). Assessing the potential of an algorithm based on mean climatic data to predict wheat yield. *Precision Agriculture, 15*, 255–272.

Dzotsi, K., Basso, B., & Jones, J., (2015). Parameter and uncertainty estimation for maize, peanut and cotton using the SALUS crop model. *Agricultural Systems, 135*, 31–47.

Fischer, G., & Sun, L., (2001). Model based analysis of future land-use development in China. *Agriculture, Ecosystems & Environment, 85*, 163–176.

Freschet, G. T., Valverde-Barrantes, O. J., Tucker, C. M., Craine, J. M., McCormack, M. L., Violle, C., Fort, F., et al., (2017). Climate, soil, and plant functional types as drivers of global fine-root trait variation. *Journal of Ecology, 105*, 1182–1196.

Gessesse, B., Bewket, W., & Bräuning, A., (2015). Model-based characterization and monitoring of runoff and soil erosion in response to land use/land cover changes in the Modjo watershed, Ethiopia. *Land Degradation & Development, 26,* 711–724.

Gobin, A., (2010). Modeling climate impacts on crop yields in Belgium. *Climate Research, 44*, 55–68.

Goudriaan, J., & Van, L. H., (1994). Radiation in crops. *Modeling Potential Crop Growth Processes* (pp. 95–119). Springer.

Goudriaan, J., (1986). A simple and fast numerical method for the computation of daily totals of crop photosynthesis. *Agricultural and Forest Meteorology, 38*, 249–254.

Gowindaknshnan, P., Singh, J., Lal, S., & Panigrahy, S., (2007). *A Methodology for Pre-harvest Prediction of Mean Potato Yield at Regional Scale using INFOCROP*. POTATO model.

Hammer, G., Kropff, M., Sinclair, T., & Porter, J., (2002). Future contributions of crop modeling—From heuristics and supporting decision making to understanding genetic regulation and aiding crop improvement. *European Journal of Agronomy, 18*, 15–31.

Han, J. C., (2016). *Development of CornSoyWater – a Web-based Irrigation App for Corn and Soybean.* The University of Nebraska-Lincoln. https://iopscience.iop.org/article/10.1088/1748-9326/ab17eb/meta#fnref-erlab17ebbib14 (accessed on 30 November 2023).

Hansen, J. W., (2005). Integrating seasonal climate prediction and agricultural models for insights into agricultural practice. *Philosophical Transactions of the Royal Society B: Biological Sciences, 360*, 2037–2047.

Hartkamp, A. D., White, J. W., & Hoogenboom, G., (1999). Interfacing geographic information systems with agronomic modeling: A review. *Agronomy Journal, 91*, 761–772.

Healy, R. W., (2008). Simulating water, solute, and heat transport in the subsurface with the VS2DI software package. *Vadose Zone Journal, 7*, 632–639.

Heng, L. K., Hsiao, T., Evett, S., Howell, T., & Steduto, P., (2009). Validating the FAO AquaCrop model for irrigated and water deficient field maize. *Agronomy Journal, 101*, 488–498.

Himanshu, S. K., Pandey, A., Yadav, B., & Gupta, A., (2019). Evaluation of best management practices for sediment and nutrient loss control using SWAT model. *Soil and Tillage Research, 192*, 42–58.

Hsiao, T. C., Heng, L., Steduto, P., Rojas-Lara, B., Raes, D., & Fereres, E., (2009). AquaCrop—The FAO crop model to simulate yield response to water: III. Parameterization and testing for maize. *Agronomy Journal, 101*, 448–459.

Jalota, S., Singh, S., Chahal, G., Ray, S., Panigraghy, S., & Singh, K., (2010). Soil texture, climate and management effects on plant growth, grain yield and water use by rainfed maize–wheat cropping system: Field and simulation study. *Agricultural Water Management, 97*, 83–90.

Jennings, S. A., Koehler, A. K., Nicklin, K. J., Deva, C., Sait, S. M., & Challinor, A. J., (2020). Global potato yields increase under climate change with adaptation and CO_2 fertilization. *Frontiers in Sustainable Food Systems, 4*, 519324.

Kamruzzaman, M., Shahid, S., Islam, A., Hwang, S., Cho, J., Zaman, M., Uz, A., et al., (2021). Comparison of CMIP6 and CMIP5 model performance in simulating historical precipitation and temperature in Bangladesh: A preliminary study. *Theoretical and Applied Climatology, 145*, 1385–1406.

Kassaye, K. T., Boulange, J., Saito, H., & Watanabe, H., (2021). Soil water content and soil temperature modeling in a vadose zone of andosol under temperate monsoon climate. *Geoderma, 384*, 114797.

Keating, B. A., Carberry, P. S., Hammer, G. L., Probert, M. E., Robertson, M. J., Holzworth, D., Huth, N. I., et al., (2003). An overview of APSIM, a model designed for farming systems simulation. *European Journal of Agronomy, 18*, 267–288.

Kropff, M. J., & Van, L. H., (1993). Modeling crop-weed interactions. *Int. Rice Res. Inst.*

Kropff, M., Cassman, K., Peng, S., Matthews, R., & Setter, T., (1994). Quantitative understanding of yield potential. *Breaking the Yield Barrier*, 21–38.

Kropff, M., Van, K. D., & Van, L. H., (1993). Program structure of the model INTERCOM. *Modeling Crop-Weed Interactions* (pp. 201–246).

Lindgren, E., & Gabrielsson, E., (2021). *Design of a Rugged Intercom System with an Interaction Design Approach.* http://lup.lub.lu.se/student papers/record/9043749 (accessed on 30 November 2023).

Liu, D., & Bull, T., (2001). Simulation of biomass and sugar accumulation in sugarcane using a process-based model. *Ecological Modeling, 144*, 181–211.

Liu, L., Xu, S., Tang, J., Guan, K., Griffis, T. J., Erickson, M. D., Frie, A. L., et al., (2022). KGML-ag: A modeling framework of knowledge-guided machine learning to simulate agroecosystems: A case study of estimating N_2O emission using data from mesocosm experiments. *Geoscientific Model Development, 15*, 2839–2858.

Loewer, O. J., Smith, E., Benock, G., Bridges, T. C., Wells, L., Gay, N., Burgess, S., Springate, L., & Debertin, D., (1981). A simulation model for assessing alternate strategies for beef production with land, energy, and economic constraints. *Transactions of the ASAE, 24*, 164–0173.

Lopez-Jimenez, J., Vande, W. A., & Quijano, N., (2022). Dynamic modeling of crop–soil systems to design monitoring and automatic irrigation processes: A review with worked examples. *Water, 14*, 889.

Lowe, M., Qin, R., & Mao, X., (2022). A review on machine learning, artificial intelligence, and smart technology in water treatment and monitoring. *Water, 14*, 1384.

Lowenberg-Deboer, J., Krause, M., Deuson, R., & Reddy, K., (1991). Simulation of yield distributions in millet-cowpea intercropping. *Agricultural Systems, 36*, 471–487.

Macias, F. M., Michaletz, S. T., & Johnson, E. A., (2011). Predicting climate change effects on wildfires requires linking processes across scales. *Wiley Interdisciplinary Reviews: Climate Change, 2*, 99–112.

McCown, R. L., Hammer, G. L., Hargreaves, J. N. G., Holzworth, D. P., & Freebairn, D. M., (1996). APSIM: A novel software system for model development, model testing and simulation in agricultural systems research. *Agricultural Systems, 50*, 255–271.

Montoya, F., Camargo, D., Ortega, J., Córcoles, J., & Domínguez, A., (2016). Evaluation of Aquacrop model for a potato crop under different irrigation conditions. *Agricultural Water Management, 164*, 267–280.

Neumann, A., Dong, F., Shimoda, Y., Arnillas, C. A., Javed, A., Yang, C., Zamaria, S., Mandal, S., Wellen, C., Paredes, D., et al., (2021). A review of the current state of process-based and data-driven modeling: Guidelines for Lake Erie managers and watershed modelers. *Environmental Reviews, 29*, 443–490.

Ntawubizi, M., Shyaka, A., Mukantwali, C., Niyonzima, E., Ndahimana, J., & Ndahetuye, J. B., (2020). *Situational Analysis of the Food Safety Control System in Rwanda*. https://cgspace.cgiar.org/bitstream/handle/10568/117934/Food%20safety%20control%20Rwanda.pdf?sequence=5 (accessed on 30 November 2023).

O'Neill, E. G., Martinez-Feria, R. A., Basso, B., & Maravelias, C. T., (2022). Integrated spatially explicit landscape and cellulosic biofuel supply chain optimization under biomass yield uncertainty. *Computers & Chemical Engineering, 160*, 107724.

Obidiegwu, J. E., Bryan, G. J., Jones, H. G., & Prashar, A., (2015). Coping with drought: Stress and adaptive responses in potato and perspectives for improvement. *Frontiers in Plant Science, 6*, 542.

Overduin, P. P., Kane, D. L., & Van, L. W. K., (2006). Measuring thermal conductivity in freezing and thawing soil using the soil temperature response to heating. *Cold Regions Science and Technology, 45*, 8–22.

Pan, J., & Wen, Y., (2014). Estimation of soil erosion using RUSLE in Caijiamiao watershed, China. *Natural Hazards, 71*, 2187–2205.

Powers, S., Ascough, II. J., Nelson, R., & Larocque, G., (2011). Modeling water and soil quality environmental impacts associated with bioenergy crop production and biomass removal in the Midwest USA. *Ecological Modeling, 222*, 2430–2447.

Prodhan, F. A., Zhang, J., Hasan, S. S., Sharma, T. P. P., & Mohana, H. P., (2022). A review of machine learning methods for drought hazard monitoring and forecasting: Current research trends, challenges, and future research directions. *Environmental Modeling & Software*, 105327.

Querol, L. S., (2021). *The SoMus Project: Close-up on Innovative Participatory Management Models in European Museums*. Les Cahiers de Muséologie.

Ramirez-Villegas, J., Koehler, A. K., & Challinor, A. J., (2017). Assessing uncertainty and complexity in regional-scale crop model simulations. *European Journal of Agronomy, 88*, 84–95.

Raza, A., Ahrends, H., Habib-Ur-Rahman, M., & Gaiser, T., (2021). Modeling approaches to assess soil erosion by water at the field scale with special emphasis on heterogeneity of soils and crops. *Land, 10*, 422.

Requena-Mesa, C., Benson, V., Reichstein, M., Runge, J., & Denzler, J., (2021). EarthNet2021: A large-scale dataset and challenge for Earth surface forecasting as a guided video prediction task. *Proceedings of the IEEE/CVF Conference on Computer Vision and Pattern Recognition*, (pp. 1132–1142).

Romain, G., Philippe, P., Julien, B., Odalric-Ambrym, M., & David, E., (2022). *Gym-DSSAT: A Crop Model Turned into a Reinforcement Learning Environment*. arXiv preprint arXiv:2207.03270.

Ruffin, T., Steele, J., Acquaah, Y., Sharma, N., Sarku, E., Tesiero, R., & Gokaraju, B., (2021). Non-invasive low-cost fever detection systems, *SoutheastCon, 2021* (pp. 1–7). IEEE.

Saljnikov, E., Lavrishchev, A., Römbke, J., Rinklebe, J., Scherber, C., Wilke, B. M., Tóth, T., et al., (2022). Understanding and monitoring chemical and biological soil degradation. *Advances in Understanding Soil Degradation* (pp. 75–124). Springer.

Salminen, H., Lehtonen, M., & Hynynen, J., (2005). Reusing legacy FORTRAN in the MOTTI growth and yield simulator. *Computers and Electronics in Agriculture, 49*, 103–113.

Sarkar, S., & Lupi, F., (2022). Modelling mid-western corn yield response to phosphorus fertilizer in Michigan. In: *2022 Annual Meeting, Anaheim, California* (No. 322478). Agricultural and Applied Economics Association. https://ideas.repec.org/p/ags/aaea22/322478.html; https://ageconsearch.umn.edu/record/322478/files/AAEA2022_SarkarLupiBasso.pdf (accessed on 30 November 2023).

Shan, Y., Li, G., Su, L., Zhang, J., Wang, Q., Wu, J., Mu, W., & Sun, Y., (2022). Performance of AquaCrop model for maize growth simulation under different soil conditioners in Shandong Coastal Area, China. *Agronomy, 12*, 1541.

Shewmake, T. A., Solis, F. J., Gillies, R. J., & Caplan, M. R., (2008). Effects of linker length and flexibility on multivalent targeting. *Biomacromolecules, 9*, 3057–3064.

Shuai, G., & Basso, B., (2022). Subfield maize yield prediction improves when in-season crop water deficit is included in remote sensing imagery-based models. *Remote Sensing of Environment, 272*, 112938.

Singh, A., Singh, G., & Kaur, P., (2020). Comparison of CERES-wheat and infocrop model to predict the phenology and yield of wheat. *Indian Journal of Ecology, 47*, 690–696.

Soualiou, S., Wang, Z., Sun, W., De Reffye, P., Collins, B., Louarn, G., & Song, Y., (2021). Functional–structural plant models mission in advancing crop science: Opportunities and prospects. *Frontiers in Plant Science, 12*.

Spitters, C., Toussaint, H., & Goudriaan, J., (1986). Separating the diffuse and direct component of global radiation and its implications for modeling canopy photosynthesis part I. Components of incoming radiation. *Agricultural and Forest Meteorology, 38*, 217–229.

Stockle, C. O., (1996). GIS and simulation technologies for assessing cropping systems management in dry environments. *American Journal of Alternative Agriculture, 11*, 115–120.

Stott, P. A., Christidis, N., Otto, F. E., Sun, Y., Vanderlinden, J. P., Van, O. G. J., Vautard, R., et al., (2016). Attribution of extreme weather and climate-related events. Wiley interdisciplinary reviews. *Climate Change, 7*, 23–41.

Swain, A., Behera, D., Karmakar, S., Dash, M., Dash, B. P., Swain, P., Molla, K. A., & Baig, M. J., (2021). Morphophysiological alterations in transgenic rice lines expressing PPDK and ME genes from the C4 model *Setaria italica. Journal of Plant Physiology, 264*, 153482.

Townsend, J. T., (2021). Issues and models concerning the processing of a finite number of inputs, 1. *Human Information Processing* (pp. 133–185). Routledge.

Trenberth, K. E., Dai, A., Van, D. S. G., Jones, P. D., Barichivich, J., Briffa, K. R., & Sheffield, J., (2014). Global warming and changes in drought. *Nature Climate Change, 4*, 17–22.

Trybula, E. M., Cibin, R., Burks, J. L., Chaubey, I., Brouder, S. M., & Volenec, J. J., (2015). Perennial rhizomatous grasses as bioenergy feedstock in SWAT: Parameter development and model improvement. *GCB Bioenergy, 7*, 1185–1202.

Van, D. C. V., Wolf, J. V., Van, K. H., & Rappoldt, C., (1989). WOFOST: A simulation model of crop production. *Soil Use and Management, 5*, 16–24.

Van, D. G. R., Van, L. T., Gubbens-Stibbe, J. M., Boddé, H. E., & Danhof, M., (1997). Lontophoretic delivery of apomorphine II: An *in vivo* study in patients with Parkinson's disease. *Pharmaceutical Research, 14*, 1804–1810.

Van, D. J. C., Groenendijk, P., Hendriks, R. F., & Kroes, J. G., (2008). Advances of modeling water flow in variably saturated soils with SWAP. *Vadose Zone Journal, 7*, 640–653.

Van, I. M. K., Leffelaar, P. A., Van, K. H., Kropff, M. J., Bastiaans, L., & Goudriaan, J., (2003). On approaches and applications of the Wageningen crop models. *European Journal of Agronomy, 18*, 201–234.

Van Kraalingen, D. W. G., (1991). *The FSE System for Crop Simulation* (No. 23). https://library.wur.nl/WebQuery/wurpubs/fulltext/216728 (accessed on 30 November 2023).

Van, L. H., Goudriaan, J., & Van, K. H., (1992). *Simulation of Crop Growth for Potential and Water-Limited Production Situations: As Applied to Spring Wheat.* CABO-DLO.

Viver, I. A., (2022). *Translation and Evaluation of the LINTUL2-Banana Crop Growth Model in R.* Doctoral dissertation, MSc Thesis, Wagenigen University & Research, Wagenigen. https://edepot.wur.nl/573896 (accessed on 30 November 2023).

Watson, J., Challinor, A. J., Fricker, T. E., & Ferro, C. A., (2015). Comparing the effects of calibration and climate errors on a statistical crop model and a process-based crop model. *Climatic Change, 132*, 93–109.

Whisler, F., Acock, B., Baker, D., Fye, R., Hodges, H., Lambert, J., Lemmon, H., et al., (1986). Crop simulation models in agronomic systems. *Advances in Agronomy, 40*, 141–208.

Williams, A. S., Kim, S., & Kiniry, J. R., (2021). Advances in application of a process-based crop model to wetland plants and ecosystems. *Wetlands, 41*, 1–17.

Yang, H., Dobermann, A., Lindquist, J. L., Walters, D. T., Arkebauer, T. J., & Cassman, K. G., (2004). Hybrid-maize—A maize simulation model that combines two crop modeling approaches. *Field Crops Research, 87*, 131–154.

Zahraee, S. M., Shiwakoti, N., & Stasinopoulos, P., (2022). Application of geographical information system and agent-based modeling to estimate particle-gaseous pollutant emissions and transportation cost of woody biomass supply chain. *Applied Energy, 309*, 118482.

Zeleke, K. T., (2019). AquaCrop calibration and validation for Faba bean (*Vicia Faba* L.) under different agronomic managements. *Agronomy, 9*, 320.

CHAPTER 9

Novel Disruptive Technology Innovations for Smart Sustainable Agricultural Practices

RAZEEF MOHD,[1] ABEENA MOHIUDDIN,[2] NAZIR AHMAD GANIE,[3] OWAIS BASHIR,[4] AND KHALID REHMAN HAKEEM[5]

[1]*NAHEP, Sher-e-Kashmir University of Agricultural Sciences and Technology of Kashmir, Kashmir, India*

[2]*Department of Computer Sciences, Mewar University Rajasthan, Chittorgarh, Rajasthan, India*

[3]*Division of Animal Genetics and Breeding, Sher-e-Kashmir University of Agricultural Sciences and Technology of Kashmir, Kashmir, India*

[4]*Division of Soil Science, Sher-e-Kashmir University of Agricultural Sciences and Technology of Kashmir, Kashmir, India*

[5]*Department of Biological Sciences, Faculty of Science, King Abdulaziz University, Jeddah, Saudi Arabia*

ABSTRACT

Adopting disruptive technologies in agriculture and allied sectors has brought a revolution in agriculture and allied sectors. Technologies like machine learning (ML), artificial intelligence (AI), IoT, drone, etc., has promising result in agricultural activities from specie selection to harvesting and logistics. These technologies help in controlled irrigation, smart pesticide spraying, and smart herbicide eradication, uphold soil fertility, and support the well-organized use of the workforce to enhance sustainable

Soil, Agriculture, and Ecosystem Modelling: Smart Technologies for Sustainable Solutions.
Owais Bashir, Khalid Rehman Hakeem, Surinder S. Kukal, and Razeef Mohd (Eds.)

smart agriculture practices. In this work, we will try to provide useful and important applications of ML and AI approaches to help and provide awareness to farmers and other stakeholders about the potential benefits of using disruptive technologies in agricultural activities and their allied sectors.

9.1 INTRODUCTION

We live in an interesting era which is a volatile, uncertain, complex, and ambiguous era, yet we also live in an era with wonderful technological opportunities and better life perspectives. Information technologies and communication have revolutionized every walk of life, and modern digital computer machines are becoming more intelligent, robust, and resilient. Artificial intelligence (AI) has made computer machines intelligent to perform tasks like humans with more accuracy and reliability. The term "Artificial Intelligence" was coined by John McCarthy in 1956 with the philosophy. Can machines think and behave like humans do? (McCarthy & Patrick, 1981).

It provides them with knowledge through various learning mechanisms. It is developed to be used by computers and machines. In terms of AI, learning is a broad term that denotes gaining knowledge, skill, and understanding from instruction, experience, data, or reflection. We establish AI systems by making machines to learn from them.

AI is one of the intriguing and universal fields of computer science which has a great scope in the future. AI holds a tendency to cause a machine to work as a human. Whether at home or at work, AI has changed our everyday lifestyles and transformed how businesses operate and bring value to customers. AI is already present in our everyday life in the form of various products and services like smart cars, smart homes, smartphones, smart businesses, smart agriculture, and in many more ways that we are using knowingly or unknowingly. AI is the world's future, which will drive tomorrow's economy into value and opportunity.

Data is the most indispensable item and oil for AI, Ml, and other disruptive technologies. We are sitting on a mountain of gold (data), but we are not mining it as effectively as we could. As data is considered —new-oil‖ and the main function of AI and ML is to unlock the concealed information/knowledge available in the data (Lujala, Jan, & Nadja, 2007). Data is oil for any data industry that is producing and consuming more data than ever before. Data is increasing every second exponentially with an astounding rate and frequency. If we do not harness this data efficiently and effectively, we will remain behind in the AI revolution transforming the world and human perception.

Data storage and collection abilities have allowed the accumulation of enormous amounts of data. To handle this huge accumulation of data, tools and techniques have been developed to analyze and manage this huge amount of data into meaningful information. AI works by combining large volumes of data with fast, iterative processing and intelligent algorithms, allowing the software to learn by itself automatically with time from patterns or features present in the data.

In today's world, AI has a wide range of applications that transform every walk of life. AI is becoming indispensable for the human race as it is used in multiple industries, such as Agriculture, Healthcare, entertainment, manufacturing, finance, education, IT, etc. AI is making our lives and business more comfortable, profitable, and highly productive. Figure 9.1 shows some of the main applications of AI in various sectors.

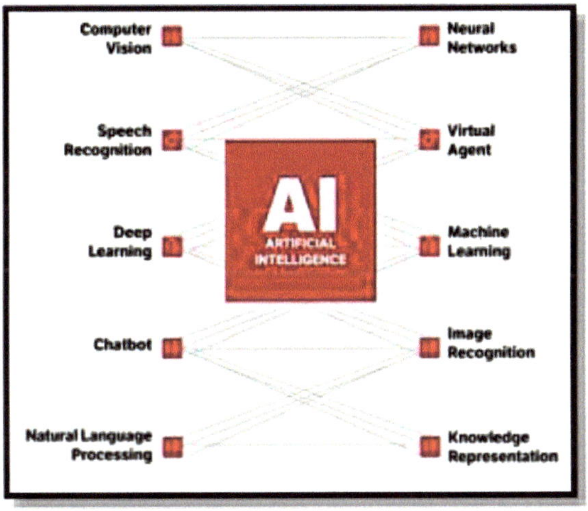

FIGURE 9.1 Application of artificial intelligence.

9.2 ARTIFICIAL INTELLIGENCE (AI) IN AGRICULTURE

Through agricultural AI and cognitive technologies, farms across the world can run more efficiently to produce the fundamental staples of our dietary lifestyles. Agriculture plays a significant role in the economic sector. For countries, like India, the agricultural sector accounts for 18% of GDP and provides employment to 50% of the country's workforce. Growth and expansion in agriculture will enhance and uplift the rural development and quality of life for farmers, leading to rural transformation and eventually resulting in structural transformation (Talaviya et al., 2020). Agriculture has been the backbone of India's economy; surprisingly, it is the least digitized sector that needs to be revolutionized and infused with new disruptive technologies for development and commercialization. Agricultural automation is the main concern and the emerging technology that farmers and agriculturalists adopt worldwide. As the population is increasing tremendously and this increases the demand for food, the traditional methods the farmers used were insufficient to fulfill these requirements. Thus, new automated and disruptive technologies and methods were introduced. AI in agriculture has brought an agriculture revolution (Zambon et al., 2019). This technology has protected crop yield from climate changes, population growth, employment issues, and food security problems. Implementing AI and ML enables agriculturists to access increasingly sophisticated data and analytics tools, which provides efficient decisions for better and quality crop yield. Due to the advanced goals of AI and ML and its cross-disciplinary approach, many tech businesses are also investing in AI.

According to Neo-Schumpeterian Economics: innovation, knowledge, and entrepreneurship are the three major driving forces of economic dynamics wherein the emergence of new industries is driven by technology, innovation, supported by knowledge and tested by entrepreneurial action (Dosi, Giorgio, & Andrea, 2010). Traditional methods used by farmers aren't sufficient enough to serve the increasing demand, so the use of high-end technology to boost the agriculture industry and the overall economy is the need of the hour. Agri-entrepreneurs and farmers can use machine learning (ML) and AI techniques to overcome increasing demand and provide a better yield of crops. This technological advancement, attributed to AI and ML, would facilitate the farm-to-table process, change the logic of agri-business models, and transform the approach and implementation of agri-entrepreneurs and farmers. The emerging technologies viz. ML, the internet of things (IoT), data science, big data, cloud computing, AI, and blockchain are changing how the agriculture industry works. Today, the majority of start-ups in agriculture are adopting an AI-enabled approach to increase the efficiency of agricultural production and quality. The Market study report stated that the global AI and ML in Agriculture market size is expected to reach 1,550 million US$ by the end of 2025 (Xie & Zhenhong, 2017).

Agriculture and its allied sectors are going through a rapid transformation driven by new innovative, disruptive technologies like AI, **ML**, bock-chain, IoT, etc., enabling the agriculture sector to move to the next level of crop yield and profitability. Realize the power of AI in agriculture will involve the integration of diverse disruptive technologies across the entirety of the agricultural domain and the adoption of improved and innovative systems-based approaches also known as the system of systems-based approach (SoS) to ensure it is applicable, robust, scalable, adaptive, and sustainable agriculture (Panetto et al., 2019). AI has begun to play a major role in our everyday activities, extending our perceptions and ability to transform our lives (Talaviya et al., 2020). Agriculturists and farmers around the globe are adopting and implementing disruptive technologies and AI-driven farming to harness various benefits like improved crop quality, better production, Reduction of costs with minimal efforts, more eco-friendly practices, and increased profitability (Malo, 2020).

9.2.1 IMPLEMENTATION OF AI IN AGRICULTURE

- Provides more efficient ways to produce, harvest, and sell essential crops.
- Helps check defective crops and improve the potential for healthy crop production.
- Provides solutions and has the potential to solve the challenges that most farmers face, such as climate variation and an infestation of pests and weeds that reduces yields.Provides advanced ways to help the farmer to remain updated with the data related to weather forecasting.
- Provides an efficient way to conduct, monitor, and identify possible defects and nutrient deficiencies in the soil.
- The growth in artificial intelligence technology has strengthened agro-based businesses to run more efficiently.

9.2.2 SPECIES MANAGEMENT USING ARTIFICIAL INTELLIGENCE (AI) AND MACHINE LEARNING (ML)

Looking for specific genes that determine the effectiveness of water and nutrient use, adaptation to climate change, disease resistance, as well as nutrient content, or a better taste is a cumbersome task (Tripodi, Nicastro, Pane, & Cammarano, 2022). Data science helps us to collect historical field data, and ML algorithms give insights and provide predictive analytics about the crop's performance and production. ML involves using ML algorithms to build prediction models (Lalwani, Mishra, Chadha, & Sethi, 2022). A predictive model is developed on training data to aid decisions without being explicitly programmed. Similarly, based on data, a ML analytical model is built to predict which genes will most likely contribute to quality crop production and disease resistance to a plant (Gu, 2022).

9.2.3 WEATHER FORECASTING USING ARTIFICIAL INTELLIGENCE (AI) AND MACHINE LEARNING (ML)

Predicting precise weather conditions, especially rainfall prediction, is very important for agriculture and allied sectors (Nyika, 2022). Rainfall is important for yield production, water resource management, humidifying the atmosphere, producing streams and rivers, replenishing the water table and redistribution of fresh water in the water cycle (Tsai, Lee, Zou, & Chu, 2022). The prolonged dry period or heavy rain at the critical stages of crop growth and development may significantly reduce crop yield (Sha, Zhao, Li, Li, Zhu, Wang, & Gao, 2022). The prolonged and continuous rainfall can also lead to floods that destroy crops, livestock, orchards, etc., and decimate housing and business infrastructure. Thus, forecasting weather is a significant factor in agricultural countries like India (Amwata, 2020). ML and deep learning (DL) algorithms can address the problems of changing weather conditions and pollution increase, making it difficult for farmers to know the exact time for sowing seeds (Ramdinthara, Bala, & Gowri, 2022). With the help of AI and other disruptive technologies, agriculturalists can examine climate conditions by using weather prediction systems which helps them to plan the type of crop that can be grown and when seeds should be sown. Some commonly used ML techniques for weather prediction are decision trees (DT), artificial neural networks (ANN), Naive Bayes networks, support vector machines (SVMs), fuzzy logic, and rule-based techniques, which include memory-based reasoning techniques and genetic algorithms (Geetha & Nasira, 2014).

9.2.4 SOIL AND CROP HEALTH MONITORING SYSTEM USING ARTIFICIAL INTELLIGENCE (AI) AND MACHINE LEARNING (ML)

Artificial intelligence systems help recognize fertile agricultural areas, soil type, seed selection, evaluate crop health status, etc. (Sarkar, Saha, Maitra,

& Mondal, 2022). Identifying crop stress factors is critical to correct crop growth simulation models and properly estimating the expected crop yield. Soil type and soil nutrition significantly affect crop growth and quality crop yield (Kadiyala et al., 2015). ML algorithms and DL algorithms are used to predict the appropriate soil type for a particular crop with good accuracy for the quality yield produced. Analysis of soil properties such as organic matter (OM), essential plant nutrients, and micronutrients that affect crops' growth is used to develop a prediction model (Pelegrino et al., 2022). Traditional evaluation approaches of soil nutrients are quite hard to operate, creating great difficulties in practical applications. Various machine-learning models play an essential role in resolving different real-world problems across multiple domains. These models are designed and developed to learn patterns from data, make forecasts, and automate decision-making processes.

For soil, nutrients are supporting vector machine (SVM), multiple linear regression (MLR), and ANNs. Training Neural networks are essential in increasing the model's accuracy. The soil properties dataset is analyzed and processed using an ANN approach. The first step is sampling where different soil with the same properties and parameters is done, then the second is Neural Network Algorithm, and the third is Weight updating. It has been seen that the ANN model evaluated using a test dataset shows that a certain number of neurons in the hidden layer had better performance in predicting soil properties. A German-based tech startup, PEAT, has developed an AI-based application called Plantix that can identify the nutrient deficiencies in soil, including plant pests and diseases, by which farmers can also get an idea to use fertilizer which helps to improve harvest quality (Rani, Dharshini, & Sonadevi, n.d.). This app uses image recognition-based technology. The farmer can capture images of plants using smartphones. Similarly, Trace Genomics is another ML-based company that helps farmers analyze soil (Garzetti, Bouabe, Heesemann, & Rakin, 2012). The app helps farmers monitor soil and crop health conditions and produce healthy crops with higher productivity.

9.2.5 PRECISION FARMING AND PREDICTIVE ANALYTICS USING ARTIFICIAL INTELLIGENCE (AI) AND MACHINE LEARNING (ML)

Precision agriculture, also known as data-driven and data-enabled agriculture or smart agriculture, is a technology-based data dependent on a sustainable farm management system (Mohapatra & Rath, 2022). It involves adopting and implementing modern information and communication technologies

(ICT) tools, along with IoT devices and sensors for smart decision support in agriculture. Implementing Precision agriculture helps minimize the manufacturing cost and environmental effects to increase yield and profitability. Modern technologies such as ML, AI, drones, IoT, information analytics, cloud computing, and blockchain technology play a vital role in precision agriculture. Smart sensors are installed on the farm to gather data about climate, soil nutrients, manures, weeds, and water necessities and to analyze crop growth (Sharma, Jain, Gupta, & Chowdary, 2020). The data is then used for predictive analytics for smart and optimized decision-making.

By implementing ML and DL algorithms. Various types of drones and robots are also used for identifying weeds and diseases in plants using computer vision techniques. Satellite images are also used in precision agriculture to monitor the field and identify the plants' diseases. Precision agriculture is an umbrella concept for IoT-based methods that make agricultural activities more controlled and accurate (Bacco, Barsocchi, Ferro, Gotta, & Ruggeri, 2019).

9.2.6 CROP YIELD PROGNOSIS USING ARTIFICIAL INTELLIGENCE (AI) AND MACHINE LEARNING (ML)

ML and AI technology help farmers predict crop yield and how it can be increased (Bali & Singla, 2022). Various parameters like pH value, soil type, quality, climate pattern: temperature, precipitation, moisture, sunlight hours, manures, etc., play a significant role in crop yield prediction. An important piece of information for any farmer is the prediction of crop yield and how the yield can be increased. pH value, soil type, and quality, and weather pattern: temperature, rainfall, humidity, sunshine hours, fertilizers, and harvesting schedules are some of the parameters which play an important role in predicting crop yield. Development and implementations of ML and Artificial intelligent models help farmers in optimal utilization of the resources when analytically implemented to a DL neural network model. With the help of correct predictive analytical models, we can predict the factors that affect crop production so that we can take corrective action accordingly to boost quality yield production (Sharma, Shekhawat, & Mehta, 2022). Thus, implementing disruptive technologies like ML and DL for smart farming by developing systems and tools which help farmers in intelligent farming practices with proper guidance and assurance so that they will have a quality of life.

9.3 DRONES IN AGRICULTURE

Nano-aircrafts initially only used for military purposes are unmanned aerial vehicles (UAVs), also known as drones, which are robotic aerial vehicles nowadays used for many agricultural purposes to help farmers with crop monitoring and yield optimization in real-time (Ayaz, Ammad-Uddin, Sharif, Mansour, & Aggoune, 2019). A drone is an aircraft without a human pilot on board; a —flying robot.

Initially, UAVs were simply remotely piloted aircraft, but autonomous control is increasingly being employed. Its flight is controlled either autonomously by onboard computers or by the remote control of a pilot on the ground or in another vehicle. Drone technology is used for surveillance in various industries and has shown remarkable progress and application in agricultural and allied sectors (Sridhar et al., 2022). According to Watts et al. (2012), drones are classified into seven categories which are micro-air vehicles (MAV), vertical take-off and landing (VTOL), low altitude-short endurance (LASE), low altitude short endurance close, low altitude low endurance, medium altitude, and high-altitude long endurance (HALE) (Schuyler & Guzman, 2017).

The aerial view provided by a drone help farmer to take quick actions against threats like weeds, pests, rodents, and early detection of crop diseases and also aid in crop scouting which saves time and enables them to take in-time action to save crops and increase efficiency in agricultural production (Sinha & Dhanalakshmi, 2022). Drones are also used in field and soil assessment, precision crop spraying, plant establishment, irrigation management, crop health assessment, and livestock monitoring. Drone technology, in combination with computer vision and AI approaches, uses high-quality cameras, radar, LiDar, and infrared technologies to see and analyze visual data captured by drones are used for identification and control of weeds, soil fertility analysis, animal health monitoring, pests, and disease identification and management (Jenssen & Roverso, 2018). Drone AI technology can be generally implemented with benefits such as location-specific precision information on spatial variation of various management practices, which can lead to improved quality yield production (Meena et al., n.d.).

9.4 ARTIFICIAL INTELLIGENCE (AI) IN AGRI-BUSINESS

ML and neural network models can aid in dealing with the alarming increase in worldwide demand for food and support (Bali & Singla, 2022). AI enables smart agriculture to help in providing a more inclusive and sustainable food system by enhancing the resilience of farming methods, slashing the cost of quality inputs and facilities to underserved farmers and enhanced and transparent market access to enable smallholder growers' integration into regional and global supply chains (Prasad et al., n.d.). AI methods are progressively developing as part of the agriculture industry's technological revolution. AI-enable approaches will not only empower farmers to improve productivity, but they will also improve the quantity, and quality and ensure faster go-to-market for crops. Achieving food challenges that the world will be facing, and agriculture-related sustainable development goals (SDGs) sustainably and inclusively is difficult. AI-based Agtech inventions have substantial development potential (Lim, Jørgensen, & Wyborn, 2018). AI-based Agtech can help farmers to access advanced technologies and methods to increase output and allow them to trace the origin and quality of their produce. The implementation and dissemination of AI expertise and precision agriculture into agtech business models help in addressing these challenges (Marshall, Turner, Richards, Foth, & Dezuanni, 2022).

AI helps in Decision support systems, on-farm analytical support, software for farm operations, and Marketplaces that help farmers and other stakeholders access online marketplaces for buying farming inputs and selling crops (Rao, 2022). Logistics and infrastructure using content management and optimization of crop produce to markets. Furthermore, financial services like Credit scoring enable giving and insurance for farmers (Orwothwun & Qutieshat, 2022).

9.5 AUGMENTED REALITY (AR) AND VIRTUAL REALITY (VR) IN AGRICULTURE

Augmented reality and virtual reality (AR and VR) are advanced technologies that replicate a computer-generated artificial environment to visualization or replace a real-life environment with a simulated one realized by hardware such as a head-mounted display (HMD) with implantations of 3D processing technologies that seize the user from the real-world environment and provides visuals and images of a virtual environment through the combination of lightweight, high-resolution screens and lenses (Joonatan, n.d.). The main difference between AR and VR is that AR uses a real-world setting, and users can control their presence in the real world, while VR is completely virtual, and users are controlled by the system (Xi, Chen, Gama, Riar, & Hamari, 2022). Areas, where AR/VR is extensively adopted, are military, e-commerce, medicine, entertainment, manufacturing, education, and agriculture (Patel, Chesmore, Legner, & Pandey, 2022). AR in agriculture and allied sectors is new hope for dealing with growing food demand (Chander, Pal, De, & Buyya, 2022). AR help farmers with constant field monitoring and provide weather updates.

For precise and real-time weather-related situations, which allows them to take proactive steps (Ayaz, Ammad-Uddin, Sharif, Mansour, & Aggoune, 2019). AR helps understand pest and crop conditions by looking at the application and may assist agriculturalists in making better Crop Selection decisions. Besides, it helps determine the soil properties and crop selection for specific land. AR in agriculture is new hope for managing the challenge of growing food demand. Implementation of AR in farming will help in improving farm production. The technological wave of AR in agriculture will revolutionize farming. There will be a continuous surge in the human population in future decades (Reardon & Timmer, 2007).

9.6 BLOCKCHAIN IN AGRICULTURE

Due to its vast complexity and intricacy, transparency, and traceability are the main challenges that Food supply chains face. Many Laws and regulations are made and implemented worldwide regarding the traceability of agricultural food products. Traceability, also known as the principle of —one step back, one step forward‖, is the ability to recall all the information about the origin of a food product (Gunawan, Vanany, & Widodo, 2019). Today, people consuming food are more concerned about food provenance and quality, which are high, and are ready to spend more money on foodstuffs whose source is certified (Tsoukas, Gkogkidis, Kampa, Spathoulas, & Kakarountas, 2022). Despite the technologies that are at present in practice, in numerous cases, most of the traceability systems are generally centralized, asymmetric, and obsolete in terms of data sharing and interoperability (Behal & Pattabiraman, 2022). Already existing systems lack transparency and customers' reliance because of the non-availability of a fast and reliable means of data retrieval and information on the product's provenance. Traditional tracking and traceability systems had a manual transfer of data collected from fields, farms, and other places to computer systems with high risks of faulty and non-reliable information and inefficient resource usage (Rai, n.d.). Agriculture traceability demands a huge volume of data that needs to be collected across the supply chain. Over the decades, Blockchain was only a topic of theoretical research, but it is only in the past few years that researchers have found its usage in agricultural traceability systems with monetary benefits in supply food chain management by providing better transparency and traceability in the agri-food domain (Bhat, Huang, Sofi, & Sultan, 2021). Blockchain technology provides a very secure technique for storing, managing, and retrieving data, enabling the implementation and use of advanced data-driven innovations, and adopting disruptive technologies for smart agricultural practices and smart index-based agriculture insurance. In Europe, the traceability of food products is obligatory as per European Directive 178/2002 since 1 January 2005, together in compliance with the HACCP (Hazard Analysis and Critical Control Points) principles. Similarly, worldwide regulations have been developed targeting reducing food epidemic incidents (Humphrey, 2006).

9.7 3D PRINTING IN AGRICULTURE

Fabrication of three-dimensional objects/items from Computer Aided Design or digital 3D model is known as 3D printing or additive manufacturing. 3D printing is a technology in which material such as plastics, liquids or powder grains is deposited, fused, or solidified through specific algorithm control by a computer system in a stratified manner (Valino et al., 2019). 3D printing is an engineering process of producing a physical object/structure from a digital model file by adding layer upon layer of plastic material/filament to construct a complete object/article. The major advantage of 3D printing is that it aids in developing very complex shapes and contours with various symmetries that would otherwise be difficult to make by hand, including hollow parts or parts with internal truss structures to decrease mass (Nicoloso, n.d.). 3D printing technology is gradually finding applications in practically all industries and sectors across the globe like the industrial sector, health sector, business sector, startups, fashion, food, agriculture, and manufacturing sectors. Recently 3D printing has also been used in the charitable and development sector to produce various medical items, prosthetics, spares, and repairs (Paunov & Planes-Satorra, n.d.). 3D printing or additive manufacturing enables and assists the agricultural and allied sectors by providing numerous agricultural models, prototypes, fabrication parts, etc. These tools and techniques aid the agriculture sector and researchers in moving rapidly (Revathi & Aithal, n.d.). 3D printing presents an image of reality to agricultural industry experts concerning their products which are robust and fast assess diverse designs and at the same time impose different modifications and adjustments at a cheaper price (Lupton, 2017). Agriculturalists and scientists across the globe are always looking, exploring, and adopting advanced and innovative technologies to increase crop production, crop quality, and self-reliance to provide food to the increasing population. 3D printing plays a significant role in helping farmers, agriculturalists, scientists, and agri-startups tackle the food problem and increase agri-business significantly.

3D printing supports the agriculture and allied sectors by providing various agricultural prototypes, production parts, etc. 3D printing facilitates replicating many objects for agriculture, including chains, gears, shock absorbers, seeder parts, and harvester attachments (Griepentrog, n.d.). Various parts and components can be printed from the new robust filament (plastics) to metal and alloy combinations (Kalender, Kılıç, Ersoy, Bozkurt, & Salman, 2019). The worldwide 3D printing market was valued at USD 13.84 billion in 2021 and is estimated to increase at a compound annual growth rate (CAGR) of 20.8% from 2022 to 2030. Across the globe, 2.2 million 3D printers were shipped in 2021, and the shipments are expected to reach 21.5 million units by 2030. 3D printing has been leveraged to develop processes and sustainability within the agricultural and allied sectors in many ways. The concept of a circular economy is increasing, and 3D printing plays a significant role in achieving circularity within it (Schröder, Albaladejo, Ribas, MacEwen, & Tilkanen, 2020).

A startup like Dragontech is using 3D printing to develop a new accessible drone-based data gathering platform to help commercial farmers to find potential problems in their farms and crops. There are many 3D food printing startup companies that are either building the technology behind 3D food printers or are making 3D-printed food, such as plant-based meat (Portanguen, Tournayre, Sicard, Astruc, & Mirade, 2019). Very soon, 3D food printers will soon be a mass-adopted consumer product that will become a common household product. 3D food printer technology remains mainly for commercial use (Sun, Zhou, Huang, Fuh, & Hong, 2015).

9.8 CONCLUSION

Adopting Information Communication Technology with ML and AI seems to constitute one of our best hopes to meet the emerging challenges of food and the world economy. As per a recent study, these technologies can potentially add 16% or about $13 trillion to the global economy. It could also boost the global gross domestic product (GDP) by up to 26%. Farmers need to use these advanced skills and technologies in their crop management practices having continuous, sustainable, and tangible benefits. AI enables systems to help build new cost-effective business models to deliver the information needed to reach small farmers in emerging markets. However, administrations, investors, agriculturists, and industry ought to collaborate to address data, infrastructure, and human capital gaps before commercial AI integration is feasible on a larger scale. It is expected that this study is going to provide a useful guide to academics, students, researchers, and farmers.

KEYWORDS

- **3D printing**
- **agricultural drones**
- **artificial intelligence**
- **blockchain**
- **disruptive technologies**
- **internet of things**
- **machine learning**

REFERENCES

Amwata, D. A., (2020). *Situational Analysis of the Agriculture Sector in Kenya.* Ministry of Agriculture, Livestock and Fisheries. Nairobi, Kenya.

Ayaz, M., Ammad-Uddin, M., Sharif, Z., Mansour, A., & Aggoune, E. H., (2019). Internet-of-Things (IoT)-based smart agriculture: Toward making the fields talk. *IEEE Access, 7*, 129551–129583.

Bacco, M., Barsocchi, P., Ferro, E., Gotta, A., & Ruggeri, M., (2019). The digitization of agriculture: A survey of research activities on smart farming. *Array, 3*, 100009.

Bali, N., & Singla, A., (2022). Emerging trends in machine learning to predict crop yield and study its influential factors: A survey. *Archives of Computational Methods in Engineering, 29*(1), 95–112.

Behal, R., & Pattabiraman, K., (2022). Applications and potential of blockchain in modern technological systems. In: *Blockchain Technologies for Sustainable Development in Smart Cities* (pp. 65–86). IGI Global.

Bhat, S. A., Huang, N. F., Sofi, I. B., & Sultan, M., (2021). Agriculture-food supply chain management based on blockchain and IoT: A narrative on enterprise blockchain interoperability. *Agriculture, 12*(1), 40.

Chander, B., Pal, S., De, D., & Buyya, R., (2022). Artificial intelligence-based internet of things for industry 5.0. In: *Artificial Intelligence-Based Internet of Things Systems* (pp. 3–45). Springer, Cham.

Dosi, G., Giorgio, F., & Andrea, R., (2010). Schumpeter meeting Keynes: A policy-friendly model of endogenous growth and business cycles. *Journal of Economic Dynamics and Control, 34*(9), 1748–1767.

Garzetti, D., Bouabe, H., Heesemann, J., & Rakin, A., (2012). Tracing genomic variations in two highly virulent yersinia enterocolitica strains with unequal ability to compete for host colonization. *BMC Genomics, 13*(1), 1–5.

Geetha, A., & Nasira, G. M., (2014). Data mining for meteorological applications: Decision trees for modeling rainfall prediction. In: *2014 IEEE International Conference on Computational Intelligence and Computing Research* (pp. 1–4). IEEE.

Griepentrog, H. W., (2019). *Crop Plant Reconstruction and Feature Extraction Based on 3-D Vision, 10*, 2.

Gu, J., (2022). Ecophysiological crop modeling combined with genetic analysis is a powerful tool for ideotype design. *Agronomy, 12*(1), 215.

Gunawan, I., Vanany, I., & Widodo, E., (2019). Cost-benefit model in improving traceability system: Case study in Indonesian bulk-liquid industry. In: *Supply Chain Forum: An International Journal* (Vol. 20, No. 2, pp. 145–157). Taylor & Francis.

Humphrey, J., (2006). Policy implications of trends in agribusiness value chains. *The European Journal of Development Research, 18*(4), 572–592.

Jenssen, R., & Roverso, D., (2018). Automatic autonomous vision-based power line inspection: A review of current status, and the potential role of deep learning. *International Journal of Electrical Power & Energy Systems, 99*, 107–120.

Joonatan, M., (2014). *Evaluation of Intuitive VR-based HRI for Simulated Industrial Robots*, p. 14.

Kadiyala, M. D., Nedumaran, S., Singh, P., Chukka, S., Irshad, M. A., & Bantilan, M. C., (2015). An integrated crop model and GIS decision support system for assisting agronomic decision making under climate change. *Science of the Total Environment, 521*, 123–134.

Kalender, M., Kılıç, S. E., Ersoy, S., Bozkurt, Y., & Salman, S., (2019). Additive manufacturing, and 3D printer technology in aerospace industry. In: *2019 9th International Conference on Recent Advances in Space Technologies (RAST)* (pp. 689–694). IEEE.

Lalwani, P., Mishra, M. K., Chadha, J. S., & Sethi, P., (2022). Customer churn prediction system: A machine learning approach. *Computing, 104*(2), 271–294.

Lim, M. M., Jørgensen, P. S., & Wyborn, C. A., (2018). Reframing the sustainable development goals to achieve sustainable development in the Anthropocene—A systems approach. *Ecology and Society, 23*(3).

Lujala, P., Jan, K. R., & Nadja, T., (2007). Fighting over oil: Introducing a new dataset. *Conflict Management and Peace Science, 24*(3), 239–256.

Lupton, D., (2017). Download to delicious: Promissory themes and sociotechnical imaginaries in coverage of 3D printed food in online news sources. *Futures, 93*, 44–53.

Malo, M., (2020). Artificial intelligence in agriculture: Strengthening the future of farming. *Agriculture & Food e-Newsletter, 71*.

Marshall, A., Turner, K., Richards, C., Foth, M., & Dezuanni, M., (2022). Critical factors of digital AgTech adoption on Australian farms: From digital to data divide. *Information, Communication & Society, 25*(6), 868–886.

McCarthy, J., & Patrick, J. H., (1981). Some philosophical problems from the standpoint of artificial intelligence. *Readings in Artificial Intelligence* (pp. 431–450). Morgan Kaufmann.

Meena, B. R., Jatav, H. S., Dudwal, B. L., Kumawat, P., Meena, S. S., Singh, V. K., Khan, M. A., & Sathyanarayana, E., (2022). *Fertilizer Recommendations by Using Different Geospatial Technologies in Precision Farming or Nanotechnology, 14*, 241–257.

Mohapatra, H., & Rath, A. K., (2022). IoE based framework for smart agriculture. *Journal of Ambient Intelligence and Humanized Computing, 13*(1), 407–424.

Nicoloso, L. G., (2021). *Data-Driven Systems Engineering for Bioinspired Integrative Design.* Doctoral dissertation, University of California, San Diego.

Nyika, J. M., (2022). Climate change situation in Kenya and measures towards adaptive management in the water sector. In: *Research Anthology on Environmental and Societal Impacts of Climate Change* (pp. 1857–1872). IGI Global.

Orwothwun, C., & Qutieshat, A., (2022). Agriculture value chain finance in Uganda: A brief review of literature. *Open Access Library Journal, 9*(4), 1–8.

Panetto, H., et al., (2019). Challenges for the cyber-physical manufacturing enterprises of the future. *Annual Reviews in Control, 47*, 200–213.

Patel, V., Chesmore, A., Legner, C. M., & Pandey, S., (2022). Trends in workplace wearable technologies and connected-worker solutions for next-generation occupational safety, health, and productivity. *Advanced Intelligent Systems, 4*(1), 2100099.

Paunov, C., & Planes-Satorra, S., (2019). *How are Digital Technologies Changing Innovation?: Evidence from Agriculture, the Automotive Industry and Retail.*

Pelegrino, M. H., Silva, S. H., De Faria, Á. J., Mancini, M., Teixeira, A. F., Pelegrino, M. H., Silva, S. H., et al., (2022). Prediction of soil nutrient content via pXRF spectrometry and its spatial variation in a highly variable tropical area. *Precision Agriculture, 23*(1), 18–34.

Portanguen, S., Tournayre, P., Sicard, J., Astruc, T., & Mirade, P. S., (2019). Toward the design of functional foods and biobased products by 3D printing: A review. *Trends in Food Science & Technology, 86*, 188–198.

Prasad, R. P., Gill, R., Gupta, V., Bordoloi, P., Ahmed, M., & Rao, R. K., (2022). *Recent Advances in Agricultural Science and Technology for Sustainable India.*

Rai, N. K., (2018). *Investigation into Various IoT (Internet of Things) Architectures, Standards, Alliances.*

Ramdinthara, I. Z., Bala, P. S., & Gowri, A. S., (2022). AI-based yield prediction and smart irrigation. In: *Internet of Things and Analytics for Agriculture* (Vol. 3, pp. 113–140). Springer, Singapore.

Rani, M., Dharshini, T., & Sonadevi, A., (2022). *Crop Suggestion Based on Soil Pattern and Rainfall Ratio, 04*(02).

Rao, E. P., (2022). Digital agriculture–a future disruption in India. *Indian Journal of Fertilizers, 18*(4), 334–342.

Reardon, T., & Timmer, C. P., (2007). Transformation of markets for agricultural output in developing countries since 1950: How has thinking changed? *Handbook of Agricultural Economics, 3*, 2807–2855.

Revathi, R., & Aithal, P. S., (2019). *A Review on Impact of Information Communication & Computation Technology (ICCT) on Selected Primary, Secondary, and Tertiary Industrial Sectors*, pp. 106–127.

Sarkar, D., Saha, S., Maitra, M., & Mondal, P., (2022). Site suitability for aromatic rice cultivation by integrating geo-spatial and machine learning algorithms in Kaliyaganj CD block, India. *Artificial Intelligence in Geosciences.*

Satapathy, S., Mishra, D., & Vargas, A. R., (2022). *Innovation in Agriculture with IoT and AI.* Springer Nature.

Schröder, P., Albaladejo, M., Ribas, P. A., MacEwen, M., & Tilkanen, J., (2020). *The Circular Economy in Latin America and the Caribbean.* The Royal Institute of International Affairs, Chatham House, London.

Schuyler, T. J., & Guzman, M. I., (2017). Unmanned aerial systems for monitoring trace tropospheric gases. *Atmosphere, 8*(10), 206.

Sha, S., Zhao, X., Li, Y., Li, C., Zhu, L., Wang, Y., & Gao, Q., (2022). Nutrient expert system optimizes fertilizer management to improve potato productivity and tuber quality. *Journal of the Science of Food and Agriculture, 102*(3), 1233–1244.

Sharma, A., Jain, A., Gupta, P., & Chowdary, V., (2020). Machine learning applications for precision agriculture: A comprehensive review. *IEEE Access, 9*, 4843–4873.

Sharma, M. K., Shekhawat, R. S., & Mehta, R., (2022). Functional framework for IoT-based agricultural system. In: *Internet of Things and Analytics for Agriculture* (Vol. 3 pp. 1–27). Springer, Singapore.

Sinha, B. B., & Dhanalakshmi, R., (2022). Recent advancements and challenges of internet of things in smart agriculture: A survey. *Future Generation Computer Systems, 126*, 169–184.

Sridhar, A., Balakrishnan, A., Jacob, M. M., Sillanpää, M., & Dayanandan, N., (2022). Global impact of COVID-19 on agriculture: Role of sustainable agriculture and digital farming. *Environmental Science and Pollution Research*, 1–7.

Sun, J., Zhou, W., Huang, D., Fuh, J. Y., & Hong, G. S., (2015). An overview of 3D printing technologies for food fabrication. *Food and Bioprocess Technology, 8*(8), 1605–1615.

Talaviya, T., et al., (2020). Implementation of artificial intelligence in agriculture for optimization of irrigation and application of pesticides and herbicides. *Artificial Intelligence in Agriculture, 4*, 58–73.

Tripodi, P., Nicastro, N., Pane, C., & Cammarano, D., (2022). Digital applications and artificial intelligence in agriculture toward next-generation plant phenotyping. *Crop and Pasture Science.*

Tsai, S., Lee, S., Zou, Z., & Chu, T., (2022). The competitions, negotiations, and collaborations of regional integration: A perspective on sustainable management of water resources in Pingtung plain, Taiwan. *Sustainability, 14*(5), 3040.

Tsoukas, V., Gkogkidis, A., Kampa, A., Spathoulas, G., & Kakarountas, A., (2022). Enhancing food supply chain security through the use of blockchain and TinyML. *Information, 13*(5), 213.

Valino, A. D., Dizon, J. R., Espera, Jr. A. H., Chen, Q., Messman, J., & Advincula, R. C., (2019). Advances in 3D printing of thermoplastic polymer composites and nanocomposites. *Progress in Polymer Science, 98*, 101162.

Xi, N., Chen, J., Gama, F., Riar, M., & Hamari, J., (2022). The challenges of entering the metaverse: An experiment on the effect of extended reality on workload. *Information Systems Frontiers*, 1–22.

Xie, F., & Zhenhong, L., (2017). Market-driven automotive industry compliance with fuel economy and greenhouse gas standards: Analysis based on consumer choice. *Energy Policy, 108*, 299–311.

Zambon, I., et al., (2019). Revolution 4.0: Industry vs. agriculture in a future development for SMEs. *Processes, 7*(1), 36.

Index

D

E

F

G

H

I

N

O

P

Q

R

S